T0326377

WAKING AND THE RETICULAR ACTIVATING SYSTEM IN HEALTH AND DISEASE

WAKING AND THE RETICULAR ACTIVATING SYSTEM IN HEALTH AND DISEASE

Edited by

EDGAR GARCIA-RILL

AMSTERDAM • BOSTON • HEIDELBERG • LONDON
NEW YORK • OXFORD • PARIS • SAN DIEGO
SAN FRANCISCO • SINGAPORE • SYDNEY • TOKYO
Academic Press is an imprint of Elsevier

Academic Press is an imprint of Elsevier
32 Jamestown Road, London NW1 7BY, UK
525 B Street, Suite 1800, San Diego, CA 92101-4495, USA
225 Wyman Street, Waltham, MA 02451, USA
The Boulevard, Langford Lane, Kidlington, Oxford OX5 1GB, UK

Notices
Knowledge and best practice in this field are constantly changing. As new research
and experience broaden our understanding, changes in research methods, professional
practices, or medical treatment may become necessary.

Practitioners and researchers must always rely on their own experience and knowledge
in evaluating and using any information, methods, compounds, or experiments described
herein. In using such information or methods they should be mindful of their own safety
and the safety of others, including parties for whom they have a professional responsibility.

To the fullest extent of the law, neither the Publisher nor the authors, contributors, or
editors, assume any liability for any injury and/or damage to persons or property as a
matter of products liability, negligence or otherwise, or from any use or operation of any
methods, products, instructions, or ideas contained in the material herein.

British Library Cataloguing in Publication Data
A catalogue record for this book is available from the British Library

Library of Congress Cataloging-in-Publication Data
A catalog record for this book is available from the Library of Congress

ISBN: 978-0-12-801385-4

For information on all Academic Press publications
visit our website at elsevierdirect.com

Typeset by SPi Global, India

Printed and bound in The United States
14   15   16   17   10   9   8   7   6   5   4   3   2   1

Working together
to grow libraries in
developing countries

www.elsevier.com • www.bookaid.org

Contents

4. Wiring Diagram of the RAS
SUSAN MAHAFFEY AND EDGAR GARCIA-RILL

5. Development and the RAS
PAIGE BECK AND EDGAR GARCIA-RILL

6. Ascending Projections of the RAS
JAMES HYDE AND EDGAR GARCIA-RILL

7. Descending Projections of the RAS
CHARLOTTE YATES AND EDGAR GARCIA-RILL

8. The 10 Hz Fulcrum
EDGAR GARCIA-RILL

9. Gamma Band Activity
NEBOJSA KEZUNOVIC, JAMES HYDE, FRANCISCO J. URBANO, AND EDGAR GARCIA-RILL

10. Preconscious Awareness
EDGAR GARCIA-RILL

11. Psychiatric Disorders and the RAS
STASIA D'ONOFRIO, ERICK MESSIAS, AND EDGAR GARCIA-RILL

12. Neurological Disorders and the RAS

BRENNON LUSTER, ERICA PETERSEN, AND EDGAR GARCIA-RILL

13. Drug Abuse and the RAS

FRANCISCO J. URBANO, VERONICA BISAGNO, AND EDGAR GARCIA-RILL

14. The Science of Waking and Public Policy

EDGAR GARCIA-RILL

Acknowledgments

A life in science is built not only on the shoulders of giants but also on the brains of friends and coworkers. I have been privileged to work with wonderful collaborators like Bernardo Dubrovsky at McGill University, Chester Hull and Nat Buchwald at UCLA, Bob Skinner at UAMS, and Francisco Urbano from Buenos Aires. Sharing ideas with colleagues has helped shape the concepts in this book, with people like Shigemi Mori and a host of excellent Japanese postdoctoral trainees, with Subimal Datta and Carlos Schenck, and definitely with Rodolfo Llinas. I have been fortunate in training and learning from graduate students like David Heister and Meijun Ye, Christen Simon and Neso Kezunovic, Nancy Reese and Charlotte Yates, and Paige Beck and James Hyde, among others.

But the kind of science that we perform takes support, a lot of it. Keith Murray at NIH gathered incredibly intelligent groups of thinkers for his supportive review committees, and it was a pleasure and an honor to work with him for many years. Merrill Mitler at NINDS was one of the most helpful program officers you could have for your individual awards. My research was supported for over 25 years by award R01 NS020246. Fred Taylor and Yanping Liu in the IDeA program at NIGMS were instrumental in helping me build the Center for Translational Neuroscience (CTN) and its core facilities, which were critical for the discoveries described herein. Their support was through awards P20 GM103425 and P30 GM110702. At UAMS, Gwen Childs and Al Reece provided me the essential opportunities and institutional resources to implement my work. Without Linda Luster's gracious and dedicated management, the administration of the CTN would be a burden instead of a pleasure. I am indebted to all of them. Thank you.

Melanie Tucker and Kathy Padilla at Elsevier were extremely helpful and supportive in the publication of this book, my sincere thanks to them. Susan Mahaffey was stellar in drafting and editing many of the figures for this book, and I am extremely grateful for her help.

Dedication

This book is dedicated to my students and trainees. They have performed the studies that allowed us to understand the RAS and how it affects brain function in health and disease. I am forever grateful and wish them long and productive careers. Today, I am as or even more excited about the next experiment, the next mystery to be resolved. I wish them the same fortune in finding a career that has brought me so much satisfaction and joy.

I also dedicate this book to the wonderful women in my life, Aracelis, Catherine, Jennifer, Sarah, and Thais (in alphabetical order). Their love and support has been integral to my successes. I love you.

Preface

There are few if any books that deal with waking proper. There are a number of books that describe the control of sleep and the mechanisms behind our sleeping hours. Considering that we spend one-third of our lives asleep, this is an important and essential endeavor. In addition, a number of thorough texts on the etiology and manifestation of a number of sleep disorders are available. These books and others on sleep medicine provide the latest on diagnosis and treatment of many of these debilitating conditions. However, we spend two-thirds of our lives awake, and it is during these hours that we accomplish such things as writing books. It is during waking that we perform great feats and create wonderful things. A number of texts describe attention, selective attention, higher cognitive abilities, and learning and memory, but almost none deals with the concept of being awake and what it takes to stay awake.

This gap in knowledge is not surprising since it has only been in the last 5–7 years that some of the most critical information on the mechanisms behind waking has been revealed. This information forms the core of this book. When we awaken, blood flow to the thalamus and the brain stem increases over the first 15 minutes. It is only later that there are increases in blood flow to the frontal cortex. Yet, we wake up as ourselves; it does not take 15–20 min to figure out who we are. We are immediately ourselves when we awaken. This suggests that subcortical regions have much to do with our sense of self. In addition, our highest functions are thought to be mediated by fast cortical oscillations, the 40 Hz rhythm, or gamma band oscillations. Yet gamma band activity has been discovered not only in the cortex but also in the hippocampus, cerebellum, basal ganglia, and, now, the reticular activating system (RAS). Not only that, but these are not independent oscillations, they are coherent depending on the task at hand. That is, the activity in cortical and subcortical regions is in synchrony. Under some conditions, gamma activity in subcortical areas even precedes cortical gamma activity.

What is the role of gamma band activity in the cortex? It is thought to participate in the binding of perception and in the unification of different aspects of a sensory event. For example, a visual image has color, depth, structure, movement, etc., and these properties are processed by different cortical regions. It is the coherent activation by gamma oscillations across regions of the cortex that has been proposed to provide the unification of all of these features in order to accomplish perception, the "binding" of the sensory event.

What is the role of gamma band activity in the RAS? We know that the RAS receives a constant stream of information from the senses and also receives ongoing activity from within the brain. What is the unifying function of gamma band activity in the RAS? We proposed that the maintenance of gamma band activity in the RAS provides information for the process necessary to support a state capable of reliably assessing the world around us on a continuous basis. That is, it provides the process of preconscious awareness.

The simple act of waking up now gains a much more complex role. It needs to integrate our world with ourselves, while we use other parts of our brains to formulate our plans and desires. As we will see, we may not be paying attention to some of these plans and desires, that is, we are not consciously paying attention to a mass of information that we process preconsciously. As such, the RAS is involved in anonymously formulating movements and actions of which we are not consciously (but only preconsciously) aware. This expands the purview of the background of activity in the RAS as not only allowing afferent information to flow into the brain but also establishing the background of activity on which we superimpose volition and free will.

These ideas have been explored before and been elucidated by some of the giants of neuroscience research. These are novel ideas only in that we have discovered mechanisms within the RAS that are similar to the mechanisms behind the generation of gamma band activity in other brain regions. More importantly, dysregulation of function in the RAS will impact the manifestation of numerous neurological and psychiatric diseases. In fact, wake–sleep disturbances are common in these disorders and may presage the onset of the disease by years. By understanding the function, specifically, the physiology, of this region, we can make more informed decisions on the treatment of these devastating diseases. One goal of this book is to improve the treatment of many, many patients with psychiatric and neurological disorders who are being treated in the absence of complete information.

This book deals mainly with disorders of waking rather than sleep. Below is a list of diseases and the chapters in which they appear, provided for rapid access by clinician scientists. However, we suggest that it is the underlying physiological mechanisms that will lead to more informed and effective treatments for these disorders. We discuss thalamocortical dysrhythmia in Chapters 1 and 10; coma, the vegetative state, and minimally conscious state in Chapter 3; sleep problems in obesity and Kleine–Levin syndrome in Chapter 5; posture, locomotion, and some aspects of spinal cord injury, as well as narcolepsy, and REM sleep behavior disorder in Chapter 7; anxiety disorder with emphasis on PTSD, autism, bipolar disorder, childhood trauma, major depression, and schizophrenia in Chapter 11; Alzheimer's disease, Huntington's disease, insomnia,

Parkinson's disease, spatial neglect, and stroke in Chapter 12; and drug and alcohol abuse in Chapter 13.

The simple act of waking and staying awake is not so simple. And it is very hard work, as we will see. The unseen hand of the RAS guides our lives, promotes our survival, and gives us identity. It is time we gave it some credit.

1

Governing Principles of Brain Activity

Edgar Garcia-Rill, PhD

Center for Translational Neuroscience, Department of Neurobiology and Developmental Sciences, University of Arkansas for Medical Sciences, Little Rock, AR, USA

WHY WAKING?

When we awaken, the world is instantaneously in harmony. We are ourselves once again, no matter what nonsense we have been dreaming. We pick up exactly where we left off the day before, the memories flooding in about the day's tasks. What does it take to wake up and, more importantly, to stay awake? This is an underappreciated faculty, yet when we lose it, we disappear. When we fall asleep, are anesthetized, or become comatose, our sense of self vanishes. There is nobody home. We have no memory when we are in slow-wave sleep (SWS), when we are anesthetized, or when we are comatose. There is a neurological condition in which this happens gradually. In Alzheimer's disease, patients seem to slowly, inexorably, lose their sense of self. On the other hand, we have some memory of our dreams. During sleep, every 90 min or so, we transition from SWS to rapid eye movement (REM) sleep. It is during this time that we dream. When we dream, we are mostly ourselves, but sometimes someone else, and we are in a distorted world, at times enjoyable and at times fearsome. Only our eyes live out our dreams because our extraocular muscles are not paralyzed, but the rest of the body is thankfully paralyzed, this is the atonia of REM sleep. Our frontal lobes have low blood flow so we are not exactly the sharpest tacks in the box. That is, we have little critical judgment, so that our dreams run the gamut of recalled and manufactured experiences. We believe the surrealistic collage of feelings and situations in dreams and accept them at face value no matter how crazy or unreal. We basically are suffering from a hallucination, but thankfully, we cannot act dreams out because of the atonia of REM sleep. There is a neurological condition in which our bodies are not paralyzed

Waking and the Reticular Activating System in Health and Disease
http://dx.doi.org/10.1016/B978-0-12-801385-4.00001-X

when we dream. If our bodies are not paralyzed while we dream, we begin to act out our dreams. We begin to fight perceived enemies, although they may be our unrecognized bed partners. We can cause them physical harm and we can persist until awakened by force. This occurs when we suffer from REM sleep behavior disorder, a rare disease discovered by the brilliant psychiatrist Carlos Schenck and his equally eminent neurologist codiscoverer Mark Mahowald (Schenck et al., 1987).

We spend a third of our lives asleep, about 80–85% of that in SWS, and the rest in REM sleep. However, we spend two-thirds of our lives awake. Waking is when we develop ideas, create objects, develop relationships, interact with other people, and earn a living, basically, when we do the really important things in life. Despite the importance of waking, there are very few books about waking and the process of staying awake. There are many books about sleep and about sleep dysregulation, about what happens when we have abnormal sleep, and when our vigilance interrupts or pushes aside our sleeping hours. The complaints patients tender are usually "problems sleeping," hardly ever do they say they have "problems waking." But the fact is that most psychiatric and neurological disorders involve just that, "problems waking." That is, hypervigilance and increased REM sleep drive cut down on our sleep, waking us early, often when we suffer from these diseases. We will address this issue in Chapter 2. In a number of other chapters, we will see how the process of waking has begun and how it is maintained. Without a firm understanding of the mechanisms behind waking, treating and controlling the so-called sleep disorders become more difficult. A firm grasp of the regulation of waking will also allow much better comprehension of the hyperarousal problems in a number of psychiatric and neurological disorders. In the following chapters, we will see how this process is disturbed by disease. Part of the problem is that we hardly ever study the process of waking per se.

THE GORILLA IN THE ROOM

There is a major difference between the real world and our studies of the human brain. In the real world, we are under continuous sensory load and always processing varied information. In the lab, our brain is studied under "controlled" conditions, that is, we are only subjected to a single stimulus at a time and we are required to elicit a response in a fixed manner. The real world is messy and complex, while the lab is neat and sterile. An animal in nature is exploring the environment, is bombarded by sensations, is exposed to multiple perceived threats, is repeatedly calculating fight-or-flight responses, and is planning the next meal. A laboratory subject is sitting still, probably listening to a background of white noise, and waiting for a stimulus to which to respond. The highest threat is the fear of falling asleep. Under these conditions, most sensory and motor physiologists ignore the gorilla in the room. They ignore the level of arousal, the stream of afferent information that establishes the background of activity upon which we *normally* superimpose our sensations, responses, and desires. Similarly, studies in animals do not take into account the excitability level of the subject. While real life is equivalent to listening to a concert played by many instruments emitting a myriad of notes, the laboratory setting is like sitting in a silent room and listening to a single note.

We have used such "controlled" conditions to dissect many basic principles and have acquired a host of information. Our need to determine cause and effect has led us to reduce experimental conditions to the smallest common denominator. This has been a productive approach to date, but we now need to study not only the "content" of sensory information but also the "context" under which we perceive that information. We have been very successful at determining very basic principles of brain function and have made great inroads into how the brain perceives simple stimuli and events. That is, we have been productive at discerning the responses to the "content" of sensory experience. However, we hardly address the other half of conscious perception, the "context" under which we perceive and act. While this may seem a lesser endeavor, the point will be made that, experimentally speaking, we ignore the majority of our waking world. How does the background of activity change the perception of sensory events? Are the responses the same when we are sedentary compared to when we are active? It would seem that the stimulus may be the same, but the consequent responses are markedly different when superimposed on a complex, continuously varying background. Are these consequences the same after being exposed again to the same stimulus, or do they change over time as the background of activity changes? The responses are probably considerably different, but we have little idea in what way they are different.

When we are awake, a mild auditory stimulus such as a voice produces a response that can be recorded in the auditory cortex. That same response is produced in the cortex if we are asleep. That is, the evoked response at the level of the cortex to the same sound stimulus is similar, yet we do not respond when we are asleep and we do respond when we are awake. If the sensory input produces a similar signal at the level of the cortex whether we are asleep or awake, what is the difference? The difference is the "context" under which we perceive the input during waking. While asleep, low-frequency activity in the electroencephalogram (EEG) does not lead to significant recognition or meaning to the sound received, despite the activation of the cortex by the sensory event. During waking, the background of activity, of high-frequency oscillations, leads to the recognition and integration of the stimulus into ongoing brain activity (Llinás and Paré, 1991). We can think of this like two gears, an internal gear that rotates in concert with the ongoing activity of the brain and an external gear that rotates with the input coming from the external world. When we are awake, the internal and external gears mesh and the internal and external worlds are merged seamlessly. They both rotate at gamma band frequency (~40 Hz) and provide unity to perception. When we are asleep, especially during SWS, the internal gear does not mesh with the external gear and is in fact rotating more slowly, at delta-band frequency (~4 Hz), than the external gear. External inputs do not force the gears to mesh unless the external input is stronger, strong enough to activate the reticular activating system (RAS). Only then does the external world intrude enough to make the gears mesh and rotate at the same rate. Only then does the arousing input awaken us to help drive the internal gear at gamma band frequency. We will see in Chapter 10 how the "context" of our sensory experience even drives our voluntary movements (Figure 1.1).

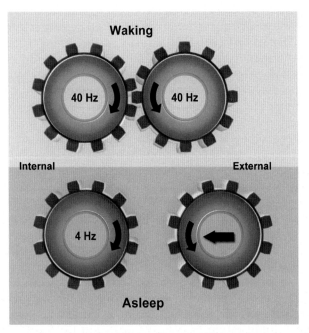

FIGURE 1.1 **Internal and external gears during waking and sleep.** Top: During waking, internal brain circuits generate high-frequency (beta/gamma) activity. These are in synchrony with the high-frequency activity generated by sensory afferents to the brain. The internal and external worlds are coherent. Bottom: During sleep, the internal gear slows to delta in the absence of high-frequency activity and the external gear is uncoupled, making awakening more difficult. During REM sleep, the internal gear generates high-frequency activity in the absence of a link to the external world. Coupled with decreased frontal lobe blood flow during REM sleep, critical judgment is absent and the brain is free to "manufacture" plots without realistic anchor. Fortunately, our muscles are paralyzed by the atonia of REM sleep that keeps us from acting out these plots. Only our eye muscles have sufficient muscle tone to live out our dreams, thus the presence of REMs.

What happens during REM sleep? The sensory input does not reset the ongoing gamma band oscillations of REM sleep so that the afferent information is not placed in "context" with the internal state of the system (Llinás and Paré, 1991). That is, the internal gear is rotating at gamma frequencies but totally disconnected from the external gear. We typically awaken during the last REM sleep episode of the night. Our RAS is already activated (the internal gear is rotating at ~40 Hz but disconnected from the external gear), and it is only a slight additional activation that shifts us from REM sleep to waking. If we are in SWS, it typically requires a strong stimulus to awaken us. Our brains are not as activated so that it seems that we are disoriented when we are awakened from a SWS episode. These considerations do not simply apply to normal, ongoing waking function, but they are involved in the disease process. Imagine a disease in which the two gears are rotating at slightly different speeds. The perceptual distortions would be disturbing. Imagine a disease in which a cog is missing and meshing "skips" on a regular basis. The dissonance between the internal world and the external environment is the basis for a number of neurological and psychiatric disorders. Many of the disorders we will consider are characterized by decreased or interrupted gamma band activity. But before we consider such concepts, we need to understand some basic principles of brain function.

CELL ASSEMBLIES

Donald Hebb synthesized certain ideas on how the environment and experience can influence brain structure and function (Hebb, 1949). He postulated the presence of cell assemblies in the brain that would fire as a consequence of sensory input to produce a sensation and that the activity would persist in the assembly and would represent the concept of the input. If the input occurred often, the circuit would be reinforced by the strengthening of synapses within that circuit and establish what he termed a "phase sequence." He proposed, "Any frequently repeated, particular stimulation will lead to the slow development of a "cell assembly," a diffuse structure comprising of cells in the cortex and diencephalon (and also, perhaps, in the basal ganglia of the cerebrum), capable of acting briefly as a closed system, delivering facilitation to other such systems and usually having a specific motor facilitation. A series of such events constitutes a 'phase sequence'—the thought process" (Hebb, 1949).

This concept became the bedrock of experimental investigation and the starting point for discussion of perception, attention, intelligence, learning, memory, and mental illness, among other subjects. The evidence for cell assemblies, or neural nets that behave the same way, is considerable, with results showing that the electrical activity of groups of neurons corresponds to the process of perception. We will see later that this involves gamma band activity or the "40 Hz rhythm." Hebb proposed (1) that the reverberation of activity around the cell assembly depended only on excitation and (2) that the synchronized activity throughout the circuit depended only on the synaptic connectivity. These have been recognized as limitations to Hebb's theory. Buzsaki pointed out that with only excitatory synapses across the assembly, there would be no segregation of assemblies (Buzsaki, 2006): "Without inhibition, activity can spread from one assembly to the next, and the whole brain would be synchronized by an external stimulus every time the stimulus is applied." Buzsaki correctly pointed out that in the absence of inhibition, any external input would generate the same pattern and soon involve the whole population. It is in the presence of excitatory and especially inhibitory interneurons that the network can self-organize and generate complex properties. These neurons are involved in feedback inhibition, feed-forward inhibition, and lateral inhibition, each of which provides a synchronizing signal no matter what the frequency of the ingoing activity. Moreover, we know that lateral inhibition is essential to the perceptual process, by which, for example, a cortical column that is activated by the most appropriate stimulus for that column sends inhibitory signals to surrounding columns to increase the contrast between the induced response in that column and the activity of those columns around it.

In addition, synchronized activity across cell assemblies or networks can be maintained at high frequencies only with the rapid oscillations provided by intrinsic subthreshold membrane oscillations (Llinas, 2001). Most cortical circuits begin to fail at frequencies around 20 Hz, so that synaptic connectivity cannot hope to maintain gamma band (~40 Hz) frequencies in a circuit. In the case of a cell assembly that includes cortical and subcortical regions, the likelihood that a 40 Hz oscillation will be accurately maintained, even in the presence of both excitatory and inhibitory neurons, is minimal. For example, an auditory stimulus elicits an evoked response in the primary auditory cortex every time until it is applied more often than $20 s^{-1}$. At higher frequencies, the synapses begin to fail, as does "perception." A visual stimulus can be detected as long as it is not changed more often than $35 s^{-1}$, after which we see the repeated images as a "movie." This is the idea behind television, which presents a series of rapid images that are perceived as moving pictures, at which point there is "flicker fusion." If our cortex could perceive at frequencies above 40 Hz, we would see a series of static pictures instead of a movie. Our cortical circuits, therefore, are not sufficiently faithful to maintain gamma band (30–60 Hz) oscillations for any length of time. The fact that television works is evidence to that effect.

A circuit that has several intervening synapses, both excitatory and inhibitory, can maintain gamma band oscillations (at or above 40 Hz) only if cells in the circuit manifest subthreshold oscillations. These oscillations ensure that action potentials at each synapse will most likely fire at the peak of the oscillations. No portion of the circuit needs to fire at every peak, as long as some of the cells fire at every peak, thus maintaining the rhythm of the cell assembly. Without these membrane oscillations, any failure in a single synapse would alter the frequency of the resonance unpredictably. *Therefore, the synchronization provided by inhibitory signals and the timing provided by subthreshold oscillations are the keys to maintaining a high-frequency rhythm for any length of time in a circuit such as a cell assembly.* That is, it is the *coherence* provided by synchronized inhibition and the *frequency* response of subthreshold oscillations that give complexity, plasticity, and endurance to high-frequency brain rhythms.

COHERENCE AND FREQUENCY

The RAS controls sleep and waking and fight-or-flight responses. While this system provides signals that modulate our wake-sleep states, it also serves to help us respond to the world around us. For example, strong stimuli simultaneously activate ascending RAS projections to the thalamus and then the cortex and cause arousal and also activate descending projections that influence the spinal cord in the form of postural changes in tone resulting from the startle response, as well as trigger locomotor events in fight-or-flight responses. During sleep, the same system is responsible for the relative lack of ascending sensory awareness during SWS, as well as the descending atonia of REM sleep. This system also modulates the activity of virtually every other system in the brain. Growing evidence suggests that the control of sleep and waking is a fundamental property of neuronal networks and prior activity within each network (Kueger et al., 2008) and that intrinsic properties of neurons in multiple regions modulate sleep autoregulation, that is, suggesting that sleep is neither a passive nor an active phenomenon (Kumar, 2010). As we will see below, it is the RAS that supplies the "context" of sensory experience during waking.

Two major elements determining the activity of large assemblies of neurons such as in the EEG are coherence and frequency. Coherence is the term for how groups of neurons, firing in coordination, can create a signal that is mirrored instantaneously and precisely by other groups of neurons across the brain. These transient episodes of coherence across different parts of the brain may be an electrical signature of thought and actions. Our recent discovery demonstrated the presence of electrical coupling in three nuclei of the RAS, a mechanism that allows groups of neurons to fire synchronously. That mechanism is addressed in Chapter 4, which describes the presence of electrical coupling in the RAS and how that mechanism is modulated by the stimulant modafinil, which increases electrical coupling to drive coherence and disinhibits a number of systems to drive higher frequencies and induce arousal (Garcia-Rill et al., 2008). Briefly, modafinil increases electrical coupling, and since most coupled neurons in the RAS are GABAergic, the coupling decreases input resistance, decreasing activity and GABA release, thus disinhibiting many other cell types. This disinhibition leads to overall higher frequency in activity, that is, during sleep and arousal, in the RAS (Garcia-Rill et al., 2007, 2008; Heister et al., 2007) and thalamocortical systems (Urbano et al., 2007). In other words, because increased coupling in GABAergic neurons will lead to decreased GABA release, the tendency will be to increase coherence and also disinhibit most other transmitter systems, leading to increased excitation, especially during waking. That is, if modafinil increases electrical coupling, it should enable better coherence at all frequencies, during waking and even after its effects are waning, during sleeping. That is why modafinil is also useful in regulating coherence during sleep. Conversely, we will see that the most fast-acting anesthetics known, inhaled halothane and injected propofol, both block gap junctions and put us to sleep very rapidly. That is, the control of gap junctions can determine if we are asleep or awake.

The other face of large-scale activity is the frequency of firing, especially of ensemble activity, which is also essential to the neural encoding process. Chapter 9 is based on another major discovery, the presence of gamma band activity in the same RAS nuclei. Recent data suggest that many, perhaps all, of the neurons in these three RAS regions fire at gamma band frequency when maximally activated, but no higher. These results now suggest that brain stem regions not only can generate but also plateau at such frequencies, which is surprising because gamma band activity was first described in the cortex and is presumably involved in consciousness, learning, and memory. This is less surprising when one considers that gamma band activity has been described in other subcortical regions like the thalamus, hippocampus, basal ganglia, and cerebellum. The goal then becomes one to identify the mechanisms behind gamma band activity in the RAS and determine their function. In Chapter 9, we will first address the classical role of gamma band activity and the presence and mechanisms behind gamma band activity in the subcortical brain regions, then turn to the mechanisms behind gamma band activity in the RAS, and finally speculate on the potential role of such activity appearing at brain stem levels, in a very old, phylogenetically speaking, region such as the RAS.

CONTENT AND CONTEXT

If groups of neurons such as cell assemblies oscillate in phase or resonate, a global activity pattern results that provides the components for a transient construct of the external world (Llinas, 2001). This coherence in time is believed to form the neurological mechanism for perceptual binding, that is, to bring together the separate sensory components of a stimulus, say, color, size and motion, that are processed in different regions of the cortex. Not only does this process serve to provide binding of sensory events, but also there are coherent oscillations that subserve the precise temporal activation of muscles needed to execute a movement accurately. There is ample evidence for the coherent activation of neurons during sensory activation. For example, a visual stimulus produces coherent gamma band oscillations in the visual cortex (Eckhorn et al., 1988; Gray and Singer, 1989). These oscillations are evident in widely separated cortical regions but they are coherent in time and frequency. That is, they are ideally manifested to perform a binding function.

Both the cortex and the thalamocortical system are endowed with intrinsic membrane properties such as sodium-dependent subthreshold oscillations and high-threshold voltage-sensitive calcium channels, both of which can resonate at gamma frequencies (Llinas et al., 1998; Pedroarena and Llinás, 1997; Steriade, 1999). This means that sensory information arriving in an awake cortex and thalamus is superimposed on ongoing gamma band oscillations. Figure 1.2 illustrates a model for the temporal

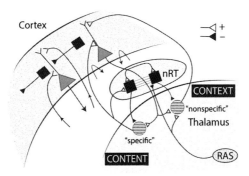

FIGURE 1.2 **Thalamocortical oscillations and the role of the RAS.** At the level of the thalamus, the "specific" thalamic nuclei receive the CONTENT of sensory experience that will be necessary for sensory discrimination. The "nonspecific" thalamus receives the CONTEXT of sensory experience mainly through RAS input, the arousal-related element essential to perception. At the level of the cortex, "specific" thalamic input terminates in layer IV, while the "nonspecific" thalamic input terminates in layers I and II. These coincident inputs summate to project back to the thalamus and thus maintain thalamocortical reverberation. This is facilitated by the continuous activation of the "nonspecific" thalamus by the RAS, along with RAS inhibition of the nRT. If RAS input to the thalamus is reduced, the electrically coupled GABAergic neurons (black squares with horizontal lines) are released. The coherent inhibition by the syncytium of electrically coupled nRT neurons induces slow inhibitory oscillations that lead to EEG slowing and ultimately sleep. At the level of the cortex, the lack of selective activation of cortical columns removes the lateral inhibition (black squares projecting laterally) necessary for sensory discrimination (after Llinas and Steriade, 2006). Excitatory inputs are designated by open triangles; inhibitory inputs by filled triangles.

binding of conscious experience developed by Rodolfo Llinas (Llinas, 2001). Basically, "specific" thalamic nuclei receive the "content" of sensory experience from primary sensory pathways and relay it to layer IV of the cortex, while "nonspecific" thalamic nuclei receive the "context" of sensory experience from the RAS and relay it to layer I of the cortex. Temporal coherence is due to the coincidence in firing of the two pathways that, when activated together, will summate to trigger cortical output to the thalamus. The cortical output flows back down to "specific" and "nonspecific" thalamic nuclei to set up a thalamocortical resonance.

The other components of this model include neurons in the reticular thalamic nucleus (nRT) that are GABAergic and electrically coupled. These cells provide feedback inhibition to both "specific" and "nonspecific" thalamic nuclei, which are themselves driven by input from these nuclei and by descending input from the cortex. When nRT neurons are active, they provide rhythmic inhibition to the circuit that results in the low-frequency oscillations of SWS (Steriade, 1999). However, when the RAS is active, it inhibits nRT cells, blocking slow oscillations, and excites the "nonspecific" intralaminar thalamus. This drives the "context" of sensory experience that will travel to layer I of the cortex. The coincidence of the "nonspecific" input to layer I and the "specific" thalamic input to layer IV of the cortex summates to drive pyramidal cells in the cortex to send descending projections back to the thalamus. For every thalamic afferent fiber to the cortex, 10 efferent fibers from the cortex will travel to the thalamus. That is, each cycle of resonance between the cortex and thalamus produces a tenfold amplification.

However, now that we determined that the RAS manifests gamma band frequencies in every one of its cells, afferent information arriving in the RAS is superimposed on an ongoing beehive of activity within its member nuclei. This activity summates with sensory tone and provides rich and powerful input to the intralaminar (nonspecific) thalamus, establishing the "context" onto which is superimposed the "content" of sensory experience (provided by primary afferent pathways and "specific" thalamic nuclei). Damage to the intralaminar thalamus that receives ascending RAS projections leads patients to ignore sensory inputs conveyed by the "specific" thalamic nuclei (Llinás and Paré, 1991). Although the input is received, the individual cannot perceive the afferent information. This suggests that the RAS and its projections to the "nonspecific" system are required to achieve binding, to place the "content" of sensory experience in its proper "context" (Llinas, 2001; Llinas and Steriade, 2006).

CORTICAL COLUMNS

How does the cortex process the "content" of sensory experience? Vernon Mountcastle first described the presence of vertically oriented cortical columns extending from the pial surface to the white matter.

These columns have connections that mainly extend vertically within the column and respond to a specific property of a limited sensory receptive field. Cells in a given column responded to cutaneous (skin and hair) or proprioceptive (joint and muscle) inputs in a common receptive field location. Mountcastle described cells extending through all cortical layers that had the same physiological properties compared with different properties in adjacent columns. He initially described sensation-specific columns in the somatosensory cortex (Mountcastle et al., 1957), but these were soon described in the visual cortex by David Hubel and Torsten Wiesel (Hubel and Wiesel, 1969). Neurons within a column are connected in the vertical direction, share inputs from other regions, and act as a functional unit. The columnar organization hypothesis is currently the most widely accepted concept to explain cortical signal processing. These vertical processing units are segregated from adjacent columns by laterally projecting inhibitory connections. Lateral inhibition is denoted in Figure 1.2 as GABAergic projections laterally to adjacent columns. This mechanism provides lateral inhibition to the optimal stimulus, decreasing activity in adjoining columns to increase contrast (Helmstaedter et al., 2009).

The concept of cortical columns as the building block of cortical function is not without its doubters (Horton and Adams, 2005), with alternative explanations for such a fixed location in the cortex but one that involves the cortex as a spatiotemporal map, rather than multiple independent columns. The facts that some species have, for example, ocular dominance columns, but other species do not and that, even within species, some members have ocular dominance columns but others do not call for a reassessment of the theory. Admittedly, the less rigid idea of a flexible spatiotemporal map is more harmonious with the idea of traveling waves of coherent oscillations determining the principles behind cortical processing.

Regardless of the role of cortical columns, what does the "context" add to the "content" of sensory experience? There has been little work addressing what we believe is the most critical feature of sensory experience, the "context" under which we assess a stimulus. In general, stimulation of the RAS potentiates sensory stimulus-induced responses, in particular increasing lateral inhibition and focusing the responses to specific sensory inputs, whether visual, auditory, somatosensory, etc. (Castro-Alamancos, 2002; Singer, 1977; Steriade, 1969). Studies on awake animals, as well as on anesthetized animals in which the RAS is stimulated, demonstrate focusing of sensory inputs (Castro-Alamancos and Oldford, 2002). These results suggest that the background of activity provided by arousal via the RAS embodies a mechanism for improving spatial processing of sensory input. This provides a more discrete representation of the external world, with cortical representations and receptive fields becoming more focused.

THALAMOCORTICAL RESONANCE

How long does it take us to perceive a sensory stimulus? The answer is about 200 ms or more. Studies by Benjamin Libet used stimulation of the somatosensory cortex in patients undergoing neurosurgery for clinical conditions. Brief stimuli to the periphery, for example, the thumb, induced a response in the corresponding region of the contralateral somatosensory cortex in 100 ms or less. At this latency, there was no conscious sensation. However, stimulation of the same region (thumb representation) of the cortex had to be applied for 500 ms in order for the patient to report a similar sensation as that following peripheral stimulation (Libet, 2004; Libet et al., 1964). These results suggested that sensory input had to persist for a prolonged period of time before it could be consciously perceived. Much more recent studies used threshold visual stimuli and measured the neuromagnetic signals elicited by perceptual events. The magnetic correlate of stimulus perception was in the order of 240 ms (Sekar et al., 2013). The earliest evoked response at the level of the cortex was in the order of 100 ms, but did not indicate conscious visual experience. Such early responses have been related to "unconscious encoding of stimulus features" (Heekeren et al., 2004); however, as we will see in Chapter 10, we take issue with the use of the term "unconscious." We suggest that these early responses represent "preconscious" detection of stimuli that will be used to build "context" for the perception of the stimulus. In any case, the implication is that a sensory event leaves a trail or, more specifically, a recurring signal, a "ringing," that must persist in order for it to ultimately be perceived. The mechanism proposed to carry out this task is thalam ocortical resonance. The idea is that, once the initial cortical response is elicited, the cortex projects back to the thalamus that, in turn, projects back to the cortex in a recurrent loop, that is, the stimulus sets up a resonance between the cortex and thalamus that lasts until the stimulus is perceived (Llinas, 2001).

What are the indicators of such a resonance? We can record evoked responses following sensory stimulation by placing electrodes on the scalp and averaging responses following repetitive stimulation. An auditory stimulus, for example, elicits a brain stem auditory evoked response (BAER) that occurs during the first 10 ms following a simple click stimulus. However, the first indicator of a cortical response is observed at the primary auditory cortex in the posterior temporal lobe, the Pa wave, which has a latency of 25 ms. This is followed by the P50 potential that is localized over the vertex at a latency of 50 ms. We will discuss the P50 potential, which is sleep state-dependent and a measure of RAS output through the intralaminar thalamus, in Chapter 6. There follows a negative shift, the N1 response; another positive wave, the P2 response; and then the N2 response. That is, there are a series of positive and negative waves recorded over the scalp, as if the stimulus induced a bouncing waveform that lasts hundreds of milliseconds. It is this rhythmic oscillation that is thought to represent thalamocortical resonance. It is not clear if the peaks of the responses represent a return to the same cortical region with each "bounce," but it is likely that the resonance engages different cortical regions. In any case, thalamocortical resonance is thought to be necessary for conscious perception.

What if the timing of the "specific" and "nonspecific" pathways is not coincident? That is, what if the timing of the "content" versus that of the "context" of sensory experience is altered? We know that eliminating arousal (context) altogether prevents the appreciation of sensory events, but what happens if only the timing between the primary or "specific" afferent pathways and the reticular or "nonspecific" inputs does not summate?

CLINICAL IMPLICATIONS: THALAMOCORTICAL DYSRHYTHMIA

Why do we need to consider both the "content" and "context" of conscious perception? Because that may be the only way to help those with a host of neurological and psychiatric disorders that involve distortions in conscious perception. If either the "specific" or "nonspecific" thalamus is overinhibited or undergoes deafferentation, one side of the two ascending pathways will be slowed. Such a decrement in thalamic input can occur after peripheral damage that reduces afferent input or similar conditions that results in a decrease in the level of input to the thalamus. As we will see in Chapter 2, thalamic relay neurons in the "specific" thalamic nuclei have a bistable mode of firing, either tonically when depolarized or in bursts when hyperpolarized. The latter is due to the presence of low-threshold spike (LTS) calcium currents in these neurons. If there is lasting deafferentation or increased inhibition, as the model proposes, the cells will express additional LTS bursting, which will slow their firing (Llinas et al., 2005). The slowed firing generates a slower frequency, in the theta band, in one of the arms of the thalamocortical projection system. At rest, one arm will undergo resonance at alpha or ~10 Hz frequency, while the other arm will resonate at slower theta or ~8 Hz frequency. This creates an imbalance between the two pathways and leads to thalamocortical dysrhythmia (TCD). The slowing may also occur due to a decrease in the flow of information through the "nonspecific" pathway, in which intralaminar thalamic neurons also begin expressing additional LTS channels, slowing their oscillations.

The consequences of such dissonance in this circuit are considerable. Figure 1.2 can be used to illustrate the effects of TCD. The lack of input to one arm of the cortical column on the right reduces the perceived responses, causing negative symptoms due to reduced oscillation frequency of one of the afferent pathways. On the other hand, the decrease in lateral inhibition (due to lack of activation of cortical inhibitory cells) will induce aberrant high-frequency oscillations, leading to misperceptions in adjacent columns (left side) that induce positive symptoms. We will discuss positive and negative symptoms in regard to neurological and psychiatric disorders in Chapters 11 and 12. That is, the same mechanism that is responsible for consciousness can generate abnormal neurological and psychiatric events when timing is disrupted (Llinas et al., 2005).

Clinically, the presence of TCD can be diagnosed and treated. Neuromagnetic recordings can detect the effects of TCD on thalamocortical resonance as slowing during eyes-closed recordings at rest peaks at 8 Hz rather than at 10 Hz (Jeanmonod et al., 2001). Once the condition is diagnosed, microelectrode recordings in the intralaminar thalamus have revealed excessive numbers of neurons manifesting high-frequency bursts attributed to LTS currents. Electrolytic lesions have been placed in a series of patients diagnosed with TCD and showing abundant LTS-like bursting. Patients with neurogenic pain, epilepsy, obsessive-compulsive disorder, major depression, tinnitus, dystonia, spasticity, and Parkinson's disease have all been diagnosed, recorded, and surgically treated (Jeanmonod et al., 2001). The lesions, mainly located in the centrolateral region of the intralaminar thalamus, lead to a speeding up of neuromagnetic recordings to the normal alpha frequency (~10 Hz), along with resolution of symptoms. These highly selected patients have been followed for only short periods (1–5 years), their numbers are still rather low, and their postsurgical state is psychiatrically fragile. Nevertheless, some success in the resolution of symptoms has been reported.

The question is why does TCD occur in so many conditions? It should be remembered that certain medications can be used successfully to treat more than one symptom, for example, carbamazepine can be used for psychiatric disorders, pain, and epilepsy. The implication of TCD is that in each disorder, there is decreased thalamic input due to deafferentation or overinhibition of the thalamus, leading to excessive expression of LTS channels, which induces thalamic slowing due to bursting instead of tonic firing. In tinnitus, hair cell damage presumably deafferents the thalamus and, in some individuals, induces LTS overexpression over time. This generates "phantom sound." In neurogenic pain, the original painful injury resolves, but dissonance is induced by additional bursting that leads to the "manufacture" of pain signals by the erroneous signaling, that is, leads to neurogenic, or brain-generated, pain. Similar mechanisms have been proposed for the other disorders that lead to TCD in at least some patients with those disorders. The key is to determine using neuromagnetic recordings if the individual manifests TCD, but if not, then the cause of the symptoms may lie elsewhere. The use of neuromagnetic recordings can identify patients who do not have TCD to good advantage. For example, a patient with chronic pain who exhibits TCD would not be a good candidate for epidural spinal cord implantation of a TENS unit for pain relief. That is, the "pain" is probably centrally (thalamic) generated and spinal cord stimulation may have little effect. However, if the patient does not have TCD, then the source of the pain may be in the spinal cord, so that the success rate of TENS unit therapy (implantation of epidural spinal cord stimulator) could improve significantly.

The RAS, then, provides the "context" for our daily lives, the background against which we evaluate the world around us, and the repository of essential information we will need to exert our will on the world. If the "context" is distorted, our perceptions are skewed; if our reference system is unreliable, our world is off kilter; and if the stream of preconscious information we rely on is altered, we cannot count on the most basic foundations for making decisions or imposing our will. It is no wonder that virtually all psychiatric and neurological disorders involve dysregulation of wake-sleep control and the RAS. The following chapters will provide us with novel information that suggests that the RAS does not simply control waking and sleep, but rules the "context" of our lives. That is why dysregulation of the RAS is intrinsic to so many neurological and psychiatric disorders and why we must address its regulation if we are to make an impact on these devastating diseases.

References

Buzsaki, G., 2006. Rhythms of the Brain. Oxford University Press, Oxford.

Castro-Alamancos, M.A., 2002. Role of thalamocortical sensory suppression during arousal: focusing sensory inputs in neocortex. J. Neurosci. 22, 9651–9655.

Castro-Alamancos, M.A., Oldford, E., 2002. Cortical sensory suppression during arousal is due to the activity-dependent depression of thalamocortical synapses. J. Physiol. 541, 319–331.

Eckhorn, R., Bauer, R., Jordan, W., Brosch, M., Kruse, W., Munk, M., Reitboeck, H.J., 1988. Coherent oscillations: a mechanism of feature linking in the visual system? Biol. Cybern. 60, 121–130.

Garcia-Rill, E., Heister, D.S., Ye, M., Charlesworth, A., Hayar, A., 2007. Electrical coupling: novel mechanism for sleep-wake control. Sleep 30, 1405–1414.

Garcia-Rill, E., Charlesworth, A., Heister, D., Ye, M., Hayar, A., 2008. The developmental decrease in REM sleep: the role of transmitters and electrical coupling. Sleep 31, 673–690.

Gray, C.M., Singer, W., 1989. Stimulus-specific neuronal oscillations in orientation columns of cat visual cortex. Proc. Natl. Acad. Sci. USA 86, 1698–1702.

Hebb, D.O., 1949. The Organization of Behavior: A Neuropsychological Theory. Lawrence Erlbaum Associates, London, 335 pp.

Heekeren, H.R., Marrett, S., Bandettini, P.A., Ungerleider, L.G., 2004. A general mechanism for perceptual decision-making in the human brain. Nature 431, 859–862.

Heister, D.S., Hayar, A., Charlesworth, A., Yates, C., Zhou, Y., Garcia-Rill, E., 2007. Evidence for electrical coupling in the SubCoeruleus (SubC) nucleus. J. Neurophysiol. 97, 3142–3147.

Helmstaedter, M., Sakmann, B., Feldmeyer, D., 2009. Neuronal correlates of local, lateral, and translaminar inhibition with reference to cortical columns. Cereb. Cortex 19, 926–937.

Horton, J.C., Adams, D.L., 2005. The cortical column: a structure without a function. Philos. Trans. R. Soc. B 360, 837–862.

Hubel, D.H., Wiesel, T.N., 1969. Anatomical demonstration of columns in the monkey striate cortex. Nature 221, 747–750.

Jeanmonod, D., Magnon, M., Morel, A., Siegemund, M., Cancro, A., Lanz, M., Llinas, R., Ribary, U., Kronberg, E., Schulman, J., Zonenshayn, M., 2001. Thalamocortical dysrhythmia II. Clinical and surgical aspects. Thalamus Relat. Syst. 1, 245–254.

Kueger, J.M., Rector, D.M., Roy, S., van Dongen, H.P., Belenky, G., Panksepp, J., 2008. Sleep as a fundamental property of neuronal assemblies. Nat. Rev. Neurosci. 9, 910–919.

Kumar, V.M., 2010. Sleep is neither a passive nor an active phenomenon. Sleep Biol. Rhythms 8, 163–169.

Libet, B., 2004. Mind Time. The Temporal Factor in Consciousness. Harvard University Press, Cambridge, MA, 248 pp.

Libet, B., Alberts, W.W., Wright, E.W., Delattre, L.D., Levin, G., Feinstein, B., 1964. Production of threshold levels of conscious sensation by electrical stimulation of human somatosensory cortex. J. Neurophysiol. 27, 546–578.

Llinas, R.R., 2001. I of the Vortex, From Neurons to Self. MIT Press, Cambridge, 303 pp.

Llinás, R.R., Paré, D., 1991. Of dreaming and wakefulness. Neuroscience 44, 521–535.

Llinas, R.R., Steriade, M., 2006. Bursting of thalamic neurons and states of vigilance. J. Neurophysiol. 95, 3297–3308.

Llinas, R.R., Ribary, U., Contreras, D., Pedroarena, C., 1998. The neuronal basis for consciousness. Philos. Trans. R. Soc. Lond. B Biol. Sci. 353, 1841–1849.

Llinas, R.R., Urbano, F.J., Leznik, E., Ramirez, R.R., van Marle, H.J.f., 2005. Rhythmic and dysrhythmic thalamocortical dynamics: GABA systems and the edge effect. Trends Neurosci. 28, 325–333.

Mountcastle, V.B., Davies, P.W., Berman, A.L., 1957. Response properties of neurons of cats somatic sensory cortex to peripheral stimuli. J. Neurophysiol. 20, 374–407.

Pedroarena, C., Llinás, R.R., 1997. Dendritic calcium conductances generate high-frequency oscillation in thalamocortical neurons. Proc. Natl. Acad. Sci. 94, 724–728.

Schenck, C.H., Bundle, S.R., Patterson, A.L., Mahowald, M.W., 1987. Rapid eye movement sleep behavior disorder—a treatable parasomnia affecting older adults. J. Am. Med. Assoc. 257, 1786–1789.

Sekar, K., Findley, W.M., Poeppel, D., Llinas, R.R., 2013. Cortical response tracking the conscious experience of threshold duration visual stimuli indicates visual perception is all or none. Proc. Natl. Acad. Sci. USA 110, 5642–5647.

Singer, W., 1977. Control of thalamic transmission by corticofugal and ascending reticular pathways in the visual system. Physiol. Rev. 57, 386–420.

Steriade, M., 1969. Alteration of motor and somesthetic thalamo-cortical responsiveness during wakefulness and sleep. Electroencephalogr. Clin. Neurophysiol. 26, 334.

Steriade, M., 1999. Cellular substrates of oscillations in corticothalamic systems during states of vigilance. In: Lydic, R., Baghdoyan, H.A. (Eds.), Handbook of Behavioral State Control. Cellular and Molecular Mechanisms. CRC Press, New York, pp. 327–347.

Urbano, F.J., Leznik, E., Llinas, R., 2007. Modafinil enhances thalamocortical activity by increasing neuronal electrotonic coupling. Proc. Natl. Acad. Sci. USA 104, 12554–12559.

2

The EEG and the Discovery of the RAS

Edgar Garcia-Rill, PhD and Christen Simon, PhD†*

*Center for Translational Neuroscience, Department of Neurobiology and Developmental Sciences, University of Arkansas for Medical Sciences, Little Rock, AR, USA
†Scientist, Study Director, Northern Biomedical Research Inc., Spring Lake, MI, USA

THE EEG AND THE RAS

Sleep and dreaming have been studied for thousands of years, notably by the Egyptians and Greeks. In 350 BCE., Aristotle wrote "On Sleep and Sleeplessness," where he stated that an awake person is exercising "sense-perception," and "all organs which have a natural function must lose power when they work beyond the natural time-limit of their working period." Sleep occurs when the sense-perception special organ "continues beyond the appointed time-limit of its continuous working period, it will lose its power and will do its work no longer." Therefore, sleep was essentially considered to be a passive event, a turning off of the "sense-perception" organ due to exhaustion (Kirsch, 2011).

Modern sleep science began with the invention of the electroencephalogram (EEG) in 1924 by Hans Berger, whose motivation for developing the technique was the study of telepathy (see a wonderful description by Buzsaki, 2006). Later, the physiologist Frédéric Bremer was one of the first to measure EEG activity following transections of brain stem regions (Figure 2.1). Transection of the brain stem at the junction of the pons and midbrain (*cerveau isolé*) in the cat anterior to the reticular activating system (RAS) produced a state with cortical high-amplitude, low-frequency EEG patterns similar to that in slow-wave sleep (SWS) (T3 in Figure 2.1). However, transection of the brain stem just rostral to the spinal cord (*encéphale isolé*) posterior to the RAS produced fluctuating EEG patterns, with spontaneous patterns of low-frequency, high-amplitude activity (as

http://dx.doi.org/10.1016/B978-0-12-801385-4.00002-1

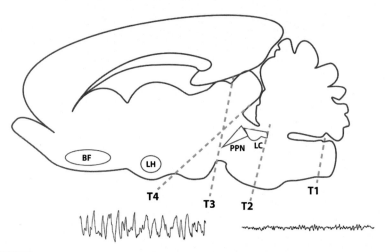

FIGURE 2.1   **Effects of acute brain stem transections on the EEG.** T1. Transections at the medullary-spinal junction allowed the manifestation of high-frequency, low-amplitude cortical EEG immediately after transection. T2. Transections at the midpontine level also permitted the manifestation of high-frequency, low-amplitude EEG activity in the EEG. T3. Midcollicular posthypothalamic transections blocked the manifestation of high-frequency activity, leading to high-amplitude, low-frequency EEG. This occurred despite the fact that this transection left the hypothalamus and basal forebrain connected to the thalamus and cortex. T4. Postcollicular prehypothalamic transections leaving the basal forebrain connected to anterior structures also failed to maintain low-amplitude, high-frequency activity. BF, basal forebrain; LC, locus coeruleus; LH, lateral hypothalamus; PPN, pedunculopontine nucleus.

observed during SWS) as well as high-frequency, low-amplitude activity (as observed during waking and rapid eye movement (REM) sleep) (T1 in Figure 2.1) (Bremer, 1973; Kerkhofs and Lavie, 2000). Therefore, the region of the RAS was considered critical for the induction and maintenance of REM sleep and the high-frequency cortical activity that is required for vigilance and consciousness during waking (Gottesmann, 1988).

The Italian Giuseppe Moruzzi and the American Horace Magoun also greatly contributed to waking and sleep research through the use of the EEG. These scientists discovered that stimulation of the brain stem reticular formation abolished low-frequency activity and induced high-frequency activity in the cortical EEG (Moruzzi and Magoun, 1949). This high-frequency activity was identical to that observed by Bremer in the *encéphale isolé* preparation. Moruzzi and Magoun concluded, "the possibility is considered that a background maintained activity within the ascending brain stem activating system may account for wakefulness, while reduction of its activity either naturally, by barbiturates, or by experimental injury and disease, may respectively precipitate normal sleep, contribute to anesthesia or produce pathological somnolence" (Moruzzi and Magoun, 1949).

Further transection studies concluded that the "maintained influence of the ascending brain stem activating system underlies wakefulness, while absence of this influence precipitates sleep" (Lindsley et al., 1949). In later studies, Moruzzi transected the brain stem at the pontomidbrain junction, a few millimeters caudal to the original *cerveau isolé* preparation. These transections at midpontine pretrigeminal levels produced spontaneous EEG patterns and eye movements, like those observed for the *encéphale isolé* preparation (T2 in Figure 2.1) (Moruzzi, 1972). Similar midpontine transections were performed by Steriade that led to waking EEG signs, while postcollicular-premammillary transections (T4 in Figure 2.1) led to SWS EEG (Steriade et al., 1969). Therefore, nuclei near the pons-midbrain junction were implicated in the generation of high-frequency EEG patterns.

Nathaniel Kleitman was a prominent sleep researcher at the University of Chicago and, along with two of his graduate students Eugene Aserinsky and William Dement, correlated dreams, increased respiration, heart rate, and eye movements to high-frequency EEG patterns during REM sleep (Aserinsky and Kleitman, 1953, 1955). They also proposed the dual nature of sleep: REM sleep is a completely different state than SWS, even though they both occur while asleep. Michel Jouvet expanded on these results to show that REM sleep, termed "paradoxical sleep," is accompanied by muscle atonia and rostropontine transections preserved muscle atonia during REM sleep (Jouvet, 1962).

The above studies led to the formulation of three distinct arousal states in the human and other mammals: waking, SWS, and REM sleep. Furthermore, they established the importance of the brain stem reticular formation in generating these states. However, later studies were required to more carefully analyze the role of specific brain stem nuclei in modulating wake–sleep states. It should be noted that the transections used in the above studies assessed EEG features soon after the lesion, that is, acutely. However, when animals were transected and maintained for days or weeks after the transection, the Chilean neuroscientist Jaime Villablanca found that prethalamic transections disconnected the RAS from the forebrain and cats did indeed ultimately recover some waking-like EEG activity (Villablanca, 2004). These results suggest that partial recovery of low-amplitude, high-frequency activity can take place in the forebrain but after chronic transections. Villablanca, however, did conclude that "true waking behavior depended on the cholinergic reticular core" (Villablanca, 2004). In Chapter 3, we will discuss some of the regions anterior to the RAS that could be mediating the recovery of some of the fast activity after such chronic lesions.

LIMITATIONS OF THE EEG

Gold cup electrodes with conducting paste are applied to the scalp in a designated system to record the EEG. The electrodes pick up electrical signals from the brain, which are somewhat distorted by the intervening bone and hair. The signals originate from thousands of neurons in the underlying cortex in which the largest cells, the pyramidal cells, are arranged radially and have apical dendrites extending to the most superficial layers with axons extending into the white matter. Since the cortex has a columnar organization, most cells are oriented perpendicularly to the surface, making current flow calculations simple for the gyri, although the presence of sulci creates a complex problem for calculating current flow. In general, the activity of as many as 500,000 neurons over a range of 3–5 mm may be measured by a single electrode. EEG amplifiers, however, typically measure activity that is filtered. The typical high-pass filter settings are at 1 Hz and in some cases 0.1 Hz. This eliminates very slow brain activity and drift in the electrodes. The typical low-pass settings are at 70 Hz and in some cases as high as 200 Hz. This eliminates high-frequency activity. That means that the EEG amplifier has inappropriate band-pass filters for detecting events as fast as action potentials, which occur in the 1–2 ms range (requiring band pass >1000 Hz).

On the other hand, EEG amplifiers more faithfully reflect the activity of slow or graded potentials such as dendritic postsynaptic potentials, which occur in the 10–20 m range. That is, the EEG is a measure of dendritic potentials and membrane oscillations, not of action potential frequencies (Murakami and Okada, 2006). That means that EEG recordings exhibit high-amplitude, low-frequency waveforms when large ensembles of cells are receiving graded potentials rather than separately, as opposed to when low-amplitude, high-frequency activity is evident in the EEG. The simplest way to think about high-amplitude, low-frequency waves is that large groups of cortical columns are receiving postsynaptic potentials and manifesting membrane oscillations in unison. This may also mean that when groups of cortical columns are active simultaneously, they have little, if any, lateral inhibition separating the activity of individual columns. On the other hand, low-amplitude, high-frequency EEG recordings reflect the activity of many different columns acting independently. This suggests that the lack of sensory perception seen during slower rhythms is a reflection of a lack of individuality of columnar organization and thus a lack of sensory perception. As the number of independent frequencies increases, more localized activity generates lower-amplitude waves, and the higher EEG frequencies suggest disparate activity patterns across these smaller groups of cells exercising lateral inhibition and, of course, sensation.

In Chapter 8, we will discuss the full range of waking EEG frequencies (alpha, beta, and gamma), but the EEG is used mostly for the determination of sleep characteristics in lower-frequency ranges (theta and delta) and especially for the detection of epileptic foci that seize in large ensembles in unison. Based on EEG recordings, sleep has been divided into nonrapid eye movement (NREM) and REM sleep. NREM sleep has been divided into stages I–IV, which mark progressively deeper stages of sleep.

That is, stage I is characterized by theta activity (4–8 Hz), and stage II by sleep spindles and K-complexes. The deeper stages represent true SWS with delta waves (1–4 Hz) of increasing amplitude. Muscle tone decreases and rolling eye movements are typical of SWS, but the transition to REM sleep is marked by further reduction of muscle tone (although myoclonic twitches occur in bursts) and by saccadic eye movements that also occur in bursts (REMs). Despite the onset of low-amplitude, high-frequency activity that led to referring REM sleep as "active" or "paradoxical" sleep, the arousal threshold is increased, making this the most difficult stage from which to awaken the individual.

EARLY THEORIES OF SLEEP

The early theory about sleep was that it occurred passively (Bremer, 1973; Moruzzi and Magoun, 1949). In fact, Kleitman was a proponent of the passive theory of sleep, suggesting that what needs to be explained is not sleep, but rather wakefulness (Kleitman, 1953). With the description of REM sleep and "paradoxical" sleep arose the suggestion that there were mechanisms for the active generation of sleep stages (Aserinsky and Kleitman, 1953; Jouvet, 1962). As we will see in Chapter 3, the presence of a number of regions that can modulate sleep and waking gave rise to the concept of a distributed system in charge of sleep control. The idea that sleep is controlled by a single region has been questioned based on these results (Kumar, 2012). Moreover, the facts that (a) single lesions cannot completely eliminate sleep, (b) single transmitter systems do not regulate sleep, and (c) sleep is integrated with a number of other regulatory mechanisms such as appetite, circadian cycles, and temperature control have prompted the suggestion that sleep is an autoregulatory global phenomenon, neither passive nor active (Kumar, 2010). The idea is that sleep is determined by cell assemblies with slow oscillating membrane properties, and it is the algebraic summation of multiple factors that leads to sleep. It is obvious that sleep, in particular SWS, is essential and probably has a number of functions; however, regardless of the mechanism(s) behind the generation of sleep, the slowing of EEG frequencies brings little in the way of perception, attention, volition, or the faculties that mark our waking state. The factors influencing sleep are multiple and probably local and global, making assignation to a single region, transmitter, or event difficult.

To complicate matters, there are transitions between waking, SWS, and REM sleep. The beginning of sleep takes time, in the order of minutes, that is, it is not instantaneous. Moreover, transitions between stages within SWS are not immediate, but can take seconds to minutes. The beginning of REM sleep is also not immediate but appears to be recruited by increasing bursts of ponto-geniculo-occipital (PGO) waves (Steriade, 1999). In the absence of strong stimuli that induce waking, therefore, we can say that waking and sleep are "recruited" and are part of a stepwise process, in which frequencies decrease from gamma, to beta, to alpha and lower or increase from delta, to theta, to alpha and higher. We will discuss this process, in which 10 Hz or alpha waves mark the fulcrum between waking and sleep, in Chapter 9. This stepwise process also suggests that the modulation of waking and sleep is progressive but piecemeal and that it is the coherence at particular frequencies that determines the particular transition state. That is, the new state may be "recruited" to another level rather than gradually formed. In Chapter 1, we dealt with two major determinants governing overall brain function, especially waking, in coherence and frequency. We saw that it is the coherence provided by synchronized inhibition and the frequency response of subthreshold oscillations that give complexity, plasticity, and endurance to high-frequency brain rhythms. Since our concern here is the control of waking and not sleep, we will concentrate on the region that Moruzzi and Magoun (Moruzzi and Magoun, 1949) first identified as part of the "ascending brain stem activating system." The effective regions that induced changes in the EEG were located in the bulbar reticular formation, the pontine and mesencephalic tegmentum, and the dorsal hypothalamus and subthalamus. The large area described, along with the potential for stimulating fibers of passage, all generated considerable debate.

Considerable time elapsed before studies that showed specific neurons with correlated activity to states of vigilance and behavior were performed. Steriade first described the firing patterns in relation to EEG arousal of reticular formation neurons that were located in the "central tegmental field and dorsal and caudal extension of the cuneiform nucleus" (the lateral but not the medial portion of the cuneiform nucleus) (Steriade et al., 1980). Increased firing in this region during waking and REM sleep was described by multiple groups (Sakai et al., 1990; Steriade et al., 1990). This region was later identified as the mesopontine cholinergic pedunculopontine tegmental nucleus (PPN) (Steriade et al., 1991), whose dorsal and posterior *pars compacta* lies embedded in the lateral portion of the cuneiform nucleus. In Chapter 7, we will describe in depth the relationship between these nuclei and electrically induced locomotion on a treadmill. Stimulation of the lateral, but not the medial cuneiform nucleus, in which the PPN *pars compacta* is embedded, elicited stepping movements in the decerebrate animal after 1–2 s of stimulation at 40–60 Hz (Garcia-Rill, 1991). The physiological bases for the appropriateness of these very specific stimulation parameters are discussed in Chapter 9. These studies demonstrated the link between the PPN and postural and locomotor control mechanisms. In Chapter 12, we will describe studies using deep brain stimulation of the PPN in the human to alleviate some of the symptoms of Parkinson's disease. In terms of waking and sleep, however, despite the fact that the locus coeruleus and the raphe nuclei are often considered part of the RAS, it is the PPN that shows firing patterns in relation to both waking and REM sleep, that is, the two states of low-amplitude, high-frequency EEG activity. The locus coeruleus and raphe nuclei are active during waking, are less active during SWS, and show the least activity during REM sleep (Hobson et al., 1975; Steriade and McCarley, 2005). Thus, they are not tightly related to the two states of arousal. Therefore, the PPN has become the critical cell group related to arousal mechanisms and the main candidate for the waking and REM sleep driver of the RAS.

THE PEDUNCULOPONTINE NUCLEUS

A detailed review of the PPN and its anatomy, morphology, anatomical projections, inputs, electrophysiological properties, membrane receptors, and role in arousal is available and will not be replicated here (Reese et al., 1995). Some of this information is provided in different chapters, but a brief description of its location and structure is called for. The PPN in the human brain stem was first described in 1954 by Olszewski and Baxter and has since been studied by many other investigators. In the human, the PPN is bounded on its lateral and medial sides by fibers of the medial lemniscus and the superior cerebellar peduncle (SCP), respectively (Geula et al., 1993; Olszewski and Baxter, 1982). Rostrally, the anterior aspect of the PPN contacts the posterolateral substantia nigra (SN) (Garcia-Rill and Skinner, 1991), while the most dorsal aspect of the PPN is within the lateral portion of the cuneiform and subcuneiform nuclei. Two subdivisions of the PPN have been characterized on the basis of cell density (Geula et al., 1993; Olszewski and Baxter, 1982). The *pars compacta* of the PPN is located within the caudal half of the nucleus in its dorsolateral aspect and embedded in the lateral portion of the cuneiform nucleus. Cells of the *pars dissipata* are distributed sparsely within the SCP and central tegmental field. This distribution coincides with the locations of effective stimulation sites for inducing fast EEG rhythms and of cells related to arousal (Moruzzi and Magoun, 1949; Steriade et al., 1980).

Anatomical studies used choline acetyltransferase (ChAT) immunocytochemical labeling or nicotinamide adenine dinucleotide phosphate (NADPH)-diaphorase histochemical labeling to describe cholinergic PPN neurons in the human, monkey, cat, and rat (Armstrong et al., 1983; Geula et al., 1993; Jones and Beaudet, 1987; Jones, 1990; Lavoie and Parent, 1994; Mesulam et al., 1984, 1989; Mizukawa et al., 1986; Nakamura et al., 1988; Rye et al., 1987; Satoh and Fibiger, 1985; Shiromani et al., 1988; Skinner et al., 1989; Sofroniew et al., 1985; Spann and Grofova, 1992; Sugimoto et al., 1984). The majority of these groups described cholinergic neurons within the PPN as being medium to large in size with an average diameter of 20–30 μm regardless of species, although a few investigators described cholinergic neurons with diameters as high as 60–80 μm (Mesulam et al., 1984: monkey; Spann and Grofova, 1992: rat) or as low as 15 μm (Jones, 1990: rat) within the region of the PPN. On average, the area of these cells is ~300 μm² (Skinner et al., 1989). The number of cholinergic neurons found within the PPN has been estimated from 1000 to 2500 in the rat (Jones, 1990; Rye et al., 1987), 6000 to 8500 in the cat (Jones and Beaudet, 1987), and 10,000 to 20,000 in the human (Garcia-Rill et al., 1995).

PPN neurons are generally fusiform in shape, and the nucleus extends posteroanteriorly in a wedge shape, making sagittal sections the best suited for showing the distribution of cells and their dendritic arbors. Figure 2.2 is an example of a sagittal section through the rat PPN showing cholinergic cells labeled using NADPH diaphorase. In the anterior portions of the nucleus, the cells are structurally influenced into a fusiform shape by the fibers of the SCP. Posteriorly, the neurons are located dorsal to the SCP and appear rounder, although unlike bushy interneurons in the thalamus and striatum. Depending on their location, individual PPN neurons may have 2–9 primary dendrites that extend for hundreds of microns (Reese et al., 1995). The anatomical relationships are evident in Figure 2.2, emphasizing the fact that the *pars compacta* is located ventrally to the inferior colliculus and the anterior portion of the nucleus is located ventral to the transition between inferior and superior colliculi.

Figure 2.3 contains three-dimensional reconstructions of the PPN and other nuclei in the rat and human brains. In Figure 2.3a, sagittal sections

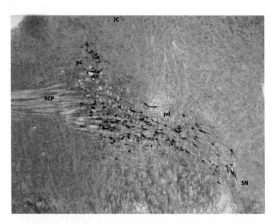

FIGURE 2.2 **The PPN in sagittal sections.** Histochemical NADPH-diaphorase labeling of only the cholinergic neurons in the PPN revealed the wedge-shaped structure of this nucleus. Posterodorsally is located the *pars compacta* (pc) of the PPN, which is located within the lateral cuneiform nucleus ventral to the inferior colliculus (IC). As the nucleus descends, its cells are intermixed with the fibers of the SCP. The body or *pars dissipata* (pd) of the PPN is located more ventrally and extends to the posterior edge of the SN. Anterior is to the right.

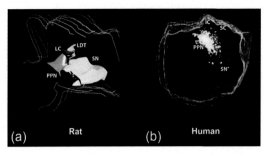

FIGURE 2.3   **(a) Three-dimensional reconstruction of four nuclei in the rat brain.** View of sagittal sections through the brain stem of the rat showing the regions encompassed by cells of the locus coeruleus (LC) in red, the laterodorsal tegmental nucleus (LDT) in green, the PPN in brown, and the SN in yellow. Note the blending of LDT with medial LC cells and of lateral LC cells with the PPN, as well as the penetration of PPN at the posterior edge of the SN. Anterior is to the right. (b) Three-dimensional reconstruction of the PPN in the human brain. Sagittal sections of the human brain stem showing the inferior colliculus (IC) and the superior colliculus (SC) dorsally and the SN anteroventrally to the PPN. Note the locations of PPN cells ventral to the IC descending toward the posterior edge of the SN. The *pars compacta* of the PPN is located posterodorsally to the *pars dissipata*. Anterior is to the right.

of the rat mesopontine region show the inferior and superior colliculi dorsally and the basis pontis ventrally. The cells of the SN (yellow), PPN (brown), locus coeruleus (red), and LDT (green) are shown as solid structures made by tiling between cells in each nucleus. The wedge shape of the PPN is evident extending from the posterior *pars compacta* to the caudal edge of the SN. Its location in relation to other critical nuclei in the region shows that the medial edge of the PPN (brown) blends into the locus coeruleus (red), which in turn blends with the lateral edge of LDT (yellow). In Figure 2.3b, sagittal sections of the human mesopontine region were reconstructed showing PPN neurons as yellow dots. Note the general wedge shape in the human as well, with the posterodorsal *pars compacta* ventral to the inferior colliculus and extending anteriorly and ventrally toward the posterior boundaries of the SN.

The relationship between the PPN and the LDT is illustrated in Figure 2.4. This semihorizontal section through the long axis of the PPN was labeled using NADPH diaphorase. The cholinergic neurons of the PPN are shown laterally curving toward its *pars compacta* posterodorsally. The LDT can be seen within the central gray medially. Between the PPN and the LDT is the locus coeruleus whose cells are not labeled. However, scattered cholinergic cells are evident between the PPN and the LDT and would be located within the boundaries of the locus coeruleus. This viewpoint demonstrates that (a) the PPN is well separated from the LDT, (b) the PPN does not have a significant medial extension except for scattered cells, and (c) scattered cholinergic cells in the transition between the PPN and the LDT are actually located within the locus coeruleus.

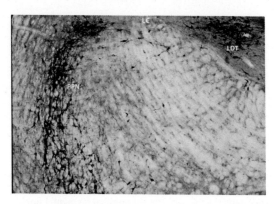

FIGURE 2.4   **Semihorizontal section through the long axis of the PPN.** This section is angled along the PPN from the posterodorsal to the medioventral direction (see Figure 2.1 for sagittal orientation). Histochemical NADPH-diaphorase labeling of cholinergic cells shows that PPN neurons of the *pars compacta* are located dorsally and are found laterally to the locus coeruleus (LC). Medial to the LC are cells of the laterodorsal tegmental nucleus (LDT) embedded within the central gray. Laterally, the PPN descends in an arc toward the posterior edge of the SN. This view demonstrates that (a) there are scattered cholinergic neurons between the PPN and the LDT that are located within the LC and (b) there is *no* medial PPN as some investigators suggest and there are only a few scattered cholinergic cells medially to the PPN *pars dissipata*.

Cholinergic PPN neurons can also be labeled using brain nitric oxide synthase (bNOS) antibody (which is equivalent to NADPH diaphorase), so that bNOS immunocytochemistry has become the label of choice. Figure 2.5 is an example of a PPN cell recorded from a rat brain stem slice. The tissue was processed for bNOS immunocytochemistry to label the cholinergic cells with a red chromophore. The cell in the center was also filled with neurobiotin by the recording microelectrode and immunocytochemically labeled with a yellow chromophore. It is evident that the recorded neuron was a cholinergic cell since it was double-labeled. In later chapters, we will describe the membrane responses and intrinsic properties of such PPN neurons.

However, cholinergic neurons are not the only cell type present in the PPN. Early work described the presence of glutamatergic, GABAergic, and noradrenergic (tyrosine hydroxylase-immunolabeled) neurons containing a number of peptides in the PPN. The controversies (localizing noradrenergic neurons mistakenly within PPN, see Figure 2.4) and disparate results, especially describing colocalization of major transmitters (GABA and acetylcholine, GABA and glutamate), have been resolved by an excellent study using triple in situ hybridization to label the message of the main transmitter types in the PPN (Wang and Morales, 2009). These authors found that the PPN contains different populations of cholinergic, GABAergic, and glutamatergic neurons, with very little colocalization

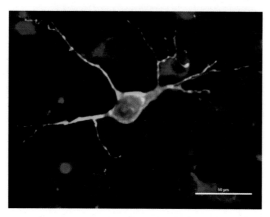

FIGURE 2.5 **Identification of recorded PPN neurons.** This 400 μm slice was processed for bNOS immunocytochemistry that is known to label cholinergic PPN cells, in this case with a red chromophore. Cells showing a red cytoplasm were identified as cholinergic neurons. The slice was also processed for neurobiotin immunocytochemistry to label the neuron recorded using a micropipette containing neurobiotin. The neurobiotin that leaked into the cell was labeled using a yellow chromophore. Therefore, the recorded neuron (yellow label) was identified as a cholinergic (red cytoplasm) PPN cell. Calibration bar 50 μm.

of transmitters. The distribution of neurons in the *pars compacta* was approximately 30% cholinergic, 50% glutamatergic, and 20% GABAergic, while the *pars dissipata* contains 20% cholinergic, 40% glutamatergic, and 40% GABAergic (Wang and Morales, 2009). The ratio of 3 cholinergic to 5 glutamatergic to 2 GABAergic cells seems a general rule across the nucleus. In Chapter 4, we will discuss the presence of mixed cell clusters that form functional units within the PPN.

Figure 2.6 is a sagittal section through the long axis of the PPN. The tissue was processed for bNOS immunocytochemistry using a red chromophore and for glutamic acid decarboxylase immunocytochemistry using a green chromophore. The section shows the *pars compacta* (pc) of the PPN dorsally and the *pars dissipata* (pd) ventrally, along with the SCP. Note the overall difference in cell size between the cholinergic neurons (red) and the GABAergic neurons (green). These labeling methods can also be combined with retrograde labeling, as shown in Figure 2.7. For example, 2 days before the slices were cut, we injected a retrogradely transported dye into the intralaminar thalamus. When the slices containing the PPN were recorded, the retrograde label was detectable using fluorescence microscopy as punctiform green dots, so that recordings could be carried out specifically from neurons projecting to the thalamus. When the tissue was processed for bNOS immunocytochemistry using a red chromophore, the thalamic-projecting cells could be identified as cholinergic (bNOS-positive with green dots) or not (bNOS-negative with green dots).

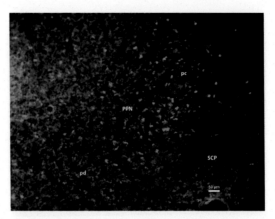

FIGURE 2.6   Labeling of cholinergic and GABAergic PPN cells. Sagittal section across the posterior PPN showing the *pars compacta* (pc) dorsally and the *pars dissipata* (pd) ventrally. Anterior is to the left, and the unlabeled fibers of the SCP enter the posteroventral edge of the PPN. Cholinergic neurons were immunocytochemically labeled for bNOS using a red chromophore, and for glutamic acid decarboxylase to label GABAergic neurons with a green chromophore. Note the intermingling of cholinergic neurons throughout the PPN but especially in the *pars compacta* (pc). Calibration bar 50 μm.

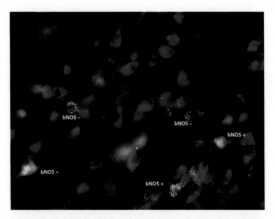

FIGURE 2.7   Labeling thalamic-projecting PPN neurons. After recording in vitro, this 400 μm slice was processed for bNOS immunocytochemistry to label PPN cholinergic neurons using a red chromophore. Three days before, fluorescent retrobeads had been injected into the intralaminar thalamus and the beads allowed to be transported retrogradely to the PPN in the intact animal. Cells projecting to the intralaminar thalamus were labeled using a green chromophore (fluorescent green beads). Neurons that showed a red cytoplasm were assumed to be cholinergic (bNOS-positive), and neurons that showed green beads were assumed to project to the thalamus. In some cases, cholinergic neurons (bNOS-positive) were also labeled with green beads, demonstrating that these cells projected to the thalamus. In other cases, cells showed no bNOS labeling (bNOS-negative) but had green beads, showing that these noncholinergic cells also projected to the thalamus.

Nomenclature of the PPN has always been inconsistent. When we began work in the region in 1979, some atlases used the contraction "PPT," others "PPTg," and others "PPN," for *nucleus tegmenti pedunculopontinus.* We chose the "PPN" contraction for this terminology because the term "tegmental" is superfluous. The main descriptor, "pedunculopontine," pinpoints the location to the body (tegmentum) of the pons near the peduncle, making the term "tegmental" unnecessary. In addition, some workers use the contraction "T" for tegmental, while others use "Tg." Since there is no other "pedunculopontine" nucleus in the brain, again, we consider the term "tegmental" or its contractions unnecessary. For example, the "laterodorsal tegmental nucleus" does require the "tegmental" since there is a laterodorsal "thalamic" nucleus. Most investigators contract the laterodorsal tegmental nucleus to "LDT," but never "LDTg." The most logical contraction for the pedunculopontine nucleus, we find, is "PPN."

DORSAL AND VENTRAL CHOLINERGIC PATHWAYS

Based on acetylcholinesterase labeling, two pathways were originally described emanating from the midbrain reticular formation ascending toward the thalamus and basal forebrain (Shute and Lewis, 1967). Later studies using ChAT immunocytochemistry identified the source of these projections as cholinergic neurons in the posterior midbrain located in the PPN and the LDT nuclei (Mesulam et al., 1983). These ascending projections from the PPN travel dorsally to the intralaminar thalamus and most medial thalamic nuclei, while ventral projections travel to the SN, subthalamic region, lateral hypothalamus, and basal forebrain. The dorsal projection system is denser, but is not limited to intralaminar targets, also terminating in every thalamic relay nucleus, but not as densely. A history of the anatomical and electrophysiological methods used to identify RAS projections is available (Steriade and McCarley, 2005). The afferent and efferent projections of the PPN specifically have also been thoroughly outlined (Garcia-Rill, 1991; Reese et al., 1995). We should note that the PPN receives ascending input from the spinal cord dorsal horn as well as medullary and pontine reticular formation. It also receives descending input from the basal forebrain, thalamus, hypothalamus, and basal ganglia (pallidum). Long projections to the cortex or spinal cord from the PPN are sparse and may be species-specific. Without a doubt, the most significant ascending projection system from the PPN is to the intralaminar thalamus and may include both cholinergic and glutamatergic fibers. We will deal with the ventral projection pathway to the lateral hypothalamus and basal forebrain in Chapter 3 but return to the dorsal projection pathway to the intralaminar thalamus and other thalamic nuclei in Chapters 6 and 9.

CLINICAL IMPLICATIONS: "SLEEP PROBLEMS" VS. "WAKING PROBLEMS"

Above, we discussed the issue of the lack of perception during states of high-amplitude, low-frequency EEG such as during delta frequencies in SWS. When we are in deep sleep, we do not even perceive our most nagging symptoms. When we are in deep sleep, we feel no pain, we hear no phantom sound such as the tinnitus we experience during our waking hours, and we lose many of the involuntary movements that frustrate our waking hours. It is only when our brain is idling at alpha frequency or higher that we perceive such neurogenic symptoms. That is, the automobile engine must be at least on idle and ready to go somewhere. It is during the transition to waking, to idling speed, that we begin to perceive pain, tinnitus, and tremor. Sleep eliminates not only our personality but also many of our psychiatric and neurological problems. Of course, there are disorders that arise during sleep such as obstructive sleep apnea, REM sleep behavior disorder, restless legs syndrome, sleep terrors, sleep walking, and bruxism. These will not be addressed here since there are a number of excellent descriptions of these frank sleep disorders called parasomnias.

However, in most cases in which patients report "sleep problems," they actually represent "waking problems." For example, insomnia, one of the most common "sleep disorders," is actually a prolongation of waking. It represents the intrusion of waking brain activity into sleep time. It is a "waking problem." In many neurological and psychiatric disorders, patients experience not only insomnia but also very vivid nightmares, which represents an additional state of increased vigilance and increased REM sleep drive. If such drive is manifested during waking, it amounts to an episode of REM sleep during waking, or a hallucination. In Chapters 11 and 12, we will discuss the mechanisms behind the elevation of waking, or hypervigilance, and of REM sleep drive that arise in psychiatric and neurological disorders. On the other hand, narcolepsy, which is characterized by sleep attacks during the day coupled with cataplexy, or loss of antigravity extensor tone, is a failure to maintain waking and in fact may be precipitated by sudden arousal. This is also a "waking problem." The most serious "waking problem," of course, is coma, a complete failure of waking on a long-term basis. We will discuss coma in Chapter 3. However, the distinctions between true "sleep problems" and "waking problems" need to be better formulated. These are important distinctions because the successful treatment of these disorders depends on understanding how the waking system has become overactive or underactive and how it needs to be calmed down or stimulated without the consequences of eliminating a normal level of arousal, disturbing attention, or draining our personality.

References

Armstrong, D.M., Saper, C.B., Levey, A.I., Wainer, B.H., Terry, R.D., 1983. Distribution of cholinergic neurons in rat brain: demonstrated by the immunocytochemical localization of choline acetyltransferase. J. Comp. Neurol. 216, 53–68.

Aserinsky, E., Kleitman, N., 1953. Regularly occurring periods of eye motility, and concomitant phenomena, during sleep. Science 118, 273–274.

Aserinsky, E., Kleitman, N., 1955. Two types of ocular motility occurring in sleep. J. Appl. Physiol. 8, 1–10.

Bremer, F., 1973. Preoptic hypnogenic area and reticular activating system. Arch. Ital. Biol. 111, 85–111.

Buzsaki, G., 2006. Rhythms of the Brain. Oxford University Press, New York, NY, 448 pp.

Garcia-Rill, F., 1991. The pedunculopontine nucleus. Prog. Neurobiol. 36, 363–389.

Garcia-Rill, E., Skinner, R.D., 1991. Modulation of rhythmic functions by the brainstem. In: Shimamura, M., Grillner, S., Edgerton, V.R. (Eds.), Neurobiological Basis of Human Locomotion. Japan Scientific Societies Press, Tokyo, pp. 137–158.

Garcia-Rill, E., Reese, N.B., Skinner, R.D., 1995. Arousal and locomotion: from schizophrenia to narcolepsy. In: Holstege, G., Saper, C. (Eds.), The Emotional Motor System. Prog. Brain Res. 107, 417–434.

Geula, C., Schatz, C.R., Mesulam, M.M., 1993. Differential localization of NADPH-diaphorase and calbindin within the cholinergic neurons of the basal forebrain, striatum and brain stem in the rat, monkey, baboon and human. Neuroscience 54, 461–476.

Gottesmann, C., 1988. What the *cerveau isolé* preparation tells us nowadays about sleep–wake mechanisms? Neurosci. Biobehav. Rev. 12, 39–48.

Hobson, J.A., McCarley, R.W., Wyzinski, P.W., 1975. Sleep cycle oscillation: Reciprocal discharge by two brainstem neuronal groups. Science 189, 55–58.

Jones, B.E., 1990. Immunohistochemical study of choline acetyltransferase-immunoreactive processes and cells innervating the pontomedullary reticular formation in the rat. J. Comp. Neurol. 295, 485–514.

Jones, B.E., Beaudet, A., 1987. Distribution of acetylcholine and catecholamine neurons in the cat brain stem: a choline acetyltransferase and tyrosine hydroxylase immunohistochemical study. J. Comp. Neurol. 261, 15–32.

Jouvet, M., 1962. Research on the neural structures and responsible mechanisms in different phases of physiological sleep. Arch. Ital. Biol. 100, 125–206.

Kerkhofs, M., Lavie, P., 2000. Frederic Bremer 1892–1982: a pioneer in sleep research. Sleep Med. Rev. 4, 505–514.

Kirsch, D.B., 2011. There and back again: a current history of sleep medicine. Chest 139, 939–946.

Kleitman, N., 1953. Sleep and Wakefulness. University of Chicago Press, Chicago, IL, 552 pp.

Kumar, V.M., 2010. Sleep is neither a passive nor an active phenomenon. Sleep Biol. Rhythms 8, 163–169.

Kumar, V.M., 2012. Sleep is an auto-regulatory global phenomenon. Front. Neurol. 3, 94 (1–2).

Lavoie, B., Parent, A., 1994. Pedunculopontine nucleus in the squirrel monkey: distribution of cholinergic and monoaminergic neurons in the mesopontine tegmentum with evidence for the presence of glutamate in cholinergic neurons. J. Comp. Neurol. 344, 190–209.

Lindsley, D.B., Bowden, J.W., Magoun, H.W., 1949. Effect upon the EEG of acute injury to the brainstem activating system. Electroencephalogr. Clin. Neurophysiol. 1, 475–486.

Mesulam, M.M., Mufson, E.J., Wainer, B.H., Levey, A.I., 1983. Central cholinergic pathways in the rat: an overview based on an alternate nomenclature (Ch1–Ch6). Neuroscience 10, 1185–1201.

Mesulam, M.M., Mufson, E.J., Levey, A.I., Wainer, B.H., 1984. Atlas of cholinergic neurons in the forebrain and upper brainstem of the macaque based on monoclonal choline acetyltransferase immunohistochemistry and acetylcholinesterase histochemistry. Neuroscience 12, 669–686.

Mesulam, M.M., Geula, C., Bothwell, M.A., Hersh, L.B., 1989. Human reticular formation: cholinergic neurons of the pedunculopontine and laterodorsal tegmental nuclei and some cytochemical comparisons to forebrain cholinergic neurons. J. Comp. Neurol. 281, 611–633.

Mizukawa, K., McGeer, P.L., Tago, H., Peng, J.H., McGeer, E.G., Kimura, H., 1986. The cholinergic system of the human hindbrain studied by choline acetyltransferase immunohistochemistry and acetylcholinesterase histochemistry. Brain Res. 379, 39–55.

Moruzzi, G., 1972. The sleep–waking cycle. Ergeb. Physiol. 64, 1–165.

Moruzzi, G., Magoun, H.W., 1949. Brain stem reticular formation and activation of the EEG. Electroencephalogr. Clin. Neurophysiol. 1, 455–473.

Murakami, S., Okada, Y., 2006. Contributions of principal neocortical neurons to magnetoencephalography signals. J. Physiol. 575 (3), 925–936.

Nakamura, S., Kawamata, T., Kimura, T., Akiguchi, I., Kameyama, M., Nakamura, N., Wakata, Y., Kimura, H., 1988. Reduced nicotinamide adenine dinucleotide phosphate-diaphorase histochemistry in the pontomesencephalic region of the human brain stem. Brain Res. 455, 144–147.

Olszewski, J., Baxter, D., 1982. Cytoarchitecture of the Human Brain Stem, second ed. Karger, New York, 199 pp.

Reese, N.B., Garcia-Rill, E., Skinner, R.D., 1995. The pedunculopontine nucleus—auditory input, arousal and pathophysiology. Prog. Neurobiol. 47, 105–133.

Rye, D.B., Saper, C.B., Lee, H.J., Wainer, B.H., 1987. Pedunculopontine tegmental nucleus of the rat: cytoarchitecture, cytochemistry, and some extrapyramidal connections of the mesopontine tegmentum. J. Comp. Neurol. 259, 483–528.

Sakai, K., El Mansari, M., Jouvet, M., 1990. Inhibition by carbachol microinjections of presumptive cholinergic PGO-on neurons in freely moving cats. Brain Res. 527, 213–223.

Satoh, K., Fibiger, H.C., 1985. Distribution of central cholinergic neurons in the baboon (Papio papio). II. A topographic atlas correlated with catecholamine neurons. J. Comp. Neurol. 236, 215–233.

Shiromani, P.J., Armstrong, D.M., Berkowitz, A., Jeste, D.V., Gillin, J.C., 1988. Distribution of choline acetyltransferase immunoreactive somata in the feline brainstem: implications for REM sleep generation. Sleep 11, 1–16.

Shute, C.C.D., Lewis, P.R., 1967. The ascending cholinergic reticular system: neocortical, olfactory, and subcortical projections. Brain 90, 497–520.

Skinner, R.D., Conrad, C., Henderson, V., Gilmore, S.A., Garcia-Rill, E., 1989. Development of NADPH diaphorase-positive pedunculopontine nucleus neurons. Exp. Neurol. 104, 15–21.

Sofroniew, M.V., Campbell, P.E., Cuello, A.C., Eckenstein, F., 1985. Central cholinergic neurons visualized by immunohistochemical detection of choline acetyltransferase. In: Paxions, G. (Ed.), The Rat Nervous System, vol. 1. Academic Press, New York, pp. 471–485.

Spann, B.M., Grofova, I., 1992. Cholinergic and non-cholinergic neurons in the rat pedunculopontine tegmental nucleus. Anat. Embryol. 186, 215–227.

Steriade, M., 1999. Cellular substrates of oscillations in corticothalamic systems during states of vigilance. In: Lydic, R., Baghdoyan, H.A. (Eds.), Handbook of Behavioral State Control. Cellular and Molecular Mechanisms. CRC Press, New York, pp. 327–347.

Steriade, M., McCarley, R., 2005. Brain Control of Wakefulness and Sleep. Springer, New York, 728 pp.

Steriade, M., Constantinescu, E., Apostol, V., 1969. Correlations between alterations of the cortical transaminase activity and EEG patterns of sleep and wakefulness induced by brainstem transections. Brain Res. 13, 177–180.

Steriade, M., Ropert, N., Kitsikis, A., Oakson, G., 1980. Ascending activating neuronal networks in midbrain reticular core and related rostral systems. In: Hobson, J.A., Brazier, M.A.B. (Eds.), The Reticular Formation Revisited. Raven Press, New York, pp. 125–167.

Steriade, M., Datta, S., Park, D., Oakson, G., Curro Dossi, R., 1990. Neuronal activities in brain stem cholinergic nuclei related to tonic activation processes in thalamocortical systems. J. Neurosci. 10, 2541–2559.

Steriade, M., Curro Dossi, R., Pare, D., Oakson, G., 1991. Fast oscillations (2040 Hz) in thalamocortical systems and their potentiation by mesopontine cholinergic nuclei in the cat. Proc. Natl. Acad. Sci. USA 88, 4396–4400.

Sugimoto, T., Mizukawz, K., Hattori, T., Konishi, A., Kaneko, T., Mizuno, N., 1984. Cholinergic neurons in the nucleus tegmenti pedunculopontinus pars compacta and the caudoputamen of the rat: a light and electron microscopic immunohistochemical study using a monoclonal antibody to choline acetyltransferase. Neurosci. Lett. 51, 113–117.

Villablanca, J., 2004. Counterpointing the functional role of the forebrain and of the brainstem in the control of the sleep-waking system. J. Sleep Res. 13, 179–200.

Wang, H.L., Morales, M., 2009. Pedunculopontine and laterodorsal tegmental nuclei contain distinct populations of cholinergic, glutamatergic and GABAergic neurons in the rat. Eur. J. Neurosci. 29, 340–358.

3

Other Regions Modulating Waking

Edgar Garcia-Rill, PhD

Center for Translational Neuroscience, Department of Neurobiology and Developmental Sciences, University of Arkansas for Medical Sciences, Little Rock, AR, USA

A MATTER OF TIME

The regions aside from the reticular activating system (RAS) implicated in the modulation of waking include the hypocretin-containing neurons of the lateral hypothalamus, the cholinergic neurons of the basal forebrain, and the histamine-containing neurons of the tuberomammillary nucleus (TMN). As mentioned in Chapter 2, Jaime Villablanca carried out a series of heroic experiments using brain stem transections and maintaining the animals for prolonged periods (Villablanca, 2004). Midcollicular or mesencephalic transections that separated the RAS from the forebrain produced animals that were awake 75% and 55% of the time, respectively. These animals were sensitive to sensory stimuli, could stand and walk, attempted to climb, and manifested low-amplitude, high-frequency activity *below* the level of the transection. Nevertheless, as mentioned in Chapter 2, Villablanca ultimately concluded that "true waking behavior depended on the cholinergic reticular core" (Villablanca, 2004). However, the permanently isolated forebrain, after high- or low-brain stem transections, led to alternating episodes of low-amplitude, high-frequency cortical electroencephalogram (EEG) activity (Batsel, 1960; Villablanca, 2004). Presumably, some or all of the regions listed above were responsible for these wake-like episodes *above* the level of the transections. In the isolated forebrain of the cat, electrical stimulation of the posterior hypothalamus and of the basal forebrain induces fast cortical EEG rhythms (Bakuradze et al., 1975; Berladetti et al., 1977), and cholinergic (Sakai et al., 1990) stimulation of these areas induces arousal, suggesting that these areas indeed modulate waking.

If the cortex and striatum were removed, leaving the thalamus, hypo-thalamus, and basal forebrain connected to the brain stem, the model was called "diencephalic." These animals became hyperactive, developing obstinate progression; were hyperreactive to sensory stimuli; and manifested low-amplitude, high-frequency activity in the thalamus. Another model used was the "athalamic" animal in which the thalamus was removed bilaterally. These animals were also hyperactive, reacted to sensory stimuli but could not localize the stimuli, and showed little awareness. Only brief periods of low-amplitude, high-frequency activity were seen in the "athalamic" animal. The implication from these chronic transection studies is that other regions in addition to the RAS can modulate at least some waking. However, it is important to note the amount of time that stimulation of these regions takes to induce waking.

Why is the latency to the induction of a waking EEG important? The implication is that short-latency effects on waking reflect more direct activation of the cortical EEG, whereas long-latency effects reflect a circuitous route for achieving high-frequency EEG activity in the cortex. Typically, stimulation of the RAS, in the region of the pedunculopontine nucleus (PPN) using either electrodes (Moruzzi and Magoun, 1949; Steriade et al., 1991a,b) or optogenetic methods (more on this technology below) activating the locus coeruleus (LC; Carter et al., 2010), will induce high-frequency EEG changes within 1–2s. (This is a similar latency as that required to induce locomotion on a treadmill following stimulation of the PPN; see Chapter 7.) However, stimulation of the basal forebrain induces high-frequency EEG but only after 15s of stimulation (Han et al., 2014), and stimulation of the lateral hypothalamus, or optogenetically activated orexin neurons, elicits high-frequency EEG activity only after 20s of stimulation (Carter et al., 2010; Carter and de Lecea, 2011). Figure 3.1

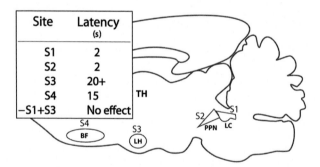

Site	Latency (s)
S1	2
S2	2
S3	20+
S4	15
–S1+S3	No effect

FIGURE 3.1   **Latency to waking following stimulation of different regions.** The sagittal diagram on the right shows the locations of the basal forebrain (BF), lateral hypothalamus (LH), locus coeruleus (LC), pedunculopontine nucleus (PPN), and thalamus (TH). Stimulation of the LC (S1) showed a 2s latency to waking, while stimulation of the mesencephalic reticular formation near the PPN (S2) showed a similar latency. However, stimulation of the LH (S3) exhibited a 20+s latency, while stimulation of the BF (S4) showed a 15s latency. Inhibition of the LC (–S1) showed that stimulation of the LH (+S3) was ineffective, suggesting that orexin neurons must activate the RAS in order to have an effect on waking.

illustrates these latencies and emphasizes the fact that stimulation of the RAS-thalamic pathway elicits cortical arousal 10 times faster than stimulation of the basal forebrain or lateral hypothalamic/orexin pathways. That is, both regions need to project elsewhere to induce a waking EEG, and, as we will see below, neither region is the final common pathway for arousal.

THE LATERAL HYPOTHALAMUS AND HYPOCRETIN

The seminal studies on this system were performed by Luis de Lecea and Thomas Kilduff, in which they discovered the mRNA that encodes hypocretin and localized the peptide to specific neurons (de Lecea et al., 1998). The cell bodies of orexin/hypocretin-containing neurons are located in the perifornical and lateral hypothalamus, they excite RAS and TMN neurons, and hypocretin is released during the low-amplitude, high-frequency EEG states of waking and rapid eye movement (REM) sleep (Siegel, 2004). Degeneration or dysfunction of these neurons is known to cause narcolepsy in human patients and animal models, and administration of hypocretin can reverse the symptoms of the disorder. Narcolepsy is characterized by excessive daytime sleepiness and bouts of cataplexy, in which affective incitement (arousal) leads to a loss of extensor muscle tone. Many patients also have hypnagogic hallucinations, a symptom that emphasizes the likely intrusion of REM sleep into the waking state. That is, both waking and REM sleep are dysregulated in narcolepsy. Almost all narcoleptic patients exhibit human leukocytic antigen (HLA) genotype expression for DQB1 (Olerup et al., 1990), which is quite similar to the HLA expression (DQW1) we found in REM sleep behavior disorder patients (Schenck et al., 1996). About 80% of patients develop the disorder after puberty, similar to patients with schizophrenia, bipolar disorder, obsessive-compulsive disorder, and panic attacks. Some of these disorders will be discussed in Chapter 11. However, a series of studies on hypocretin knockout animals led to the conclusion that hypocretin was less related to arousal than to the levels of motor activity, perhaps mediated by its link to the LC (Torterolo et al., 2003).

While much attention has been paid to hypothalamic neurons in the control of waking, de Lecea's lab has optogenetically engineered animals that have rhodopsin cation channels in orexin and in noradrenergic LC neurons (Carter et al., 2012). Their findings showed that light activation of orexin neurons requires several seconds of stimulation with a latency ~20s to induce waking, implying that its output must travel elsewhere before the animal awakens. When they light-activated LC cells, the animals awoke immediately, within 1–2s, but if LC was inactivated by anion channels, orexin neuron stimulation failed to awaken the animals. These results suggest that orexin neurons must first affect at least one of their descending RAS targets, the LC, in order to manifest a waking effect in vivo.

That is, the lateral hypothalamic system may act through the RAS to elicit arousal. In addition, the prolonged latency required to induce arousal after orexin neuron activation suggests that this region "recruits" waking along with the RAS. The result showing that optogenetic inactivation/inhibition of the LC in advance of activation of optogenetically altered orexin neurons fails to induce waking (Carter et al., 2012) suggests that, in the absence of ascending RAS, specifically the LC, orexin cells cannot induce a waking-like EEG. In other words, descending projections to the RAS may be essential for these cells to ultimately induce an effect on waking. Moreover, the role of these neurons is regulated by sleep deprivation. Optogenetic stimulation experiments found that sleep deprivation blocks the ability of orexin to activate its downstream targets and enhance waking (Carter et al., 2009), suggesting that these neurons can be kept from exercising an effect on waking by simple sleep deprivation. Therefore, rather than a specific role in arousal, hypocretin neurons have been implicated in the integration of motor, metabolic, circadian, and limbic inputs that can influence sleep to wake transitions (de Lecea and Huerta, 2014).

THE BASAL FOREBRAIN AND ACETYLCHOLINE

The basal forebrain contains a magnocellular region with cholinergic, GABAergic, and glutamatergic neurons, much like the PPN (Henny and Jones, 2008; Szymusiak, 1995; Wang and Morales, 2009). Activation of the basal forebrain induces low-amplitude, high-frequency EEG in the cortex (Metherate et al., 1992). Recordings in this region showed that most cells were related to waking and very few related to SWS (Szymusiak and McGinty, 1986). Later studies found that basal forebrain neurons did indeed fire in relation to both waking and REM sleep (Lee et al., 2005). In addition, lesions of basal forebrain neurons were found to increase SWS (Buzsaki and Gage, 1989). All of these findings suggest that basal forebrain cells help manifest the high-frequency cortical EEG. However, a number of groups have suggested that this cell group modulates waking via projections to the brain stem (McGinty and Szymusiak, 2004; Saper et al., 2001; Steininger et al., 2001) and, perhaps, also via the thalamus (Asanuma and Porter, 1990; Gritti et al., 1998). This suggestion is supported by experiments in which stimulation of mesopontine cholinergic nuclei in the brain stem resulted in cortical activation, an effect that persisted even after excitotoxic lesions of the basal forebrain (Steriade et al., 1991a,b).

Moreover, early studies showed that acute precollicular transections in which the basal forebrain is anterior to the transection eliminate fast activity related to waking and REM sleep; i.e., the basal forebrain by itself cannot drive the cortex to maintain gamma band activity in vivo (Lindsley et al., 1949; Moruzzi and Magoun, 1949). This was confirmed by studies in which transection at the level of the posterior edge of the inferior colliculus to the anterior hypothalamus, thus disconnecting the basal forebrain and cortex from the brain stem, induced profound coma (Steriade et al., 1969). Despite the fact that chronic transections did suggest that the basal forebrain modulates waking to some extent, the RAS appears to be the final common pathway for the induction of the high-frequency EEG states of waking and REM sleep by the basal forebrain.

THE TMN AND HISTAMINE

The histamine neurons of the TMN represent the only histaminergic cells in the brain. They are located in the posterior hypothalamus, and they have a reciprocal relationship with ventrolateral preoptic (VLPO) cells (McGinty and Szymusiak, 2004; Saper et al., 2001). TMN cells are known to be inhibited by VLPO cells that promote sleep (Yang and Hatton, 1997) and are excited by hypocretin cells of the lateral hypothalamus (Peyron et al., 1998). However, TMN histamine neurons do not manifest pronounced changes in activity in relation to high-frequency EEG (Lin, 2000). Moreover, knockout animals lacking the histamine-synthesizing enzyme histidine decarboxylase do not show major changes in cortical EEG and waking time (Anaclet et al., 2009; Parmentier et al., 2002). These authors proposed that the TMN system is more related to waking in novel environments, suggesting a modified role in arousal. These findings collectively imply that the TMN system is also not a final common pathway for arousal.

CLINICAL IMPLICATIONS: COMA

Coma is a state of unconsciousness lasting for prolonged periods, from which persons cannot be awakened and have no ability to initiate movement. These patients may randomly move limbs and respond to painful stimuli, but do not make defensive motions to avoid pain. However, this state is not equivalent to deep sleep or SWS, since the EEG is quite complex and variable. In fact, the EEG of the comatose patient appears to be more similar to that during general anesthesia, which has been referred to as a "drug-induced coma," and both states are characterized by burst suppression (Brown et al., 2010). The burst suppression seen in the EEG in some comatose or anesthetized patients has been proposed to be due to thalamic discharges to a cortex that is unresponsive (Steriade et al., 1994). Such discharges are manifested after anoxia (Koenig et al., 2006), hypothermia (Stecker et al., 2001), and some forms of epilepsy (Yamatogi and Ohtahara, 2002), but are not present during normal sleep. The variability across patients with coma suggests that damage by multiple insults can result in coma. For example, it is thought that widespread damage to the cortex such as can occur with carbon monoxide poisoning, anoxia, or trauma can lead to coma since there is no cortex available to be awakened. Coma may also arise from damage to the mesencephalic tegmentum (RAS) or sometimes the hypothalamus (Ingvar and Sourander, 1970; Jasper, 1966; Jefferson, 1958; Smith et al., 2002; Steriade et al., 1969; Young, 2009). Most texts list the main sites inducing coma as the cerebral hemispheres and the RAS, occasionally the hypothalamus, but not the basal forebrain or TMN. Early studies showing that the RAS was the locus of damage in most cases of coma (Chase et al., 1968; Loeb, 1958; Plum and Posner, 1980) have been confirmed using new anatomical methods (Parvizi and Damasio, 2003). Therefore, the RAS remains the primary region damaged or disconnected in coma.

Disorders of consciousness include coma, which represents a total failure of arousal, as well as the chronic vegetative state, in which there is spontaneous or stimulation-induced arousal and wake–sleep cycling. The locked-in syndrome is characterized by full consciousness and is not included here as a disorder of consciousness. However, some coma patients attain a minimally conscious state that has a partial preservation of consciousness and includes following simple commands and purposeful behavior and have better outcomes than those in the vegetative state (Giacino et al., 2002). The partial preservation of a stage of consciousness suggests that the mechanism for consciousness is not a smooth, continuous process, but may require stepping through a series of stages. So, what is the mechanism of consciousness that is disrupted in coma? The answer is that we do not know; however, a recent study used recovery from anesthesia, which, as mentioned above, is also characterized by burst suppression, to determine that the recovery passes through several discrete activity states (Hudson et al., 2014). This study showed that there was an orderly progression through intermediate states rather than a continuous and gradual recovery of consciousness from anesthesia. The implication is that this may also be the case for recovery from coma. What do we know about the process of attaining and maintaining full consciousness?

Roger Sperry, one of the giants of neuroscience research, proposed that the critical organizational features of the neural circuitry for generating conscious awareness are activated through the RAS and, once activated, become responsive to changing sensory as well as centrally generated input (Sperry, 1969). In Chapters 1 and 2, we saw how conscious perception may take place as a result of intrinsic activity rather than sensory drive (Llinas et al., 1998). That is, consciousness may be modulated rather than generated by sensory input. This process is thought to arise from coherence and specific frequency activity that helps maintain a state of continuous appreciation of the world (Garcia-Rill et al., 2013). This is the process of preconscious awareness to be discussed in Chapter 10. The mechanism thought to underlie this process has been proposed to be gamma band activity in the RAS. We will deal in depth with the mechanisms behind this gamma band activity in Chapter 9.

For the purposes of the present discussion of coma, however, gamma frequency oscillations are thought to participate in sensory perception, problem solving, and memory (Eckhorn et al., 1988; Gray and Singer, 1989; Jones, 2007; Palva et al., 2009; Philips and Takeda, 2009; Voss et al., 2009). Coherence at gamma band frequencies can occur at cortical or thalamocortical levels (Llinas et al., 1991; Singer, 1993), but we now know that gamma band activity is also present in the hippocampus (Charpak et al., 1995; Chrobak and Buzsáki, 1998), the cerebellum (Lang et al., 2006; Middleton et al., 2008), the basal ganglia (Trottenberg et al., 2006), and the RAS (Garcia-Rill et al., 2013; Kezunovic et al., 2011; Simon et al., 2010; Urbano et al., 2012). As we will see in Chapter 9, gamma activity across regions is coherent so that resonance occurs across cortical and subcortical areas. That is, gamma band oscillations appear to be a brain-wide property of the awake brain, but gamma band oscillations are not present monolithically across the entire cortex. In the awake brain, regions exhibiting gamma oscillations do so for some period locally, and the oscillations then shift to other regions of the cortex. This may also reflect a series of transitions between discrete intermediate states that ultimately gain full consciousness. Therefore, for an individual interacting with a complex environment, there must be a dizzying display of gamma oscillations throughout the cortical mantle, but no one region is always manifesting gamma band activity at all times. Moreover, the resonance of these oscillations is amplified by projections from the cortex to the thalamus, cerebellum, basal ganglia, RAS, etc. That is, a cortical region that received thalamic input sends a reciprocal 10-fold volley to a thalamic target, which in turn projects back to that same or different area of cortex. Therefore, the delayed (by conduction time) return volley represents another stepwise change in the process of sensory perception. We conclude that different regions of the cortex are repeatedly shifting from one level to another, not in a gradual manner, but in stepwise fashion. Presumably, there must be a sufficient number of regions manifesting gamma band activity to maintain full awareness. It is the attainment of a sufficiently stable process that appears to be blocked in coma. The insult or damage to the brain makes it unable to step through all the levels needed to reach full consciousness. We assume that, based on the amount or degree of damage, partial "levels" of consciousness may be achieved in different patients. For example, in the chronic vegetative state, the EEG changes may reflect stepwise shifts that do not reach higher levels of consciousness, whereas in the minimally conscious state, some degree of consciousness is attained, but the next level cannot be fully manifested.

Gamma band resonance also needs to be maintained for prolonged periods in order to realize cognitive and perceptual functions properly. Because gamma waves can occur briefly during slow-wave sleep (SWS) and anesthesia, a close relation to consciousness has been questioned (Vanderwolf, 2000). It was suggested that consciousness is associated mainly with *continuous* gamma band activity, as opposed to an interrupted pattern of activity (Vanderwolf, 2000). The original description of the RAS specifically suggested that it participates in *tonic or continuous* arousal (Moruzzi and Magoun, 1949), and lesions of this region were found to eliminate *tonic* arousal (Watson et al., 1974). Therefore, the mechanisms that help maintain gamma band activity for prolonged periods are critically important. We will discuss the mechanisms behind the *maintenance* of gamma band activity in Chapter 9. However, in terms of coma, it may be that either the ability to generate significant stepwise levels of gamma activity or the mechanism for maintaining gamma oscillations is disrupted. Moreover, it is possible that some pathological insults may affect one or both processes, which further explains the variability in the symptomatic manifestations during coma. Given the potential causes of coma, from trauma, to metabolic (e.g., glucose), to organ failure (e.g., hepatic), to toxicological, to drug-induced, to infection, it is difficult to determine how each of these insults could affect stepwise gamma band generation or maintenance mechanisms specifically.

These considerations suggest that the RAS must have the ability to generate gamma band activity in order for the rest of the brain to be fully conscious. However, if the RAS is able to generate gamma band activity but the cortical hemispheres are unable to trigger their own gamma band oscillations because of trauma or other insults, then the RAS has no substrate for launching processes related to perception and volition. The question is, if the hemispheres are inactive but the RAS is functional, is there somebody home? That is, if the RAS indeed is (a) in charge of preconscious awareness, the property that allows us to assess the environment on a continuing basis, and (b) also responsible for our sense of self immediately when we awaken (see Chapters 9 and 10), while the cortex has yet to increase blood flow, then there must be at least some awareness if not higher functions such as perception, attention, memory, and learning. This raises many questions and prompts the need for much more research to firmly establish how aware are comatose, minimally conscious, and vegetative state patients. Recent studies have attempted to determine if patients in the vegetative state can communicate in the absence of motor control. Using functional magnetic resonance imaging (fMRI) mental imagery, about 20% of patients in the vegetative state were found to respond intentionally (Cruse et al., 2011; Monti et al., 2010). Some results suggest that these techniques may be useful in predicting outcome (Vogel et al., 2013). But, on the one hand, these studies had small samples and, on the other, the testing paradigm implied that all patients were responsive and had preserved language comprehension. Nevertheless, these results do suggest that at least some patients in the vegetative state may have some awareness, but better paradigms in larger samples are required.

The treatment of coma is generally supportive or directed at proximal causes such as glucose replacement for diabetic shock, antibiotic for infection, or surgery/diuretics for swelling. Coma can last days to weeks, but many patients gradually come out of a coma, while others progress to the vegetative state. Recovery, if it occurs, is gradual, only some cases achieving full recovery. Patients coming out of a coma regain awareness for only minutes at first, which slowly lengthens to hours. A number of therapies have been found to be successful in improving arousal levels in at least some patients. For example, deep brain stimulation of the intralaminar thalamus has been used in some patients with minimally conscious state to improve function (Schiff et al., 2007). Amantadine, a drug that affects n-methyl-D-aspartic acid (NMDA) and dopamine systems, accelerated the recovery of some patients above that of standard of care (Giacino et al., 2012). One of fifteen patients responded to the hypnotic zolpidem with improved coma scores (Whyte and Myers, 2009). The use of the stimulant modafinil has improved wakefulness in some cases (Rivera, 2005; Scott and Ahmed, 2004). These varied therapies, most with only modest success or selective improvements, suggest that the patient population is mixed in terms of regions and processes affected. Unfortunately, treatment must be tailored to the individual and remains a trial-and-error process. We speculate that, for cases in which the RAS is functional, the use of the stimulant modafinil, whose mechanism of action is described in detail in Chapter 4, will become the agent of choice for restoring awareness, but large-scale studies are required.

References

Anaclet, C., Parmentier, R., Ouk, K., Guidon, G., Buda, C., Sastre, J.P., Akaoka, H., Sergeeva, O.A., Yanagisawa, M., Ohtsu, H., Franco, P., Haas, H.L., Lin, J.S., 2009. Orexin/hypocretin and histamine: distinct roles in the control of wakefulness demonstrated using knock-out mouse models. J. Neurosci. 29, 14423–14438.

Asanuma, C., Porter, L.L., 1990. Light and electron-microscopy evidence for a GABAergic projection from the caudal basal forebrain to the thalamic reticular nucleus in rats. J. Comp. Neurol. 302, 159–172.

Bakuradze, A.N., Naneishvili, T.L., Noselidze, A.G., 1975. The role of some limbic structures in the activation of the neocortex following transection of the brain stem at the colliculi level. Fiziol Zh SSSR 61, 985–990.

Batsel, H.L., 1964. Spontaneous desynchronization in the chronic cat "cerveau isole." Arch. Int. Biol. 97, 1–12.

Berladetti, F., Borgia, R., Mancia, M., 1977. Prosencephalic mechanisms of EEG desynchronization in the "cerveau isolé" of the cat. Electroencephalogr. Clin. Neurophysiol. 42, 213–225.

Brown, E.N., Lydic, R., Schiff, N.D., 2010. General anesthesia, sleep, and coma. N. Engl. J. Med. 363, 2638–2650.

Buzsaki, G., Gage, F.H., 1989. The nucleus basalis a key structure in neocortical arousal. In: Fortscher, M., Misgeld, U. (Eds.), Central Cholinergic Synaptic Transmission. Birhkauser Verlag, Basel, pp. 159–171.

Carter, M.E., de Lecea, L., 2011. Optogenetic investigation of neural circuits in vivo. Trends Mol. Med. 17, 197–206.

Carter, M.E., Adamantidis, A., Ohtsu, H., Deisseroth, K., de Lecea, L., 2009. Sleep homeostasis modulates hypocretin-mediated sleep-to-wake transitions. J. Neurosci. 29, 10939–10949.

Carter, M.E., Yizhar, O., Chikahisa, S., Nguyen, H., Adamantidis, A., Nishino, S., Deisseroth, K., de Lecea, L., 2010. Tuning arousal with optogenetic modulation of locus coeruleus neurons. Nat. Neurosci. 13, 1526–1533.

Carter, M.E., Brill, J., Bonnavion, P., Huguenard, J.R., Huerta, R., de Lecea, L., 2012. Mechanisms of hypocretin-mediated sleep-to-wake transitions. Proc. Natl. Acad. Sci. U. S. A. 109, E2635–E2644.

Charpak, S., Paré, D., Llinás, R.R., 1995. The entorhinal cortex entrains fast CA1 hippocampal oscillations in the anaesthetized guinea-pig: role of the monosynaptic component of the perforant path. Eur. J. Neurosci. 7, 1548–1557.

Chase, T.N., Moretti, L., Prensky, A.L., 1968. Clinical and electroencephalographic manifestations of vascular lesions of the pons. Neurology 18, 357–368.

Chrobak, J.J., Buzsáki, G., 1998. Gamma oscillations in the entorhinal cortex of the freely behaving rat. J. Neurosci. 18, 388–398.

Cruse, D., Chennu, S., Chatelle, C., Bekinschtein, T.A., Fernandez-Espejo, D., Pickard, J.D., Laureys, S., Owen, A.M., 2011. Bedside detection of awareness in the vegetative state: a cohort study. Lancet 378, 2088–2094.

de Lecea, L., Huerta, R., 2014. Hypocretin (orexin) regulation of sleep-to-wake transitions. Front. Phamacol. 5, 16/1–16/7.

de Lecea, L., Kilduff, T., Peyron, C., Gao, X.B., Foye, P.E., Danielson, P.E., Fukuhara, C., Battenberg, E.L., Gautvik, V.T., Bartlett, F.S., Frankel, W.N., van den Pol, A.N., Bloom, F.E., Gautvik, K.M., Sutcliffe, J.G., 1998. The hypocretins: hypothalamus-specific peptides with neuroexcitatory activity. Proc. Natl. Acad. Sci. USA 95, 322–327.

Eckhorn, R., Bauer, R., Jordan, W., Brosch, M., Kruse, W., Munk, M., Reitboeck, H.J., 1988. Coherent oscillations: a mechanism of feature linking in the visual system? Biol. Cybern. 60, 121–130.

Garcia-Rill, E., Kezunovic, N., Hyde, J., Beck, P., Urbano, F.J., 2013. Coherence and frequency in the reticular activating system (RAS). Sleep Med. Rev. 17, 227–238.

Giacino, J.T., Ashwal, S., Childs, N., Cranford, R., Jennett, B., Katz, D.I., Kelly, J.P., Rosenberg, J.H., Whyte, J., Zafonte, R.D., Zesler, N.D., 2002. The minimally conscious state. Definition and diagnostic criteria. Neurology 58, 349–353.

Giacino, J.T., Whyte, J., Bagiella, E., Kalmar, K., Childs, N., Khademi, A., Eifert, B., Long, D., Katz, D.I., Cho, S., Yablon, S.A., Luther, M., Hammond, F.M., Nordenbo, A., Novak, P., Mercer, W., Maurer-Karattup, P., Sherer, M., 2012. Placebo-controlled trial of amantadine for severe traumatic brain injury. N. Engl. J. Med. 366, 819–826.

Gray, C.M., Singer, W., 1989. Stimulus-specific neuronal oscillations in orientation columns of cat visual cortex. Proc. Natl. Acad. Sci. USA 86, 1698–1702.

Gritti, I., Mariotti, M., Mancia, M., 1998. Gabaergic and cholinergic basal forebrain and preoptic anterior hypothalamus projections to the mediodorsal nucleus of the thalamus of the cat. Neuroscience 85, 149–188.

Han, Y., Shi, Y., Xi, W., Zhou, R., Tan, Z., Wang, H., Li, M., Chen, Z., Feng, G., Luo, M., Huang, Z., Duan, S., Yu, Y., 2014. Selective activation of cholinergic basal forebrain neurons induces immediate sleep–wake transitions. Curr. Biol. 24, 693–698.

Henny, P., Jones, B.E., 2008. Projections from basal forebrain to prefrontal cortex comprise cholinergic, GABAergic and glutamatergic inputs to pyramidal cells or interneurons. Eur. J. Neurosci. 27, 654–670.

Hudson, A.E., Calderon, D.P., Pfaff, D.W., Proekt, A., 2014. Recovery of consciousness is mediated by a network of discrete metastable activity states. Proc. Natl. Acad. Sci. U. S. A. 111 (25), 9283–9288.

Ingvar, D.H., Sourander, P., 1970. Destruction of the reticular core of the brain stem. Arch. Neurol. 23, 1–8.

Jasper, H.H., 1966. Pathological studies of brain mechanisms of different states of consciousness. In: Eccles, J.C. (Ed.), Brain and Conscious Experience. Springer-Verlag, New York, pp. 256–282.

Jefferson, G., 1958. Altered consciousness associated with brain-stem lesions. Brain 75, 55–67.

Jones, E.G., 2007. Calcium channels in higher-level brain function. Proc. Natl. Acad. Sci. U. S. A. 14, 17903–17904.

Kezunovic, N., Urbano, F.J., Simon, C., Hyde, J., Smith, K., Garcia-Rill, E., 2011. Mechanism behind gamma band activity in the pedunculopontine nucleus (PPN). Eur. J. Neurosci. 34, 404–415.

Koenig, M.A., Kaplan, P.W., Thakor, N.V., 2006. Clinical neurophysiologic monitoring and brain injury from cardiac arrest. Neurol. Clin. 24, 89–106.

Lang, E.J., Sugihara, I., Llinás, R.R., 2006. Olivocerebellar modulation of motor cortex ability to generate vibrissal movements in rat. J. Physiol. 571, 101–120.

Lee, M.G., Hassani, O.K., Alonso, A., Jones, B.E., 2005. Cholinergic basal forebrain neurons burst with theta during waking and paradoxical sleep. J. Neurosci. 25, 4365–4369.

Lin, J.S., 2000. Brain structures and mechanisms involved in the control of cortical activation and wakefulness, with emphasis on the posterior hypothalamus and histaminergic neurons. Sleep Med. Rev. 4, 471–503.

Lindsley, D.B., Bowden, J.W., Magoun, H.W., 1949. Effect upon the EEG of acute injury to the brainstem activating system. Electroencephalogr. Clin. Neurophysiol. 1, 475–486.

Llinas, R.R., Grace, A.A., Yarom, Y., 1991. In vitro neurons in mammalian cortical layer 4 exhibit intrinsic oscillatory activity in the 10- to 50-Hz frequency range. Proc. Natl. Acad. Sci. USA 88, 897–901.

Llinas, R.R., Ribary, U., Contreras, D., Pedroarena, C., 1998. The neuronal basis for consciousness. Philos. Trans. R. Soc. Lond. B Biol. Sci. 353, 1841–1849.

Loeb, C., 1958. Electroencephalographic changes during the state of coma. Electroencephalogr. Clin. Neurophysiol. 10, 589–606.

McGinty, D., Szymusiak, R., 2004. Sleep-promoting mechanisms in mammals. In: Kryger, M., Roth, T., Dement, W. (Eds.), Principles and Practice of Sleep Medicine. W. B. Saunders, Philadelphia, PA, pp. 169–184.

Metherate, R., Cox, C.L., Ashe, J.H., 1992. Cellular bases of neocortical activation: modulation of neural oscillations by the nucleus basalis and endogenous acetylcholine. J. Neurosci. 12, 4701–4711.

Middleton, S.J., Racca, C., Cunningham, M.O., Traub, R.D., Monyer, H., Knöpfel, T., Schofield, I.S., Jenkins, A., Whittington, M.A., 2008. High-frequency network oscillations in cerebellar cortex. Neuron 58, 763–774.

Monti, M.M., Vanhaudenhuyse, A., Coleman, M.R., Boly, M., Pickard, J.D., Tshibanda, L., Owen, A.M., Laureys, S., 2010. Willful modulation of brain activity in disorders of consciousness. N. Engl. J. Med. 362, 579–589.

Moruzzi, G., Magoun, H.W., 1949. Brain stem reticular formation and activation of the EEG. Electroencephalogr. Clin. Neurophysiol. 1, 455–473.

Olerup, O., Schaffer, M., Hillert, J., Sachs, C., 1990. The narcolepsy-associated DRw15, DQw6, Dw2 haplotype has no unique HLA-DQA or -DQB restriction fragments and does not extend to the HLA-DP region. Immunogenet. 32, 41–44.

Palva, S., Monto, S., Palva, J.M., 2009. Graph properties of synchronized cortical networks during visual working memory maintenance. Neuroimage 49, 3257–3268.

Parmentier, R., Ohtsu, H., Djebbara-Hannas, Z., Valatx, J.L., Watanabe, T., Lin, J.S., 2002. Anatomical, physiological, and pharmacological characteristics of histidine decarboxylase knock-out mice: evidence for the role of brain histamine in behavioral and sleep-wake control. J. Neurosci. 22, 7695–7711.

Parvizi, J., Damasio, A.R., 2003. Neuroanatomical correlates of brainstem coma. Brain 126, 1524–1536.

Peyron, C., Tighe, D.K., van den Pol, A.N., de Lecea, L., Heller, H.C., Sutcliffe, J.G., Kilduff, T., 1998. Neurons containing hypocretin (orexin) project to multiple neuronal systems. J. Neurosci. 18, 9996–10015.

Philips, S., Takeda, Y., 2009. Greater frontal-parietal synchrony at low gamma-band frequencies for inefficient then efficient visual search in human EEG. Int. J. Psychophysiol. 73, 350–354.

Plum, F., Posner, J.B., 1980. The Diagnosis of Stupor and Coma, third ed. F. A. Davis, Philadelphia, PA.

Rivera, V.M., 2005. Modafinil for the treatment of diminished responsiveness in a patient recovering from brain surgery. Brain Inj. 19, 725–727.

Sakai, A., El Mansari, M., Lin, J.S., Zhang, J.G., Vanni-Mercier, G., 1990. The posterior hypothalamus in the regulation of wakefulness and paradoxical sleep. In: Mancia, M., Marini, G. (Eds.), The Diencephalon and Sleep. Raven Press, New York, pp. 171–198.

Saper, C.B., Chou, T.C., Scammell, T.E., 2001. The sleep switch: hypothalamic control of sleep and wakefulness. Trends Neurosci. 24, 726–731.

Schenck, C., Garcia-Rill, E., Segall, M., Noreen, H., Mahowald, M.W., 1996. HLA class II genes associated with REM sleep behavior disorder. Ann. Neurol. 39, 261–263.

Schiff, N.D., Giacino, J.T., Kalmar, K., Victor, J.D., Baker, K., Gerber, M., Fritz, B., Eisenberg, B., Biondi, T., O'Connor, J., Koblarz, E.J., Farris, S., Machado, A., McCagg, C., Plum, F., Fins, J.J., Rezal, A.R., 2007. Behavioural improvements with thalamic stimulation after severe traumatic brain injury. Nature 448, 600–603.

Scott, C., Ahmed, I., 2004. Medafinil in endozepine stupor. A case report. Can. J. Neurol. Sci. 31, 409–411.

Siegel, J.M., 2004. Hypocretin (orexin): role in normal behavior and neuropathology. Annu. Rev. Psychol. 55, 125–148.

Simon, C., Kezunovic, N., Ye, M., Hyde, J., Hayar, A., Williams, D.K., Garcia-Rill, E., 2010. Gamma band unit and population responses in the pedunculopontine nucleus. J. Neurophysiol. 104, 463–474.

Singer, W., 1993. Synchronization of cortical activity and its putative role in information processing and learning. Annu. Rev. Physiol. 55, 349–374.

Smith, D.H., Nonaka, M., Miller, R., Leoni, M., Chen, X.H., Alsop, D., Meaney, D.F., 2002. Immediate coma following inertial brain injury dependent on axonal damage in the brainstem. J. Neurosurg. 93, 315–322.

Sperry, R.W., 1969. A modified concept of consciousness. Psychol. Rev. 76, 532–536.

Stecker, M.M., Cheung, A.T., Pochettino, A., Kent, G.P., Patterson, T., Weiss, S.J., Bavaria, J.E., 2001. Deep hypothermic circulatory arrest: I. Effects of cooling on electroencephalogram and evoked potentials. Ann. Thorac. Surg. 71, 14–21.

Steininger, T.L., Gong, H., McGinty, D., Szymusiak, R., 2001. Subregional organization of preoptic area/anterior hypothalamic projections to arousal-related monoaminergic cell groups. J. Comp. Neurol. 429, 638–653.

Steriade, M., Constantinescu, E., Apostol, V., 1969. Correlation between alterations of the cortical transaminase activity and EEG patterns of sleep and wakefulness induced by brainstem transections. Brain Res. 13, 177–180.

Steriade, M., Curro-Dossi, R., Nunez, A., 1991a. Network modulation of a slow intrinsic oscillation of cat thalamocortical neurons implicated in sleep delta waves: cortical induced synchronization and brainstem cholinergic suppression. J. Neurosci. 11, 3200–3217.

Steriade, M., Curro Dossi, R., Pare, D., Oakson, G., 1991b. Fast oscillations (20–40 Hz) in thalamocortical systems and their potentiation by mesopontine cholinergic nuclei in the cat. Proc. Natl. Acad. Sci. USA 88, 4396–4400.

Steriade, M., Amzica, F., Contreras, D., 1994. Cortical and thalamic cellular correlates of electroencephalographic burst-suppression. Electroencephalogr. Clin. Neurophysiol. 90, 1–16.

Szymusiak, R., 1995. Magnocellular nuclei of the basal forebrain: substrates of sleep and arousal regulation. Sleep 18, 478–500.

Szymusiak, R., McGinty, D., 1986. Sleep-related neuronal discharge in the basal forebrain of cats. Brain Res. 370, 82–92.

Torterolo, P., Yamuy, J., Sampogna, S., Morales, F.R., Chase, M.H., 2003. Hypocretinergic neurons are primarily involved in activation of the somatomotor system. Sleep 26, 25–28.

Trottenberg, T., Fogelson, N., Kuhn, A.A., Kivi, A., Kupsch, A., Schneider, G.H., Brown, P., 2006. Subthalamic gamma activity in patients with Parkinson's disease. Exp. Neurol. 200, 56–65.

Urbano, F.J., Kezunovic, N., Hyde, J., Simon, C., Beck, P., Garcia-Rill, E., 2012. Gamma band activity in the reticular activating system (RAS). Front. Neurol. 3 (6), 1–16.

Vanderwolf, C.H., 2000. What is the significance of gamma wave activity in the pyriform cortex? Brain Res. 877, 125–133.

Villablanca, J.R., 2004. Counterpointing the functional role of the forebrain and of the brainstem in the control of the sleep–waking system. J. Sleep Res. 13, 179–208.

Vogel, D., Markl, A., Yu, T., Kotchoubey, B., Lang, S., Muller, F., 2013. Can mental imagery functional magnetic resonance imaging predict recovery in patients with disorders of consciousness? Arch. Phys. Med. Rehabil. 94, 1891–1898.

Voss, U., Holzmann, R., Tuin, I., Hobson, J.A., 2009. Lucid dreaming: a state of consciousness with features of both waking and non-lucid dreaming. Sleep 32, 1191–1200.

Wang, H., Morales, M., 2009. Pedunculopontine and laterodorsal tegmental nuclei contain distinct populations of cholinergic, glutamatergic and GABAergic neurons in the rat. Eur. J. Neurosci. 29, 340–358.

Watson, R.T., Heilman, K.M., Miller, B.D., 1974. Neglect after mesencephalic reticular formation lesions. Neurology 24, 294–298.

Whyte, J., Myers, R., 2009. Incidence of clinically significant responses to zolpidem among patients with disorders of consciousness: a preliminary placebo controlled trial. Am. J. Phys. Med. Rehabil. 88, 410–418.

Yamatogi, Y., Ohtahara, S., 2002. Early-infantile epileptic encephalopathy with suppression-bursts, Ohtahara syndrome: its overview referring to our 16 cases. Brain Dev. 24, 13–23.

Yang, Q.Z., Hatton, G.I., 1997. Electrophysiology of excitatory and inhibitory afferents to rat histaminergic tuberomammillary nucleus neurons from hypothalamic and forebrain sites. Brain Res. 773, 162–172.

Young, G.B., 2009. Coma. Ann. N. Y. Acad. Sci. 1157, 32–47.

4

Wiring Diagram of the RAS

Susan Mahaffey, BS and Edgar Garcia-Rill, PhD

Center for Translational Neuroscience, Department of Neurobiology and
Developmental Sciences, University of Arkansas for Medical Sciences,
Little Rock, AR, USA

WIRING DIAGRAM

There are three main nuclei in the reticular activating system (RAS): the locus coeruleus (LC) nucleus, with norepinephrine/noradrenaline (NE/NA)-containing neurons; the dorsal raphe nucleus (RN), with serotonin (5-HT)-containing neurons; and the pedunculopontine nucleus (PPN), with acetylcholine (ACh)- and glutamate (GLU)-containing neurons (Figure 4.1). These nuclei also contain neurons with the inhibitory neurotransmitter gamma-aminobutyric acid (GABA). The dorsal subcoeruleus nucleus (SubCD) is not traditionally considered part of the RAS, but is as critical for generating wake-sleep states as the major nuclei in the RAS. Because LC and RN neurons are most active during waking, have slower firing rates during slow-wave sleep (SWS), and are inhibited during rapid eye movement (REM) sleep, they are sometimes called "REM-off" nuclei. On the other hand, the PPN is most active during waking and REM sleep and the SubCD is most active during REM sleep.

LOCUS COERULEUS

LC neurons, which contain NE or GABA (Aston-Jones et al., 2004), fire action potentials (APs) spontaneously at low frequency (1–5 Hz) and are electrically coupled (Christie et al., 1989; Williams et al., 1984), which allows these neurons to fire in synchrony. The LC is considered a "REM-off" nucleus because it is most active during waking, with little activity during non-REM (NREM) and no activity during REM sleep (Aston-Jones and Bloom, 1981). Stimulation of the LC increased wakefulness, while

Waking and the Reticular Activating System in Health and Disease
http://dx.doi.org/10.1016/B978-0-12-801385-4.00004-5

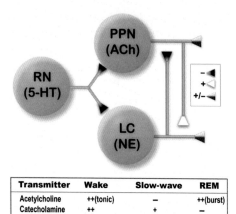

Transmitter	Wake	Slow-wave	REM
Acetylcholine	++(tonic)	−	++(burst)
Catecholamine	++	+	−

FIGURE 4.1 **Wiring diagram of the RAS.** Top. Recordings in slices have shown that the dorsal raphe nucleus (RN), mostly made up of serotonergic (5-HT) cells, generally inhibits (filled triangles) the pedunculopontine nucleus (PPN), which has cholinergic (ACh) cells among others, and the locus coeruleus (LC), which is made up mostly of noradrenergic (NE) cells. The LC inhibits the PPN, while the PPN excites LC neurons (open triangle). Both the PPN and LC project in parallel to ascending and descending sites and modulate these (half-filled triangles). Bottom. Table showing the overall firing patterns of these cell groups during different wake–sleep states. Cholinergic cells of the PPN as well as both LC and RN catecholaminergic neurons are active during waking, while the cholinergic cells decrease firing during SWS and the catecholamine cells groups still show activation. During REM sleep, the catecholamine cells groups are generally silent, but the cholinergic PPN cells fire in bursts, especially in the cat.

cooling or lesion of this region increased SWS and REM sleep (Cespuglio et al., 1982; Delagrange et al., 1993).

The LC receives excitatory cholinergic input from the PPN that is mediated by nicotinic and muscarinic M_2 receptors (Egan and North, 1985, 1986), serotonergic afferents from the RN (Pickel et al., 1977), and excitatory orexigenic input from the lateral hypothalamus (Hagan et al., 1999). LC neurons also release NE onto other LC neurons, which provides local feedback inhibition (Egan et al., 1983). Microdialysis studies have shown that GABA release into the LC is lowest during waking, increases during SWS, and is highest during REM sleep, while no changes were observed in the levels of glycine and GLU release (Nitz and Siegel, 1997b). These results suggest that GABA release in the LC increases especially during REM sleep, accounting for the cessation of its activity (Table 4.1).

The LC projects to the PPN and thalamus, but the majority of its projections are to the cortex (Jones and Yang, 1985). Noradrenergic fibers enter layer IV of the cortex and send collaterals to the other cortical layers (Levitt and Moore, 1978). Cholinergic PPN neurons are inhibited by NE through α-2 adrenergic receptors (Figure 4.1), while noncholinergic PPN neurons are

TABLE 4.1   RAS Cell Properties

Nucleus	Waking	Slow Wave Sleep	REM Sleep
Locus coeruleus	Active	Active	Inactive
Raphe nucleus	Active	Active	Inactive
Pedunculopontine	Active	Inactive	Active
	In vivo		
Pedunculopontine	Wake/REM-on		REM-on
Raphe nucleus			REM-off
Locus coeruleus			REM-off
	In vitro		
Pedunculopontine	Type I (LTS)	Type II (I_A)	Type III (I_A+LTS)

excited by NE (Kohlmeier and Reiner, 1999; Williams and Reiner, 1993). On the other hand, NE inhibits thalamic interneurons and excites thalamocortical (TC) relay and cortical neurons (Dodt et al., 1991; Kayama et al., 1982).

In conclusion, during waking, the LC is most active due to excitation by cholinergic and orexigenic afferents, resulting in increased NE release and excitation of the cortex. During SWS, excitatory input decreases and inhibitory input increases, decreasing NE release. Finally, during REM sleep, LC neurons turn off due to decreased excitatory and increased GABAergic inhibitory drive.

DORSAL RAPHE NUCLEUS

The RN, which contains 5-HT neurons (Jones and Beaudet, 1987), has similar activity patterns as the LC (tonic AP firing at low frequency with long afterhyperpolarization (AHP) (Vandermaelen and Aghajanian, 1983)), but electrical coupling has not been reported. RN neurons fire at their highest rate during waking, decrease activity during SWS, and show the lowest activity during REM sleep (Trulson and Jacobs, 1979; Trulson and Trulson, 1982). Like the LC, stimulation of the RN increased waking, and cooling or inactivation increased SWS and REM (Cespuglio et al., 1979; Nitz and Siegel, 1997a).

ACh release from the PPN excites RN neurons through nicotinic receptors (Figure 4.1) (Mihailescu et al., 2002). The RN receives noradrenergic input from the LC (Baraban and Aghajanian, 1981), which is excitatory through α-1 adrenergic receptors (Aghajanian, 1985). RN neurons are also excited by orexin and histamine (Brown et al., 2002). RN neurons contain 5-HT$_{1A}$ autoreceptors, which inhibit further 5-HT release (Bjorvatn et al.,

1997). GABA release in the RN increases during REM sleep, but there are no changes in glycine or GLU release (Nitz and Siegel, 1997a).

Like the LC, RN neurons send projections to the PPN and thalamus but primarily project to the cortex. RN neurons project to both cholinergic and noncholinergic neurons in the PPN (Steininger et al., 1997), and a 5-HT$_{1A}$ receptor agonist inhibited "REM-on" neurons in the PPN (Figure 4.1), but had little effect on "wake/REM-on" neurons (Thakkar et al., 1998). In the thalamus, 5-HT inhibited TC relay neurons through 5-HT$_{1A}$ receptors and activated local interneurons (Monckton and McCormick, 2002). Serotonergic fibers from the RN extensively innervate all regions of the cerebral cortex, and axons arborize throughout all of the cortical layers, even more extensively than NE neurons from the LC (Lidov et al., 1980), and these connections are primarily inhibitory (Olpe, 1981).

The RN has traditionally been considered a "REM-off," wakefulness-promoting nucleus. While the "REM-off" activity is supported by in vivo recordings during arousal states, the idea of wakefulness promotion has been challenged by evidence showing the inhibitory actions of 5-HT on the thalamus and cortex (Olpe, 1981). Perhaps the RN is involved in modulating waking and SWS or only SWS (Datta and Maclean, 2007).

PEDUNCULOPONTINE NUCLEUS

The PPN is most active during waking and REM sleep and is directly involved in generating the activated states of waking and REM sleep (Steriade et al., 1990). Lesion of the PPN reduced or eliminated REM sleep (Deurveilher and Hennevin, 2001; Shouse and Siegel, 1992), while stimulation of the PPN blocked SWS, increased REM sleep and waking, and potentiated the appearance of fast (20–40 Hz) oscillations in the cortical electroencephalogram (EEG) (Datta and Siwek, 1997; Steriade et al., 1991). The lack of significant effect on waking due to PPN lesions suggests that other regions driving waking can compensate for aspects of the waking state to some extent, as discussed in Chapter 3. However, it is not known if PPN lesions undermine the capacity of the animal to maintain high-frequency activity related to arousal and cognitive function. Lesions of the PPN do have effects on higher functions, as we will discuss in Chapter 10.

The presence of PPN neurons containing multiple neurotransmitters was previously reported (Jia et al., 2003; Lavoie and Parent, 1994). These studies found markers for GLU and GABA colocalized with markers for ACh, indicating that cholinergic PPN neurons may also release GLU and GABA. However, a recent study using in situ hybridization and immunohistochemistry determined that the PPN contains separate populations of GLU, GABAergic, and cholinergic cells, with little colocalization of neurotransmitters (Wang and Morales, 2009). There are some differences in the distribution of neurotransmitter cell types within the boundaries of the PPN (Mena-Segovia et al., 2009; Wang and Morales, 2009). In the more posterior *pars compacta*, GLU neurons make up the majority (~50%), followed by ACh (~30%) and GABA (~20%) cells. However, in the more anterior *pars dissipata*, GABA neurons were more numerous (~40%), followed by GLU (~35%) and ACh (~25%) cells.

Measure of the activity of PPN neurons during wake-sleep states has revealed the presence of neurons that are most active during high-frequency states. In the cat, some neurons were most active during waking ("wake-on"), others were most active during REM sleep ("REM-on"), and a third group was most active during both waking and REM sleep ("wake/REM-on") (Kayama et al., 1992; Steriade et al., 1990). However, there are serotonergic and noradrenergic neurons near the PPN of the cat (Leonard et al., 1995), which were suggested to be the "wake-on" neurons in the cat (Kayama et al., 1992). In the PPN of the rat, where there are no serotonergic and adrenergic neurons (Wang and Morales, 2009), there appear to be only "REM-on" and "wake/REM-on" neurons and no "wake-on" neurons (Datta and Siwek, 2002). Extracellular recordings of PPN neurons in vivo identified six categories of thalamic projecting PPN cells distinguished by their firing rates in relation to PGO waves (Steriade et al., 1990). Some of these neurons had low rates of spontaneous firing (<10 Hz), but most had high rates of tonic firing (20–80 Hz).

The question remains, are there "wake-on" neurons in the PPN of the rat? The answer is inconclusive and may in part be due to the characteristics of recording single neurons in vivo. That is, extracellular recordings during active waking are very difficult when the animal is engaging in motor acts, which could cause loss or damage of the neuron being studied. Moreover, these studies do not test the responsiveness of cells to sensory inputs, much less their activity during trained movements. Recordings during SWS and during REM sleep (with atonia facilitating recording stability) are the most likely cells to be acquired. Therefore, the determination of the presence or absence of "wake-on" neurons in the PPN must await better recording methods and certainly more appropriate recording protocols that involve testing for afferent sensory responsiveness.

It should be noted that there is overlap in the distribution of cells at the borders of the PPN, LC, and laterodorsal tegmental (LDT) nucleus. Cells in the PPN become less numerous medially and its cells intermingle with neurons of the LC, which then becomes dense at its core, but medially to the LC, scattered LC neurons are evident among lateral LDT neurons. That is, the distribution of PPN, LC, and LDT cells overlaps at the transition between PPN and LC and between LC and LDT. Moreover, the most anterior pole of the PPN is embedded in the posterior edge of the substantia nigra. That is, as discussed in Chapter 2, there is an anatomical alternation between cholinergic and catecholaminergic cell groups

(Garcia-Rill and Skinner, 1991). The synaptic interactions between the PPN, LC, and RN do not easily explain how these nuclei might control waking, slow wave sleep, and REM sleep. Obviously, these nuclei have more than one cell type (PPN has three cell types; LC and RN have two) and are under the influence of multiple transmitter inputs. Since our main concern is waking and high-frequency states, we will focus on intrinsic properties of PPN neurons. The following is a description of some of the intrinsic membrane properties of these neurons that help explain their firing patterns when altered by changes in state or by their inputs.

INTRINSIC PROPERTIES OF PPN NEURONS

While the more medial cholinergic partner of the PPN, the LDT nucleus, is more compact and easier to localize, the fact remains that stimulation of the LDT does not induce changes in muscle tone or locomotion like the PPN (Garcia-Rill, 1991; Reese et al., 1995a), i.e., may not participate in the motor components of adult waking and REM sleep. Moreover, its ascending projections do not help generate arousal-related evoked potentials (Reese et al., 1995b,c), and ascending LDT projections are directed at more "limbic" regions (ventral tegmental area, basal forebrain, and accumbens) than those of the PPN, which is more "motor" (substantia nigra, intralaminar thalamus, and striatum) (Garcia-Rill, 1991; Reese et al., 1995a,b,c). It is also possible that the characteristics of LDT neurons across the critical stage of the developmental decrease in REM sleep may not generalize to those of PPN neurons and may even differ in terms of ascending and descending modulation of thalamic and pontomedullary neurons, respectively. Certainly, based on its connectivity, the LDT could not be considered a target for deep brain stimulation for the alleviation of motor deficits in Parkinson's disease and has typically not been investigated for its ability to modulate intralaminar thalamus and cortical high-frequency EEG. The LDT simply does not have the same effects on arousal and waking as the PPN, and its role may be more related to limbic functions in keeping with its connectivity. Its location within the central gray and its projections to the ventral tegmental area that is a way station to the accumbens and limbic system would suggest that the LDT may respond to painful stimuli or at least afferents with "affective" properties. Assuming, just because the neuronal complements and cellular properties are similar to those of PPN neurons, that the LDT has a similar function on waking and arousal is unsubstantiated. Figure 2.4 (Chapter 2) shows a semihorizontal section along the axis of the PPN with cholinergic cells labeled with NADPH-diaphorase. Posteriorly, the PPN bends medially towards the lateral edge of the LC and anteriorly towards the posterior substantia nigra. The LDT is located within the central gray medially, although there are scattered cholinergic cells between the PPN and LDT, lying within the boundaries of the LC. This view shows the marked regional separation between the PPN and LDT.

On the other hand, physiologically, PPN neurons in vivo display both tonic and phasic activities in relation to waking and REM sleep (Steriade and McCarley, 1990) and "wake/REM-on" and "REM-on" firing patterns as described above. However, intracellular recordings of the PPN in vitro have provided somewhat conflicting results. Recordings in slices are typically performed using electrodes containing a dye such as neurobiotin, so that the tissue can be processed for immunocytochemical labeling of the recorded cell. By using immunocytochemical labeling for a transmitter type, the nature of the recorded and nonrecorded cells can be determined. Figure 4.2 shows such labeling from two different slices. In one slice, the recorded cell was also shown to be positive for bNOS immunocytochemistry, which selectively labels cholinergic PPN cells. In the other slice, the recorded cell was not labeled by bNOS, indicating that it was noncholinergic.

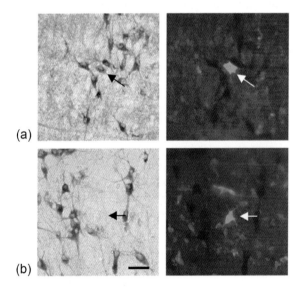

FIGURE 4.2 **Labeling of PPN neurons.** (a) Left side. Section through the PPN processed for NADPH-diaphorase histochemistry that labels cholinergic PPN cells dark. Right side. Fluorescence microscopy revealed that one of the cholinergic PPN neurons was also labeled using fluorescent dye in the recording pipette. Therefore, the recorded neuron was shown to be cholinergic. (b) Left side. Another section through the PPN showing cholinergic neurons labeled using NADPH-diaphorase, along with an arrow pointing at an unlabeled cell. Right side. Fluorescence labeling of the recorded neuron that, in this case, was found not to be cholinergic.

The presence of three types of PPN neurons was reported in the guinea pig, namely, neurons with a low-threshold spike (LTS) (type I), A current (type II), and both A+LTS (type III) (Leonard and Llinas, 1990). Most type II and III neurons were identified as cholinergic. It should be noted that even newborn guinea pigs show adult-like REM sleep percent, so that their REM sleep drive is more like that of the adult rat, and they undergo no major changes in wake–sleep control across postnatal development (Jouvet-Mounier et al., 1970). Others (Kang and Kitai, 1990) found type I (LTS) and type II (A) neurons, over half of the latter cholinergic, but their type III neurons had neither A nor LTS properties in the adult rat. On the other hand, another lab (Kamondi et al., 1992) reported data from the neonate rat similar to those of Leonard and Llinas (1990), that is, three types. Subsequent analysis confirmed the presence of type I and II neurons, with one-half to two-thirds of these being cholinergic (Takakusaki and Kitai, 1997). The discrepancies in the number and presence of type III neurons between labs in part can be explained in terms of age, since the Kitai lab recorded from animals >30 days of age, whereas the Reiner lab recorded from 9 to 15 day olds. A later study from the Kitai lab reported the presence of A+LTS neurons in young (70–150 g) rats (Takakusaki et al., 1997), but they classified them as type II. They did find a difference such that 65% of type II (A) neurons were cholinergic, but only 36% of type II (A+LTS) neurons were cholinergic. Figure 4.4 shows current clamp recordings of each type of PPN neuron, type I (LTS), type II (I_A), and type III (I_A+LTS). The three types of electrophysiologically identified PPN neurons in vitro are shown in Figure 4.3.

LTSs are important calcium-dependent currents that activate (open) on depolarization and inactivate (close) with a slow time course (Greene and Rainnie, 1999; Keele et al., 1999). This allows a burst of APs to be generated even below resting membrane potential, which is similar to the firing patterns of ponto–geniculo–occipital (PGO)-burst neurons found in PPN (Steriade et al., 1990b). The problem with this view is that PGO waves are generated by cholinergic mechanisms (Steriade and McCarley, 1990), but PPN type I (LTS) cells appear to be noncholinergic. This suggests that the small number of cholinergic type III neurons may mediate PGO bursting, but the number of type III neurons may be decreasing with age, at least in the rat. Such bursting could occur in these cells if potassium channels are blocked (Leonard and Llinas, 1990), but additional evidence is needed to resolve the issue. How and which PPN cells generate PGO bursts remains controversial. It will be critical to determine if neurons classified as cholinergic/noncholinergic or by bursting activity display unique projection patterns and, if they do, whether they are differentially affected by specific neurotransmitters. This will necessitate identification of a neuron's projection pattern and its neurochemical identity in the in vitro slice preparation (Rye, 1997). Another confounding factor may be related to

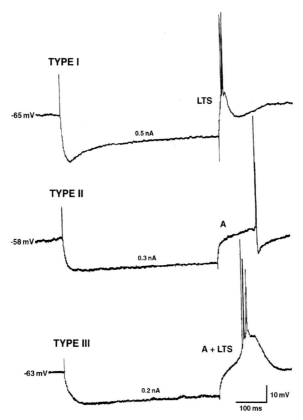

FIGURE 4.3 **Electrophysiological cell types in the PPN recorded in vitro.** Type I cells have low-threshold spike (LTS) T-type calcium channels that open when the cell is released from hyperpolarization at the end of a current step. In this case, the hyperpolarizing step also revealed the presence of I_H current in this cell, which is present in ~40% of PPN cells. Type I cells are typically noncholinergic. Type II cells have an I_A current, which delays the return to potential threshold upon release from hyperpolarization. About two-thirds of type II cells are cholinergic. Type III cells have both I_A and LTS currents, and only one-third are cholinergic.

species, as recent findings suggest that PGO bursting in the rat may not be as widespread as in other species (Datta and Siwek, 2002). As will be seen in Chapter 9, other intracellular mechanisms may dictate the kinds of activity manifested by PPN neurons during different arousal states.

The other current described above is a voltage-gated, transient (outward potassium, I_A) A current (Greene and Rainnie, 1999), which is activated when a neuron is depolarized after a hyperpolarization, and is an outward current that tends to counteract inward currents such as LTS if they occur in the same cells (e.g., type III). The A current displays a delayed return to baseline after hyperpolarization, is rapidly deinactivated, and thus serves to increase interspike interval and, therefore, slow the discharge rate of these neurons (Kamondi et al., 1992). The presence of a high-amplitude and long-duration, calcium-dependent AHP in type II (I_A) cells further limits spontaneous firing rate.

Type II neurons have other conductances, such as a persistent, TTX-sensitive, sodium conductance, which produces plateau potentials (Leonard and Llinas, 1990). This is different from the TTX-sensitive sodium conductance that produces the rapid rise of the AP. These cells also have high-threshold spikes that are calcium-dependent and are evident particularly in cells with longer-duration APs bearing a prominent inflection on the falling phase of the AP (Leonard and Llinas, 1990; Takakusaki et al., 1997). Based on these observations, it is difficult to ascribe phasic (e.g., PGO bursts) and tonic (e.g., change in state) firing patterns of individual neurons that match up exactly with what we know about the modulation of waking and REM sleep by the PPN. Table 4.1 summarizes the known firing patterns during different states and intrinsic properties of RAS neurons in vivo and in vitro. In addition to these intrinsic membrane properties, in Chapter 9, we will discuss newly discovered membrane channels in these cells that impart them the capacity for high-frequency activity.

INPUTS TO PPN NEURONS

Any transmitter that modulates the expression of either I_A or LTS currents will have a dramatic impact on firing frequency and bursting behavior. PPN neurons have excitatory amino acid and GABAergic receptors with ionotropic effectors and serotonergic, noradrenergic, cholinergic, adenosinergic, and histaminergic receptors with a common metabotropic effector (Garcia-Rill et al., 2003).

Serotonin (5-HT)—5-HT induces a tetrodotoxin-resistant hyperpolarization in most PPN neurons (Leonard and Llinas, 1990; Luebke et al., 1992) but inhibits only 25% of noncholinergic neurons (Garcia-Rill et al., 2003). In the cat, inhibition of serotonergic transmission does not increase REM sleep (Sakai and Crochet, 2001). Moreover, 5-HT neurons in the dorsal raphe cease firing during drug-induced atonia (Steinfels et al., 1983), but not during REM sleep without atonia (Trulson et al., 1981), in keeping with evidence showing that injections of 5-HT into the mesopontine region suppressed REM sleep (Horner et al., 1997). Conversely, injection of a 5-HT$_{1A}$ agonist into the PPN did not affect PGO wave induction, and the PPN has few 5-HT$_{1A}$ binding sites compared to LDT (Sanford et al., 1996). This represents yet another example of differences between PPN and LDT cells, such that the function of the LDT remains unclear. Earlier studies concluded that only about 12% of the 5-HT terminals in PPN synapse on cholinergic cells (Steininger et al., 1992). 5-HT$_2$ receptors were found on cholinergic neurons (Morilak and Ciaranello, 1993), but others suggested that 5-HT_2 receptors were found instead on noncholinergic neurons (Fay and Kubin, 2000). These findings require more evidence to be reconciled in order to understand specifically how 5-HT modulates PPN function. A later study described the inhibition of extracellularly recorded "REM-on" neurons in the mesopontine region by a 5-HT_{1A} agonist, which had minimal effect on "wake/REM-on" neurons (Thakkar et al., 1998). These authors suggested that 5-HT raphe neurons slow their firing in drowsiness and SWS, and less inhibition via 5HT_{1A} receptors on cholinergic PPN "REM-on" cells leads to increased firing to promote REM sleep (see also Strecker et al., 1999). This suggests that some cholinergic PPN neurons will be inhibited, while others will not be affected by 5-HT.

Glutamic acid (GLU)—PPN neurons may receive glutamate inputs from the reticular formation (Steininger et al., 1992; Stevens et al., 1992). Responses in guinea pig LDT neurons appear mediated by both *n*-methyl-D-aspartic acid (NMDA) and non-NMDA receptors, especially kainic acid (KA) receptors (Sanchez and Leonard, 1996). Injections of glutamate (which activates both NMDA and KA receptors) into the PPN are known to induce increased duration of both wakefulness and REM sleep in the rat (Datta and Siwek, 1997; Datta et al., 2001). NMDA receptors appear to be involved specifically in the induction of increased duration of wakefulness (Datta and Siwek, 1997), whereas KA receptors may be involved selectively in the induction of increased duration of REM sleep (Datta, 2009). A shift in the development of PPN responses to NMDA vs. KA receptors is described in Chapter 5. Briefly, responses to NMDA decrease with the developmental decrease in REM sleep, and responses to KA increase during this period (Garcia-Rill et al., 2003). It is not clear if "wake-on" neurons in the mesopontine region are preferentially driven by NMDA receptor activation, while "REM-on" neurons are preferentially driven by KA receptor activation, but this would be explained by the developmental changes reported. "Wake/REM-on" neurons may represent a subpopulation that is affected by both NMDA and KA receptor activation. As discussed in Chapter 9, the differential activity of these cells during waking compared to during REM sleep may be driven at least in part by intracellular mechanisms.

Noradrenaline (NA)—NA has been reported to hyperpolarize 7–15 day (i.e., during the first half of the developmental decrease in REM sleep) cholinergic mesopontine neurons in the LDT (Williams and Reiner, 1993), although similar studies have never been carried out in the PPN, or at later stages (15–30 days) of the developmental decrease in REM sleep. Moreover, α_2 adrenergic receptor development is known to undergo marked changes in the LC and other mesopontine regions after 15 days of age in the rat (Happe et al., 2004). Interestingly, in adult cholinergic mesopontine neurons, only about one-half showed α_2 adrenergic receptor immunocytochemical labeling, whereas one-third of cholinergic cells showed α_1 adrenergic receptor labeling (Happe et al., 2004). These authors suggested that there may be differential activation of different populations of cholinergic cells that underlie aspects of behavioral state control. However, they did not perform triple labeling studies in order to determine if different populations of cholinergic cells are differentially modulated by the two receptor subtypes. This would determine if there is indeed differential adrenergic regulation of "REM-on" vs. "wake/REM-on" cells. Our results show that the clonidine, the α_2 adrenergic receptor agonist, blocked or reduced the hyperpolarization-activated inward cation conductance Ih, so that its effects on decreasing the firing rate of a specific population of PPN neurons could be significant (Bay et al., 2006).

Acetylcholine (ACh)—The PPN is modulated by cholinergic input (Velazquez-Moctezuma et al., 1989). The cholinergic cells in the PPN have been hypothesized to be regulated, at least in part, by muscarinic (mAChR) and nicotinic (nAChR) cholinergic receptors located on either the soma and dendrites to modulate AP discharge or on presynaptic axon terminals to regulate neurotransmitter release (Tribollet et al., 2004; Vilaro et al., 1994). Studies utilizing in situ hybridization have multiply labeled cells with choline acetyltransferase mRNA with those containing mRNA for the five subtypes of mAChRs (M_1-M_5) in the PPN and found that the cholinergic cells not only expressed primarily the M_2 muscarinic receptor subtype but also had detectable levels of M_3 and M_4 mRNA (Vilaro et al., 1994). However, it is not known if these were colocalized on the same cells, if they represented different populations of cells, or if the different subtypes were differentially located on the soma, dendrites, and/or presynaptic terminals. PPN neurons receive cholinergic input from the contralateral PPN and bilaterally from the LDT (Semba and Fibiger, 1992). Previous electrophysiological studies have demonstrated muscarinic inhibition of PPN neurons, although these were carried out only on the guinea pig, which has an adult-like wake–sleep cycle at birth (Leonard and Llinas, 1994).

Later work in the adult PPN using *c-Fos* expression showed that nicotine, when administered systemically, activated primarily the noncholinergic cells, whereas only a few cholinergic cells were labeled with *c-Fos*. The authors concluded that the PPN is a primary target for the action of nicotine and hypothesized that the noncholinergic cells were directly activated by nicotine, which in turn modulates the activity of the cholinergic cells (Lanca et al., 2000). When microinjections of nicotine were made in the medial pontine reticular formation of freely moving cats, REM sleep was induced (Velazquez-Moctezuma et al., 1989). Compared to controls, these animals showed increased REM sleep at the expense of both waking and stage I SWS while also exhibiting a decrease in the time to REM sleep onset. Nicotine, injected intraperitoneally, has been shown to increase waking in wild-type mice while also causing a significant decrease in non-REM and REM sleep during the first hour post injection. However, in nicotine receptor knockout (KO) $\beta_2^{-/-}$ mice, there was no change in state following similar injections, indicating that the alerting effects of nicotine are probably mediated by the β_2-containing nicotinic receptors (Lena et al., 2004). When nicotine was administered subcutaneously in an acute model, there was a dose-dependent decrease in REM sleep such that the highest doses suppressed REM sleep the most when compared to saline or lower amounts of nicotine. However, when nicotine was given repeatedly at low doses, there was an increase in REM sleep following the third injection that lasted until the following day. It should be noted that these effects could be blocked by mecamylamine, indicative of a purely nicotinic receptor effect. These studies show that nicotine may have differing effects on sleep architecture depending on route of administration, dosing, and/or species of animal used. In Chapter 13, we will discuss the clinical implications of tobacco dependence, both in development and in adulthood. Most people do not realize the pronounced effect that nicotine has on the RAS.

The responses to PPN neurons bearing muscarinic cholinergic receptors will be described in detail in Chapter 9, since these are involved in generating high-frequency activity in PPN cells.

GABA—The GABAergic neurons in the PPN are intermingled with cholinergic neurons (Ford et al., 1995) and express *c-Fos* during carbachol-induced REM sleep in the cat (Torterolo et al., 2001). The PPN also receives GABAergic projections from the substantia nigra (Rye, 1997), which not only terminate mostly on noncholinergic neurons (Steininger et al., 1992) but also terminate on cholinergic neurons (Grofova and Zhou, 1998).

Given the variety of transmitter inputs to the PPN, it is obvious that a complex regulation of its function in modulating waking and REM sleep is present. In Chapter 5, we will discuss the changes that these transmitter inputs undergo during the developmental decrease in REM sleep. In Chapter 9, we will discuss how the cholinergic input to the PPN has a biphasic effect in regulating activity in the PPN and how there are two separate intracellular mechanisms that guide the function of PPN cells differentially during waking compared to during REM sleep. However, before we can understand how the PPN regulates these high-frequency states, we need to know more about the details of its intranuclear organization. Are different cell types active during different states? Are groups of unlike cells, such as microcircuits, acting in unison? How does the activity in the PPN synchronize or achieve coherence? The discovery of electrical coupling in the PPN, as described below, was a critical element in understanding intranuclear PPN function (Figure 4.4).

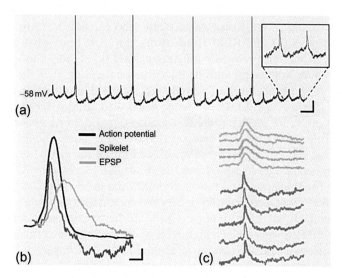

FIGURE 4.4 **Spikelets in the PPN.** (a) Occasionally (~7–10% of the time), spikelets were observed interspersed with action potentials. (b) Spikelets (dark gray) are biphasic waveforms indicative of low-amplitude action potentials (black) and not monophasic postsynaptic potentials (light gray). The assumption is that the cell being recorded and that has full amplitude action potentials is linked through gap junctions to a second neuron that is also firing. Because of the high resistance of the gap junctions, the amplitude of the spikelet is only a fraction of the intrinsic action potential. (c) Biphasic spikelets (dark gray) differ significantly from monophasic excitatory postsynaptic potentials (light gray).

ELECTRICAL COUPLING

Electrical coupling in the mammalian brain was described in the 1970s, and connexin 36 (Cx36) was identified as the main gap junction protein in neurons (Connors and Long, 2004). Connexins are proteins that oligomerize into hemichannels and are transported to the membrane, where they can be apposed or unapposed, congregate into "plaques," allow the passage of ions and molecules as large as cyclic adenosine monophosphate, and have a half-life of several hours (Bukauskas and Verselis, 2004; Laird, 2006). Cx36 gap junctions are the least voltage-dependent (Hughes et al., 2002) and are closed by low intracellular pH and low intracellular calcium (Srinivas et al., 1991). Spikelets are stereotypical, usually rhythmic, subthreshold depolarizing potentials thought to reflect firing in the coupled neurons (Long et al., 2004) (Figure 4.5). Electrical coupling (Deleuze and Huguenard, 2006; Fuentealba and Steriade, 2005; Landisman and Connors, 2005) and Cx36 labeling (Liu and Jones, 2003) are present in the reticular nucleus of the thalamus. Electrical coupling is also evident in thalamic relay neurons but only when activated by metabotropic glutamate receptor agonists (Hughes et al., 2002). Electrical synapses appear mainly between

FIGURE 4.5   **Dye-coupled PPN neurons.** Recordings in which a single PPN neuron was patched using an electrode containing neurobiotin showed upon immunocytochemical processing that more than one cell was labeled. That is, the dye was presumed to flow between neurons through gap junctions. In the *pars compacta* of the PPN, typically only two cells were labeled, appearing to exist in pairs. On rare occasions, more than two cells were labeled, as in the case shown here in which four cells were dye-coupled. The labeling was not unspecific since dye was not detected outside the cells and in cases labeled cells at a distance without labeling intervening neurons.

GABAergic neurons in the thalamic reticular nucleus and in the cortex, where they may enhance the synchrony of gamma oscillations (Beierlein et al., 2000; Hestrin and Galarreta, 2005; Traub et al., 2001). On the other hand, Cx36 KO mice show disruption of gamma frequency (Hormudzi et al., 2001). Electrical coupling in the hippocampus may promote seizure activity (Kohling et al., 2001; Yang and Michelson, 2001).

Until recently, the presence of electrical coupling in the RAS had not been reported (Garcia-Rill et al., 2007, 2008; Heister et al., 2007). Figure 4.4a shows typical spikelets (low-amplitude biphasic waves), presumably generated by APs by another, coupled cell, in a PPN neuron manifesting its own APs (high-amplitude spikes). Figure 4.4b compares the biphasic waveforms of APs (black), their similarity to spikelets (dark gray), and marked differences with monophasic excitatory postsynaptic potentials (EPSPs) (light gray). Figure 4.4c shows differences between EPSPs (light gray) and spikelets (dark gray). Injection of the dye neurobiotin into a single cell will produce labeling in other cells with which the recorded cell is electrically coupled. This is known as dye coupling since the dye flows through the gap junctions to cells coupled to the recorded, injected neuron. Figure 4.5 is an example of dye coupling in the PPN. In this case, recording and injection of a single cell led to dye coupling in three other PPN neurons. Usually, however, only one other cell was dye-coupled when a PPN neuron was injected, suggesting that couplets represent the normal distribution of electrical coupling.

Electrical coupling, arousal, and anesthesia—Some anesthetics induce their effects by blocking gap junctions, and one stimulant induces its effects by increasing electrical coupling. On the one hand, gap junctions can be blocked through membrane fluidization such as that induced by anesthetics such as halothane and propofol (Evans and Boitano, 2001; He and Burt, 2000). Oleamide promotes sleep and blocks gap junctions (Boger et al., 1998; Guan et al., 1997). Anandamide enhances adenosine levels to induce sleep (Murillo-Rodriguez et al., 2003) and blocks gap junctions (Evans and Boitano, 2001). One possibility is that a mechanism by which these agents may induce sleep and/or anesthesia is through blockade of electrical coupling in the RAS. Carbenoxolone, a gap junction blocker previously used to treat ulcers by blocking junctions between acid-secreting cells, decreases the synchronicity of gamma oscillations (Gigout et al., 2006), as well as seizure activity (Gareri et al., 2004). The glycyrrhetinic acid derivative 18-glycyrrhetinic acid blocks gap junctions (Rozental et al., 2000). Quinine and a related compound, mefloquine, used in the treatment of malaria, also block gap junctions (Cruikshank et al., 2004; Gajda et al., 2005; Pais et al., 2003; Srinivas et al., 2001). Mefloquine is particularly useful because it blocks Cx36 and Cx50 gap junctions at low concentrations but not Cx43, Cx32, or Cx26 gap junctions (Cruikshank et al., 2004). All of these agents share the property of being soporific. In fact, a common reason for the lack of compliance in their prescribed medical use is the sleep-inducing "side effects" of these agents. The issue of prescription drugs that can induce somnolence is germane to the operation of public and private transport and will be discussed in terms of public policy in Chapter 14.

On the other hand, increased electrical coupling can be induced by low calcium artificial cerebrospinal fluid and trimethylamine, a gap junction opener, perhaps by increasing intracellular pH (Pais et al., 2003; Rozental et al., 2000). The stimulant modafinil is approved for use in treating excessive sleepiness in narcolepsy, sleepiness in obstructive sleep apnea, and shift work sleep disorder. As we will discuss in Chapter 11, it is also prescribed in a number of neuropsychiatric conditions. Early publications on this agent acknowledged that the mechanism of action of modafinil was unknown, although there was general agreement that it increases glutamatergic, adrenergic, and histaminergic transmission and decreases GABAergic transmission (Ballon and Feifel, 2006). However, a definitive study found that a major mechanism of action of modafinil is to increase electrical coupling between cortical interneurons, thalamic reticular neurons, and inferior olive neurons (Urbano et al., 2007). Following pharmacological blockade of connexin permeability, modafinil restored electrotonic coupling. The effects of modafinil were counteracted by the gap junction blocker mefloquine. These authors proposed that modafinil acts in these areas by increasing electrotonic coupling in such a way that the high input resistance typical of GABAergic neurons is reduced. These authors proposed that this "shunting effect" of modafinil may activate the whole TC system by mildly downregulating inhibitory networks while increasing synchronous activation of both interneurons and noninhibitory neurons (Urbano et al., 2007).

The background behind this suggestion stems from imaging studies using voltage-sensitive dyes that have shown that inhibitory interneurons modulate cortical activation by afferent input (Contreras and Llinas, 2001). These cortical interneurons manifest gamma band (~40 Hz) oscillations (Traub et al., 2001) that are reduced by pharmacological blockade of gap junctions. The presence of both electrical coupling and chemical synapses between inhibitory interneuron networks is thought to enhance the timing of APs (Beierlein et al., 2000; Connors and Long, 2004; Galarreta and Hestrin, 2001; Laird, 2006). In the cortex, electrical coupling may contribute to AP synchronization and network oscillations and may (a) coordinate and reinforce IPSPs and (b) promote coincidence detection in inhibitory networks (Fricker and Miles, 2001; Traub et al., 2001). There is extensive electrical coupling in the cortex during development (Peinado et al., 1993), but epileptiform activity is virtually absent. Therefore, electrical coupling may result in a "shunting effect" by decreasing the input resistance of coupled cells, thereby reducing the excitability of cortical interneurons. Such a "shunting effect" has been proposed as the mechanism behind the general absence of epileptiform discharges during early postnatal development of the rat neocortex (Sutor et al., 1994). In conclusion, modafinil may elicit its stimulant effects by decreasing input resistance in GABAergic cells and thus GABA release, essentially disinhibiting networks throughout the brain. On the other hand, anesthetics such as halothane and propofol block gap junctions mainly present in GABAergic neurons, thereby increasing input resistance and thus firing and in turn increasing GABA release and essentially inhibiting networks throughout the brain.

Electrical coupling and gap junctions in the RAS—We described the presence of dye and electrical coupling in the RAS, specifically in the Pf, PPN, and SubCD (Garcia-Rill et al., 2007; Heister et al., 2007). We also found that modafinil decreased the resistance of PPN and SubCD neurons (Garcia-Rill et al., 2007), in keeping with results in the cortex, reticular thalamus, and inferior olive (Urbano et al., 2007). The effects of modafinil were evident in the absence of changes in resting membrane potential or of changes in the amplitude of induced EPSCs and were blocked by low concentrations of the gap junction blocker mefloquine and also in the absence of changes in resting membrane potential or in the amplitude of induced EPSCs. This suggests that these compounds do not act indirectly by affecting voltage-sensitive channels such as potassium channels, but rather modulate electrical coupling via gap junctions. Figure 4.6 is an example of dye coupling within the SubCD nucleus, in which one cell was recorded but the neurobiotin labeled two adjacent neurons.

FIGURE 4.6   **Dye-coupled neurons in the SubCD.** Recordings in the SubCD also showed a 7–10% incidence of dye coupling. In this case, two neurons located in the SubCD were found to be dye-coupled, presumably through gap junctions. No unspecific dye labeling was observed, and like in the PPN, dye-coupled cells were usually found in pairs. However, in order to confirm that electrical coupling is present, recording from two coupled cells is necessary.

Figure 4.7 shows the protocol used to study electrically coupled SubCD neurons. Two recording electrodes are used, in this case both filled with fluorescent dye (Lucifer yellow) so that both neurons can be visualized during recording. After adding tetrodotoxin (sodium channel blocker to prevent APs within the slice) and synaptic blockers (CNQX to block AMPA/KA receptors, APV to block NMDA receptors, and gabazine to block GABA receptors, together referred to as CAG) to block all fast synaptic transmission, then signals between the neurons can only be due to electrical coupling through gap junctions. Delivery of a current step to the first cell will result in a reduced response in the second cell, but only if they are coupled. Delivery of a current step to the second cell should result in a reduced response in the first cell. The ratio between the response in the cell to which the current step was applied and the response in the second cell is known as the coupling ratio. The coupling ratio for Pf, PPN, and SubCD cells was in the 3–4% range. This suggests that a 70 mV AP

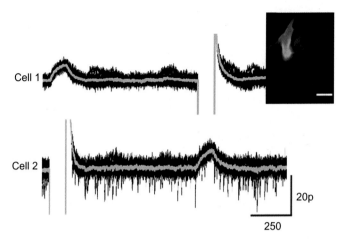

Cell 1

Cell 2

20p

250

FIGURE 4.7   **Electrophysiological confirmation of electrical coupling.** In this case, Cell 1 in the SubCD was patched, followed by patching of the nearby Cell 2. Lucifer yellow in both electrodes showed fluorescence in both cells. In the presence of fast synaptic blockers (AVP, CNQX, gabazine, and strychnine—to block glutamatergic, GABAergic, and glycinergic inputs) and tetrodotoxin (TTX—to block all action potential generation), depolarizing pulses delivered to Cell 2 (left side) induced a lower-amplitude response in Cell 1. Conversely, depolarizing steps administered to Cell 1 (right side) induced a lower-amplitude response in Cell 2. This suggests that the response in the adjacent cell could only have been mediated by electrical coupling through gap junctions. The ratio of the amplitude of the response to the stimulus step to the amplitude of the response in the adjacent cell provides the coupling ratio. In the PPN, SubCD, and Pf, the coupling ratio was in the order of 3–4%. That is, spikelets occurring in one cell will depolarize the coupled cells by a small amount in order to promote coherence. However, the final confirmation that electrical coupling is present is if the response is blocked by a gap junction blocker.

in one cell will induce a spikelet or change in membrane response in the coupled cell in the order of 2–3 mV.

Figure 4.8 shows that the cholinergic agonist carbachol-induced oscillations in SubCD cells at frequencies in the theta range (right-side power spectrum). After addition of fast synaptic blockers (CAG), the record in Figure 4.8a showed that there were no EPSPs evident but low-frequency biphasic oscillations typical of spikelets could be observed (indicating mediation by electrical coupling). Following superfusion with the cholinergic agonist carbachol (CAR), the frequency of the spikelets increased, as evident in the record in Figure 4.8b (red record). A faster time base of this record is shown in the box at bottom right in which the biphasic nature of the spikelets is evident. After addition of the gap junction blocker carbenoxolone (CBX) at the bottom (green record), the oscillations were blocked without changing resting membrane potential. The power spectrum on the right shows that the frequency of the spikelets increased to the theta range after addition of CAR. We used another gap junction blocker,

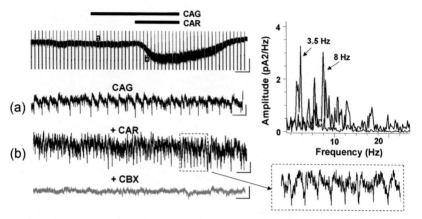

FIGURE 4.8   **Cholinergic modulation of spontaneous spikelets, indicative of electrical coupling.** Left. In the presence of fast synaptic blockers (CAG), the nonselective cholinergic receptor agonist, carbachol (CAR), induced an inward current in this PPN neuron with spontaneous oscillations (top, vertical bar 50 pA, horizontal bar 4 s). Enlarged records from points (a) and (b) are shown below. Recordings (a, red record) showed spontaneous oscillations in the presence of CAG. Recordings (b, black record) demonstrated that CAR increased the frequency of oscillations. The box on the bottom right shows enlarged 1 s record (b). Carbenoxolone, a gap junction blocker, completely blocked the oscillations (bottom, green record). Blockade did not affect membrane potential. This suggested that the oscillations were modulated by electrical coupling. Scale bars for the enlarged records are vertical 10 pA and horizontal 500 ms. Top right. Power spectrum of the oscillations in (a, red line) and (b, black line).

mefloquine, which is known to affect potassium channels, but with similar effects. Because gap unction blockers have side effects, it is important to use both in order to ensure that unspecific effects are not responsible for the changes observed.

We should emphasize that gap junctions are not necessary for generating subthreshold oscillations; rather, they are required for clustering coherent oscillatory activity (Leznik and Llinas, 2005). Oscillatory properties of single neurons, such as subthreshold gamma band oscillations (Chapter 9), endow the system with important reset dynamics, while gap junctions are mainly required for synchronized neural activity (Llinas and Yarom, 1986). For example, oscillations in single inferior olive neurons persisted over a limited range of voltages in the presence of gap junction blockers, leading these authors to conclude that gap junctions allow oscillations to persist over a wide range of membrane potentials making their frequency independent of membrane potential (Llinas and Yarom, 1986). This is in keeping with the suggestion that the output from a network of electrically coupled cells may maintain oscillations over a wide range of voltages that would otherwise inactivate rhythmic conductances in single neurons.

Cx36 in the RAS—We sampled each of the nuclei of interest for Cx36 protein using 400 μm slices like those used for recordings and punched 1 mm of Pf, PPN, and SubCD on days 10 and 30, spanning the developmental decrease in REM sleep (Chapter 5). Figure 4.9 shows the locations of punches of tissue from slices containing the SubCD (left), the Pf (middle), and the PPN (right). Cx36 protein levels in relation to tubulin were measured from these punches taken at the beginning (day 10) and at the end of the developmental decrease in REM sleep (day 30). Cx36 levels at day 30 were about 25% of those at day 10 in Pf, PPN, and SubCD. This suggests the presence of a marked developmental decrease in Cx36 protein levels, with considerable amounts of Cx36 protein still present in the adult, suggesting that this gap junction protein may participate in developmental regulation and contribute to wake-sleep control in the adult (Garcia-Rill et al., 2007; Heister et al., 2007).

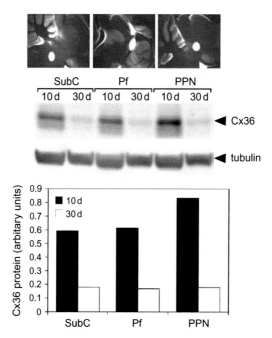

FIGURE 4.9 **Presence of connexin 36 (Cx36) protein in the SubCD, Pf, and PPN.** At the top are shown sagittal slices cut in the same manner as those used for recordings. On the left is a medial section through the region of the SubCD in which a 1 mm core was taken for Cx36 protein measure. In the middle is a section through the Pf in which a 1 mm core was taken containing the Pf. On the right is a lateral section through the PPN showing the location of the 1 mm core from the region of the PPN. In the middle are the results of Cx36 labeling relative to tubulin for each of the three regions. Cx36 levels were higher at the beginning of the developmental decrease in REM sleep at 10 days than at the end of the decrease at 30 days. At the bottom are histograms showing the relative amounts of Cx36 in each region and at 10 and 30 days.

A study of Cx36 KO mice showed that gap junctions in the inferior olive were mostly abolished but the genetically uncoupled neurons still generated subthreshold oscillations of their membrane potential at a similar amplitude and frequency as those recorded in the wild type (De Zeeuw et al., 2003). This implies that inferior olive oscillations may be generated by conductances of individual neurons; however, these authors suggested the alternative explanation that oscillations of Cx36 KO animals are qualitatively different from those in wild-type mice and that these differences are due to structural and physiological changes in the Cx36 KO mouse. That is, single-cell oscillations may occur in such Cx36 KO cells as a result of compensatory ionic mechanisms not encountered in the normal animal. Others showed instead that oscillatory properties of uncoupled inferior olive neurons are not caused by long-term compensatory changes but are due to the single-cell properties of normal inferior olive cells (Llinas and Yarom, 1986). Thus, gap junctions are required for synchronizing the oscillatory responses of neurons and for clustering of coherent rhythmic activity.

What functional significance does electrical coupling have if Cx36 KO mice survive, sleep, and walk? Several studies showed that while gross motor activity patterns appeared normal in the Cx36 KO mouse, detailed analysis of motor patterns showed a 10–20 ms degradation in coordination (Placatonakis et al., 2004) and a delay of more than 20 ms in the optokinetic reflex (Kistler et al., 2002). These differences appear vital to survival. In terms of consciousness, arousal, and sleep, two laboratories have found that cortical gamma oscillations in vitro are abnormal in Cx36 KO mice (Deans et al., 2001; Yang and Michelson, 2001). A later study on Cx36 KO mice showed that Cx36 gap junctions contributed to gamma oscillations (Buhl et al., 2003), whereas others showed that gap junctions play a role in learning and memory (Frisch et al., 2005). These studies suggest that gap junctions confer an advantage in timing, probably due to their ability to promote coherence in brain rhythms for optimal performance. The unique ability of coupled cells to maintain synchrony across a wide range of membrane potentials probably allows brain rhythms to persist for longer periods without waning. This is a critical property for the maintenance of states like attention that require prolonged activation.

Great care is needed, however, in the interpretation of findings in Cx36 KO animals, since some brain cells can compensate for the absence of cell coupling (He et al., 2000) and AMPA currents are absent in these animals (Maher and Westbrook, 2007). That is, these animals are not really "pure" Cx36 KO; the genetic manipulation alters certain basic elements of cell function. On the other hand, these animals have potential for exploring developmental changes since preliminary findings suggest that thalamic reticular neurons upregulate gap junctions in development, presumably to compensate for changes in membrane properties in order to maintain the strength of electrical coupling (Parker et al., 2009). Needless to say, research on neuronal gap junctions requires greater attention, especially given the importance they bear on one of the main determinants of nerve cell activity, coherence.

CELL CLUSTERS

Hebb advanced the concept of cell assemblies to describe a network that is repeatedly activated, and thus, the synaptic connections among members of the circuit are strengthened (Hebb, 1949). Recent studies determined that in the hippocampus, there are, according to one group, "patches" of entorhinal cortex cells that play a role in learning and memory (Ray et al., 2014). Another group described clusters of cells that may have joint function as "islands" that form a hexagonal lattice over the cortex (Kimura et al., 2014). Are there clusters of cells that act together in the PPN to modulate coherence and frequency? The answer is we do not know; however, there is intriguing evidence to suggest that the PPN may contain subgroups of functional units. Figure 4.5 shows a group of four dye-coupled PPN neurons, which is a rare event, especially since most dye coupling is evident between only two cells rather than four. For example, Figure 4.6 shows a couplet of SubCD cells that were electrically and dye-coupled, and Figure 4.10 shows a pair of dye-coupled Pf cells. That is, these couplets are evident in all three nuclei of the RAS, the Pf, PPN, and SubCD that modulate waking and REM sleep.

As discussed early in this chapter, anatomically, neurons in the PPN are scattered such that in the *pars compacta*, there are GLU, ACh, and GABA neurons in the ratio of 5:3:2, respectively (Wang and Morales, 2009). Studies using calcium imaging in the PPN *pars compacta* reveal an interesting anatomical organization within the nucleus. Figure 4.11 shows that pairs of PPN cells are labeled throughout the nucleus even in the control, unstimulated condition. The spatial separation between couplets suggests that there are clusters of cells throughout the nucleus. Since electrically coupled neurons generally represent GABAergic neurons, we speculate that there are five GLU and three ACh neurons closely associated with each GABAergic pair. That is, there may be clusters of approximately 10 neurons scattered within the *pars compacta* that may create a functional subgroup. Much additional evidence is required to support this hypothesis, but it may be possible to dissect such an organization to determine how the nucleus as a whole generates coherent activity at specific frequencies. It is also important to determine how PPN neurons respond to sensory input and how that input generates coherent activity. Similar functional clustering has been proposed for the hippocampus, especially in relation to epileptic networks (Muldoon et al., 2013). In a

FIGURE 4.10 **Dye-coupled neurons in the Pf.** Recordings in the Pf also showed a 7–10% incidence of dye coupling. In this case, two neurons located in the Pf were found to be dye-coupled, presumably through gap junctions. No unspecific dye labeling was observed, and like in the PPN, dye-coupled cells were usually found in pairs. However, in order to confirm that electrical coupling is present, recording from two coupled cells is necessary.

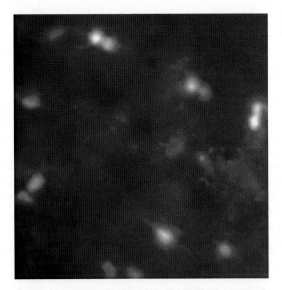

FIGURE 4.11 **Calcium imaging of cell pairs in the PPN.** View of background activity in a PPN slice after loading with Fura2. Note that the majority of cells showing high levels of calcium in the absence of stimulation were found in pairs spaced about 100 μm apart.

study of cell assemblies in the hippocampus, Buzsáki described subsets of about 10 neurons that showed repeated synchronous firing during open field exploration (Buzsáki, 2010). Interestingly, the timescale of activity between these neurons had a median of 23 ms, and the peak optimal timescale was ~16 ms, that is, most activity occurred in the 40–60 Hz range. We hypothesize that a similar temporal relationship will be evident among cell clusters in the PPN. The evidence for gamma band activity in the PPN is discussed in Chapter 9.

CLINICAL IMPLICATIONS

The clinical consequences of the characteristics of the wiring diagram of the RAS are multiple and will be discussed in relation to psychiatric disorders in Chapter 11, in relation to neurological disorders in Chapter 12, and in relation to drugs of abuse in Chapter 13.

References

Aghajanian, G.K., 1985. Modulation of a transient outward current in serotonergic neurones by alpha 1-adrenoceptors. Nature 315, 501–503.

Aston-Jones, G., Bloom, F.E., 1981. Activity of norepinephrine-containing locus coeruleus neurons in behaving rats anticipates fluctuations in the sleep-waking cycle. J. Neurosci. 1, 876–886.

Aston-Jones, G., Zhu, Y., Card, J.P., 2004. Numerous GABAergic afferents to locus ceruleus in the pericerulear dendritic zone: possible interneuronal pool. J. Neurosci. 24, 2313–2321.

Ballon, J.S., Feifel, D., 2006. A systematic review of Modafinil: potential clinical uses and mechanisms of action. J. Clin. Psychiat. 67, 554–566.

Baraban, J.M., Aghajanian, G.K., 1981. Noradrenergic innervation of serotonergic neurons in the dorsal raphe: demonstration by electron microscopic autoradiography. Brain Res. 204, 1–11.

Bay, K., Mamiya, K., Good, C., Skinner, R.D., Garcia-Rill, E., 2006. Alpha-2 adrenergic regulation of pedunculopontine nucleus (PPN) neurons during development. Neuroscience 141, 769–779.

Beierlein, M., Gibson, J.R., Connors, B.W., 2000. A network of electrically coupled interneurons drives synchronized inhibition in neocortex. Nat. Neurosci. 3, 904–910.

Bjorvatn, B., Fagerland, S., Eid, T., Ursin, R., 1997. Sleep/waking effects of a selective 5-HT1A receptor agonist given systemically as well as perfused in the dorsal raphe nucleus in rats. Brain Res. 770, 81–88.

Boger, D.L., Henriksen, S.J., Cravatt, B.F., 1998. Oleamide: an endogenous sleep-inducing lipid and prototypical member of a new class of biological signaling molecules. Curr. Pharm. Des. 4, 303–314.

Brown, R.E., Sergeeva, O.A., Eriksson, K.S., Haas, H.L., 2002. Convergent excitation of dorsal raphe serotonin neurons by multiple arousal systems (orexin/hypocretin, histamine and noradrenaline). J. Neurosci. 22, 8850–8859.

Buhl, D.L., Harris, K.D., Hormuzdi, S.G., Monyer, H., Buzsáki, G., 2003. Selective impairment of hippocampal gamma oscillations in connexin-36 knock-out mouse in vivo. J. Neurosci. 23, 1013–1018.

Bukauskas, F.F., Verselis, V.K., 2004. Gap junction channel gating. Biochim. Biophys. Acta 1662, 42–60.

Buzsáki, G., 2010. Neural syntax: cell assemblies, synapsembles, and readers. Neuron 68, 362–385.

Cespuglio, R., Gomez, M.E., Walker, E., Jouvet, M., 1979. Effect of cooling and electrical stimulation of nuclei of raphe system on states of alertness in cat. Electroencephalogr. Clin. Neurophysiol. 47, 289–308.

Cespuglio, R., Gomez, M.E., Faradji, H., Jouvet, M., 1982. Alterations in the sleep-waking cycle induced by cooling of the locus coeruleus area. Electroencephalogr. Clin. Neurophysiol. 54, 570–578.

Christie, M.J., Williams, J.T., North, R.A., 1989. Electrical coupling synchronizes subthreshold activity in locus coeruleus neurons in vitro from neonatal rats. J. Neurosci. 9, 3584–3589.

Connors, B.W., Long, M.A., 2004. Electrical synapses in the mammalian brain. Annu. Rev. Neurosci. 27, 393–418.

Contreras, D., Llinás, R., 2001. Voltage-sensitive dye imaging of neocortical spatiotemporal dynamics to afferent activation frequency. J. Neurosci. 21, 9403–9413.

Cruikshank, S.J., Hopperstad, M., Younger, M., Connors, B.W., Spray, D.C., 2004. Potent block of Cx36 and Cx50 gap junction channels by mefloquine. Proc. Natl. Acad. Sci. USA 101, 12364–12369.

Datta, S., 2009. Regulation of neuronal activities within REM sleep-sign generators. Sleep 32, 1135–1147.

Datta, S., Maclean, R.R., 2007. Neurobiological mechanisms for the regulation of mammalian sleep–wake behavior: reinterpretation of historical evidence and inclusion of contemporary cellular and molecular evidence. Neurosci. Biobehav. Rev. 31, 775–824.

Datta, S., Siwek, D.F., 1997. Excitation of the brain stem pedunculopontine tegmentum cholinergic cells induces wakefulness and REM sleep. J. Neurophysiol. 77, 2975–2988.

Datta, S., Siwek, D.F., 2002. Single cell activity patterns of pedunculopontine tegmentum neurons across the sleep–wake cycle in the freely moving rats. J. Neurosci. Res. 70, 611–621.

Datta, S., Spoley, E.E., Patterson, E.H., 2001. Microinjection of glutamate into the pedunculopontine tegmentum induces REM sleep and wakefulness in the rat. Am. J. Physiol. Regul. Integr. Comp. Physiol. 280, R752–R759.

Deans, M.R., Gibson, J.R., Sellitto, C., Connors, B.W., Paul, D.L., 2001. Synchronous activity of inhibitory networks in neocortex requires electrical synapses containing connexin36. Neuron 31, 477–485.

De Zeeuw, C.I., Chorev, E., Devor, A., Manor, Y., Van Der Giessen, R.S., De Jeu, M.T., Hoogenraad, C.C., Bijman, J., Ruigrok, T.J., French, P., Jaarsma, D., Kistler, W.M., Meier, C., Petrasch-Parwez, E., Dermietzel, R., Sohl, G., Gueldenagel, M., Willecke, K., Yarom, Y., 2003. Deformation of network connectivity in the inferior olive of connexin 36-deficient mice is compensated by morphological and electrophysiological changes at the single neuron level. J. Neurosci. 23, 4700–4711.

Delagrange, P., Canu, M.H., Rougeul, A., Buser, P., Bouyer, J.J., 1993. Effects of locus coeruleus lesions on vigilance and attentive behaviour in cat. Behav. Brain Res. 53, 155–165.

Deleuze, C., Huguenard, J.R., 2006. Distinct electrical and chemical connectivity maps in the thalamic reticular nucleus. J. Neurosci. 26, 8633–8645.

Deurveilher, S., Hennevin, E., 2001. Lesions of the pedunculopontine tegmental nucleus reduce paradoxical sleep (PS) propensity: evidence from a short-term PS deprivation study in rats. Eur. J. Neurosci. 13, 1963–1976.

Dodt, H.U., Pawelzik, H., Zieglgansberger, W., 1991. Actions of noradrenaline on neocortical neurons in vitro. Brain Res. 545, 307–311.

Egan, T.M., North, R.A., 1985. Acetylcholine acts on m2-muscarinic receptors to excite rat locus coeruleus neurones. Br. J. Pharmacol. 85, 733–735.

Egan, T.M., North, R.A., 1986. Actions of acetylcholine and nicotine on rat locus coeruleus neurons in vitro. Neuroscience 19, 565–571.

Egan, T.M., Henderson, G., North, R.A., Williams, J.T., 1983. Noradrenaline-mediated synaptic inhibition in rat locus coeruleus neurones. J. Physiol. 345, 477–488.

Evans, W.H., Boitano, S., 2001. Connexin mimetic peptides: specific inhibitors of gap-junctional intercellular communication. Biochem. Soc. Trans. 29, 606–612.

Fay, R.A., Kubin, L., 2000. 5-HT 2A receptor-like immunoreactivity is present in cells and cellular processes adjacent to mesopontine nitric oxide synthase-containing neurons. Sleep 23, A108.

Ford, B., Holmes, C.J., Mainville, L., Jones, B.E., 1995. GABAergic neurons in the rat mesencephalic tegmentum: codistribution with cholinergic and other tegmental neurons projecting to the lateral hypothalamus. J. Comp. Neurol. 363, 177–196.

Fricker, D., Miles, R., 2001. Interneurons, spike timing, and perception. Neuron 32, 771–774.

Frisch, C., De Souza-Silva, M.A., Sohl, G., Güldenagel, M., Willecke, K., Huston, J.P., Dere, E., 2005. Stimulus complexity dependent memory impairment and changes in motor performance after deletion of the neuronal gap junction protein connexin36 mice. Behav. Brain Res. 157, 177–185.

Fuentealba, P., Steriade, M., 2005. The reticular nucleus revisited: intrinsic network properties of a thalamic pacemaker. Prog. Neurobiol. 75, 125–141.

Gajda, Z., Szupera, Z., Blaszo, G., Szente, M., 2005. Quinine, a blocker of neuronal Cx36 suppresses seizure activity in rat neocortex in vivo. Epilepsia 46, 1581–1591.

Galarreta, M., Hestrin, S., 2001. Electrical synapses between GABA-releasing interneurons. Nature Neurosci. 2, 425–433.

Garcia-Rill, E., 1991. The pedunculopontine nucleus. Prog. Neurobiol. 36, 363–389.

Garcia-Rill, E., Skinner, R.D., 1991. Modulation of rhythmic functions by the brainstem. In: Shimamura, M., Grillner, S., Edgerton, V.R. (Eds.), Neurobiological Basis of Human Locomotion. Japan Scientific Press, Tokyo, Japan, pp. 137–158.

Garcia-Rill, E., Kobayashi, T., Good, C., 2003. The developmental decrease in REM sleep. Thalamus Relat. Syst. 2, 115–131.

Garcia-Rill, E., Heister, D.S., Ye, M., Charlesworth, A., Hayar, A., 2007. Electrical coupling: novel mechanism for sleep-wake control. Sleep 30, 1405–1414.

Garcia-Rill, E., Charlesworth, A., Heister, D., Ye, M., Hayar, A., 2008. The developmental decrease in REM sleep: the role of transmitters and electrical coupling. Sleep 31, 673–690.

Gareri, P., Condorelli, D., Belluardo, N., Russo, E., Loiacono, A., Barresi, V., Trovato-Salinaro, A., Mirone, M.B., Ferreri Ibbadu, G., De Sarro, G., 2004. Anticonvulsant effects of carbenoxolone in genetically epilepsy prone rats (GEPRs). Neuropharmacology 47, 1205–1216.

Gigout, S., Louvel, J., Kawasaki, H., D'Antuono, M., Armand, V., Kurcewicz, I., Olivier, A., Laschet, J., Turak, B., Devaux, B., Pumain, R., Avoli, M., 2006. Effects of gap junction blockers on human neocortical synchronization. Neurobiol. Dis. 22, 496–508.

Green, R.W., Rainnie, D.G., 1991. Mechanisms affecting neuronal excitability in brainstem cholinergic centers and their impact on behavioral state. In: Lydic, R., Baghdoyen, H.A., (Eds.), Handbook of Behavioral State Control. Cellular and molecular mechanisms. CRC Press, New York, pp. 277–309.

Grofova, I., Zhou, M., 1998. Nigral innervation of cholinergic and glutamatergic cells in the rat mesopontine tegmentum: light and electron microscopic anterograde tracing and immunohistochemical studies. J. Comp. Neurol. 395, 359–379.

Guan, X., Cravatt, B.F., Ehring, G.R., Hall, J.E., Boger, D.L., Lerner, R.A., Gilula, N.B., 1997. The sleep-inducing lipid oleamide deconvolutes gap junction communication and calcium wave transmission in glial cells. J. Cell Biol. 139, 1785–1792.

Hagan, J.J., Leslie, R.A., Patel, S., Evans, M.L., Wattam, T.A., Holmes, S., Benham, C.D., Taylor, S.G., Routledge, C., Hemmati, P., Munton, R.P., Ashmeade, T.E., Shah, A.S., Hatcher, J.P., Hatcher, P.D., Jones, D.N., Smith, M.I., Piper, D.C., Hunter, A.J., Porter, R.A., Upton, N., 1999. Orexin A activates locus coeruleus cell firing and increases arousal in the rat. Proc. Natl. Acad. Sci. USA 96, 10911–10916.

Happe, H.K., Coulter, C.L., Gerety, M.E., Sanders, J.D., O'Rourke, M., Bylund, D.B., Murrin, L.C., 2004. Alpha-2 adrenergic receptor development in rat CNS: an autoradiographic study. Neuroscience 123, 157–178.

He, D.S., Burt, J.M., 2000. Mechanism and selectivity of the effects of halothane on gap junction channel function. Circ. Res. 86, 1–10.

He, S., Weiler, R., Vaney, D.I., 2000. Endogenous dopaminergic regulation of horizontal cell coupling in the mammalian retina. J. Comp. Neurol. 418, 33–40.

Hebb, D.O., 1949. The Organization of Behavior: A Neuropsychological Theory. Wiley, New York.

Heister, D.S., Hayar, A., Charlesworth, A., Yates, C., Zhou, Y., Garcia-Rill, E., 2007. Evidence for electrical coupling in the subcoeruleus (SubC) nucleus. J. Neurophysiol. 97, 3142–3147.

Hestrin, S., Galarreta, M., 2005. Electrical synapses define networks of neocortical GABAergic neurons. Trends Neurosci. 28, 304–309.

Hormudzi, S.G., Pais, I., LeBeau, F.E., Towers, S.K., Rozov, A., Buhl, E.H., Whittington, M.A., Monyer, H., 2001. Impaired electrical signaling disrupts gamma frequency oscillations in connexin-deficient mice. Neuron 31, 487–495.

Horner, R.L., Sanford, L.D., Annis, D., Pack, A.I., Morrison, A.R., 1997. Serotonin at the laterodorsal tegmental nucleus suppresses rapid-eye-movement sleep in freely behaving rats. J. Neurosci. 17, 7541–7552.

Hughes, S.W., Blethyn, K.L., Cope, D.W., Crunelli, V., 2002. Properties and origin of spikelets in thalamocortical neurons in vitro. Neuroscience 3, 395–401.

Jia, H.G., Yamuy, J., Sampogna, S., Morales, F.R., Chase, M.H., 2003. Colocalization of gamma-aminobutyric acid and acetylcholine in neurons in the laterodorsal and pedunculopontine tegmental nuclei in the cat: a light and electron microscopic study. Brain Res. 992, 205–219.

Jones, B.E., Beaudet, A., 1987. Distribution of acetylcholine and catecholamine neurons in the cat brainstem: a choline acetyltransferase and tyrosine hydroxylase immunohistochemical study. J. Comp. Neurol. 261, 15–32.

Jones, B.E., Yang, T.Z., 1985. The efferent projections from the reticular formation and the locus coeruleus studied by anterograde and retrograde axonal transport in the rat. J. Comp. Neurol. 242, 56–92.

Jouvet-Mounier, D., Astic, L., Lacote, D., 1970. Ontogenesis of the states of sleep in rat, cat, and guinea pig during the first postnatal month. Dev. Psychobiol. 2, 216–239.

Kamondi, A., Williams, J., Hutcheon, B., Reiner, P., 1992. Membrane properties of mesopontine cholinergic neurons studied with the whole-cell patch-clamp technique: implications for behavioral state control. J. Neurophysiol. 68, 1359–1372.

Kang, Y., Kitai, S.T., 1990. Electrophysiological properties of pedunculopontine neurons and their postsynaptic responses following stimulation of substantia nigra reticulata. Brain Res. 535, 79–95.

Kayama, Y., Negi, T., Sugitani, M., Iwama, K., 1982. Effects of locus coeruleus stimulation on neuronal activities of dorsal lateral geniculate nucleus and perigeniculate reticular nucleus of the rat. Neuroscience 7, 655–666.

Kayama, Y., Ohta, M., Jodo, E., 1992. Firing of "possibly" cholinergic neurons in the rat laterodorsal tegmental nucleus during sleep and wakefulness. Brain Res. 569, 210–220.

Keele, N.B., Neugebauer, V., Shinniek-Gallagher, R., 1991. Differential effects of metabotropic glutamate receptor antagonists on bursting activity in the amygdala. J. Neurophysiol. 81, 2056–2065.

Kimura, T., Pignatelli, M., Suh, J., Kohara, K., Yoshiki, A., Abe, K., Tonegawa, S., 2014. Island cells control temporal association memory. Science 343, 896–901.

Kistler, W.M., De Jeu, M.T., Elgersma, Y., Van Der Giessen, R.S., Hensbroek, R., Luo, C., Koekkoek, S.K., Hoogenraad, C.C., Hamers, F.P., Gueldenagel, M., Sohl, G., Willecke, K., De Zeeuw, C.I., 2002. Analysis of Cx36 knockout does not support tenet that olivary gap junctions are required for complex spike synchronization and normal motor performance. Ann. N. Y. Acad. Sci. 978, 391–404.

Kohling, R., Gladwell, S.J., Bracci, E., Vreugdenhil, M., Jefferys, J.G.R., 2001. Prolonged epileptiform bursting induced by 0 Mg++ in rat hippocampal slices depends on gap junctional coupling. Neuroscience 105, 579–587.

Kohlmeier, K.A., Reiner, P.B., 1999. Noradrenaline excites non-cholinergic laterodorsal tegmental neurons via two distinct mechanisms. Neuroscience 93, 619–630.

Laird, D.W., 2006. Life cycle of connexins in health and disease. Biochem. J. 394, 527–543.

Lanca, J.A., Sanelli, T.R., Corrigall, W.A., 2000. Nicotine-induced fos expression in the pedunculopontine mesencephalic tegmentum in the rat. Neuropharmacology 39, 2808–2817.

Landisman, C.E., Connors, B.W., 2005. Long-term modulation of electrical synapses in the mammalian thalamus. Science 310, 1809–1813.

Lavoie, B., Parent, A., 1994. Pedunculopontine nucleus in the squirrel monkey: distribution of cholinergic and monoaminergic neurons in the mesopontine tegmentum with evidence for the presence of glutamate in cholinergic neurons. J. Comp. Neurol. 344, 190–209.

Lena, C., Popa, D., Grailhe, R., Escourrou, P., Changeux, J.-P., Adrien, J., 2004. β2-containing nicotinic receptors contribute to the organization of sleep and regulate putative micro-arousals in mice. J. Neurosci. 24, 5711–5718.

Leonard, C.S., Llinas, R.R., 1990. Electrophysiology of mammalian pedunculopontine and laterodorsal tegmental neurons in vitro: implications for the control of REM sleep. In: Steriade, M., Biesold, D. (Eds.), Brain Cholinergic Systems. Oxford Science, Oxford, pp. 205–223.

Leonard, C.S., Llinas, R.R., 1994. Serotonergic and cholinergic inhibition of mesopontine cholinergic neurons controlling REM sleep: an in vitro electrophysiological study. Neuroscience 59, 309–330.

Leonard, C.S., Kerman, I., Blaha, G., Taveras, E., Taylor, B., 1995. Interdigitation of nitric oxide synthase-, tyrosine hydroxylase-, and serotonin-containing neurons in and around the laterodorsal and pedunculopontine tegmental nuclei of the guinea pig. J. Comp. Neurol. 362, 411–432.

Levitt, P., Moore, R.Y., 1978. Noradrenaline neuron innervation of the neocortex in the rat. Brain Res. 139, 219–231.

Leznik, E., Llinas, R.R., 2005. Role of gap junctions in synchronized oscillations in the inferior olive. J. Neurophysiol. 94, 2447–2456.

Lidov, H.G., Grzanna, R., Molliver, M.E., 1980. The serotonin innervation of the cerebral cortex in the rat—an immunohistochemical analysis. Neuroscience 5, 207–227.

Liu, X.B., Jones, E.G., 2003. Fine structural localization of connexin-36 immunoreactivity in mouse cerebral cortex and thalamus. J. Comp. Neurol. 466, 457–467.

Llinas, R.R., Yarom, Y., 1986. Oscillatory properties of guinea-pig inferior olivary neurons and their pharmacological modulation: an in vitro study. J. Physiol. 376, 163–182.

Long, M.A., Landisman, C.E., Connors, B.W., 2004. Small clusters of electrically coupled neurons generate synchronous rhythms in the thalamic reticular nucleus. J. Neurosci. 24, 341–349.

Luebke, J.I., Greene, R.W., Semba, K., Kamondi, A., McCarley, R.W., Reiner, P.B., 1992. Serotonin hyperpolarizes cholinergic low threshold burst neurons in the rat laterodorsal tegmental nucleus in vitro. Proc. Natl. Acad. Sci. USA 89, 743–747.

Maher, B.J., Westbrook, G.L., 2007. Sensory deprivation and the development of dendrodendritic excitation in olfactory glomeruli. Neurosci. Abst. 33, 503–515.

Mena-Segovia, J., Micklem, B.R., Nair-Roberts, R.G., Ungless, M.A., Bolam, J.P., 2009. GABAergic neuron distribution in the pedunculopontine nucleus defines functional subterritories. J. Comp. Neurol. 515, 397–408.

Mihailescu, S., Guzman-Marin, R., Dominguez Mdel, C., Drucker-Colin, R., 2002. Mechanisms of nicotine actions on dorsal raphe serotoninergic neurons. Eur. J. Pharmacol. 452, 77–82.

Monckton, J.E., McCormick, D.A., 2002. Neuromodulatory role of serotonin in the ferret thalamus. J. Neurophysiol. 87, 2124–2136.

Morilak, D.A., Ciaranello, R.D., 1993. 5-HT2 receptor immunoreactivity on cholinergic neurons of the pontomesencephalic tegmentum shown by double immunofluorescence. Brain Res. 627, 49–54.

Muldoon, A.F., Soltesz, I., Cossart, R., 2013. Spatially clustered neuronal assemblies comprise the microstructure of synchrony in chronically epileptic networks. Proc. Natl. Acad. Sci. USA 110, 3567–3572.

Murillo-Rodriguez, E., Blanco-Centurion, C., Sanchez, C., Piomelli, D., Shiromani, P.J., 2003. Anandamide enhances extracellular levels of adenosine and induces sleep: an in vivo microdialysis study. Sleep 26, 943–947.

Nitz, D., Siegel, J.M., 1997a. GABA release in the dorsal raphe nucleus: role in the control of REM sleep. Am. J. Physiol. 273, R451–R455.

Nitz, D., Siegel, J.M., 1997b. GABA release in the locus coeruleus as a function of sleep/wake state. Neuroscience 78, 795–801.

Olpe, H.R., 1981. The cortical projection of the dorsal raphe nucleus: some electrophysiological and pharmacological properties. Brain Res. 216, 61–71.

Pais, I., Hormudzi, S.G., Monyer, H., Traub, R.D., Wood, I.C., Buhl, E.H., Whittington, M.A., LeBeau, F.E., 2003. Sharp wave-like activity in the hippocampus in vitro in mice lacking the gap junction protein connexin 36. J. Neurophysiol. 89, 2046–2054.

Parker, P.R.L., Cruikshank, S.J., Connors, B.W., 2009. Stability of electrical coupling despite massive developmental changes of intrinsic neuronal physiology. J. Neurosci. 29, 9761–9770.

Peinado, A., Yuste, R., Katz, L.C., 1993. Extensive dye coupling between rat neocortical neurons during the period of cortical formation. Neuron 10, 103–114.

Pickel, V.M., Joh, T.H., Reis, D.J., 1977. A serotonergic innervation of noradrenergic neurons in nucleus locus coeruleus: demonstration by immunocytochemical localization of the transmitter specific enzymes tyrosine and tryptophan hydroxylase. Brain Res. 131, 197–214.

Placatonakis, D.G., Bukovsky, A.A., Zeng, X.H., Kiem, H.P., Welsh, J.P., 2004. Fundamental role of inferior olive connexin 36 in muscle coherence during tremor. Proc. Natl. Acad. Sci. USA 101, 7164–7169.

Ray, S., Naumann, R., Burgalossi, A., Tang, O., Schmitt, H., Brecht, M., 2014. Grid-layout and theta-modulation of layer 2 pyramidal neurons in medial entorhinal cortex. Science 343, 891–896.

Reese, N.B., Garcia-Rill, E., Skinner, R.D., 1995a. The pedunculopontine nucleus—auditory input, arousal and pathophysiology. Prog. Neurobiol. 47, 105–133.

Reese, N.B., Garcia-Rill, E., Skinner, R.D., 1995b. Auditory input to the pedunculopontine nucleus. I. Evoked potentials. Brain Res. Bull. 37, 247–255.

Reese, N.B., Garcia-Rill, E., Skinner, R.D., 1995c. Auditory input to the pedunculopontine nucleus. II. Unit responses. Brain Res. Bull. 37, 257–264.

Rozental, R., Srinivas, M., Spray, D.C., 2000. How to close a gap junction channel. In: Bruzzone, R., Glaume, C., (Eds.), Methods in Molecular Biology, Vol. 154, Connexin methods and protocols. Humana Press, New Jersey, pp. 447–477.

Rye, D.B., 1997. Contributions of the pedunculopontine region to normal and altered REM sleep. Sleep 20, 757–788.

Sakai, K., Crochet, S., 2001. Role of dorsal raphe neurons in paradoxical sleep generation in the cat: no evidence for a serotonergic mechanism. Eur. J. Neurosci. 13, 103–112.

Sanchez, R., Leonard, C.S., 1996. NMDA-receptor-mediated synaptic currents in guinea pig laterodorsal tegmental neurons in vitro. J. Neurophysiol. 76, 1101–1111.

Sanford, L.D., Tejanir Butt, S.M., Ross, R.J., Morrison, A.R., 1996. Elicited PGO waves in rats: lack of 5-HT 1a inhibition in putative pontine generator region. Pharmacol. Biochem. Behav. 53, 323–327.

Semba, K., Fibiger, H.C., 1992. Afferent connections of the laterodorsal and the pedunculopontine tegmental nuclei in the rat: a retro- and anterograde transport and immunocytochemical study. J. Comp. Neurol. 323, 387–410.

Shouse, M.N., Siegel, J.M., 1992. Pontine regulation of REM sleep components in cats: integrity of the pedunculopontine tegmentum (PPT) is important for phasic events but unnecessary for atonia during REM sleep. Brain Res. 571, 50–63.

Srinivas, M., Rozental, R., Kojima, T., Dermietzel, R., Mehler, M., 1991. Functional properties of channels formed by the neuronal gap junction protein connexin 36. J. Neurosci. 19, 9848–9855.

Srinivas, M., Hopperstad, M.G., Spray, D.C., 2001. Quinine blocks specific gap junction channel subtypes. Proc. Natl. Acad. Sci. USA 98, 10942–10947.

Steinfels, G.F., Heym, J., Strecker, R.E., Jacobs, B.L., 1983. Raphe unit activity in freely moving cats is altered by manipulations of central but not peripheral motor systems. Brain Res. 279, 77–84.

Steininger, T.L., Rye, D.B., Wainer, B.H., 1992. Afferent projections to the cholinergic pedunculopontine nucleus and adjacent midbrain extrapyramidal area in the albino rat. J. Comp. Neurol. 321, 515–543.

Steininger, T.L., Wainer, B.H., Blakely, R.D., Rye, D.B., 1997. Serotonergic dorsal raphe nucleus projections to the cholinergic and noncholinergic neurons of the pedunculopontine tegmental region: a light and electron microscopic anterograde tracing and immunohistochemical study. J. Comp. Neurol. 382, 302–322.

Steriade, M., McCarley, R., 2005. Brain Control of Wakefulness and Sleep. Springer, New York, 728 pp.

Steriade, M., Datta, S., Pare, D., Oakson, G., Curro Dossi, R.C., 1990. Neuronal activities in brain-stem cholinergic nuclei related to tonic activation processes in thalamocortical systems. J. Neurosci. 10, 2541–2559.

Steriade, M., Paré, D., Datta, S., Oakson, G., Curro Dossi, R., 1990b. Different cellular types in mesopontine cholinergic nuclei related to ponto–geniculo–occipital waves. J. Neurosci. 10, 2560–2579.

Steriade, M., Dossi, R.C., Pare, D., Oakson, G., 1991. Fast oscillations (20–40 Hz) in thalamocortical systems and their potentiation by mesopontine cholinergic nuclei in the cat. Proc. Natl. Acad. Sci. USA 88, 4396–4400.

Stevens, D.R., McCarley, R.W., Greene, R.W., 1992. Excitatory amino acid-mediated responses and synaptic potentials in medial pontine reticular formation neurons of the rat in vitro. J. Neurosci. 12, 4188–4194.

Strecker, R.E., Thakkar, M., Porkka-Heiskanen, M., Dauphin, L.J., Bjorkum, A.A., McCarley, R.W., 1999. Behavioral state-related changes of extracellular serotonin concentration in the pedunculopontine tegmental nucleus: a microdialysis study in freely moving animals. Sleep Res. Online 2, 21–27.

Sutor, B., Hablitz, J.J., Rucker, F., ten Bruggencate, G., 1994. Spread of epileptiform activity in the immature rat neocortex studied with voltage-sensitive dyes and laser scanning microscopy. J. Neurophysiol. 72, 1756–1768.

Takakusaki, K., Kitai, S.T., 1997. Ionic mechanisms involved in the spontaneous firing of tegmental pedunculopontine nucleus neurons of the rat. Neurosci. 78, 771–794.

Takakusaki, K., Shiroyame, T., Kitai, S.T., 1997. Two types of cholinergic neurons in the rat tegmental pedunculopontine nucleus: electrophysiological and morphological characterization. Neurosci. 79, 1089–1109.

Thakkar, M.M., Strecker, R.E., McCarley, R.W., 1998. Behavioral state control through differential serotonergic inhibition in the mesopontine cholinergic nuclei: a simultaneous unit recording and microdialysis study. J. Neurosci. 18, 5490–5497.

Torterolo, P., Yamuy, J., Sampogna, S., Morales, F.R., Chase, M.H., 2001. GABAergic neurons of the laterodorsal and pedunculopontine tegmental nuclei of the cat express c-fos during carbachol-induced active sleep. Brain Res. 892, 309–319.

Traub, R.D., Kopell, N., Bibbig, A., Buhl, E.H., LeBeau, F.E., Whittington, M.A., 2001. Gap junctions between interneuron dendrites can enhance synchrony of gamma oscillations in distributed networks. J. Neurosci. 21, 9476–9486.

Tribollet, E., Bertrand, D., Marguerat, A., Raggenbass, M., 2004. Comparative distribution of nicotinic receptor subtypes during development, adulthood and aging: an autoradiographic study in the rat brain. Neuroscience 12, 405–420.

Trulson, M.E., Jacobs, B.L., 1979. Raphe unit activity in freely moving cats: correlation with level of behavioral arousal. Brain Res. 163, 135–150.

Trulson, M.E., Trulson, V.M., 1982. Activity of nucleus raphe pallidus neurons across the sleep-waking cycle in freely moving cats. Brain Res. 237, 232–237.

Trulson, M.E., Jacobs, B.L., Morrison, A.R., 1981. Raphe unit activity during REM sleep in normal cats and in pontine lesioned cats displaying REM sleep without atonia. Brain Res. 226, 74–91.

Urbano, F.J., Leznik, E., Llinas, R., 2007. Modafinil enhances thalamocortical activity by increasing neuronal electronic coupling. Proc. Natl. Acad. Sci. USA 104, 12554–12559.

Vandermaelen, C.P., Aghajanian, G.K., 1983. Electrophysiological and pharmacological characterization of serotonergic dorsal raphe neurons recorded extracellularly and intracellularly in rat brain slices. Brain Res. 289, 109–119.

Velazquez-Moctezuma, J., Gillin, J., Shiromani, P., 1989. Effect of specific M1, M2 muscarinic receptor agonists on REM sleep generation. Brain Res. 503, 128–131.

Vilaro, M.T., Palacios, J.M., Mengod, G., 1994. Multiplicity of muscarinic autoreceptor subtypes? Comparison of the distribution of cholinergic cells and cells containing mRNA for five subtypes of muscarinic receptors in the rat brain. Mol. Brain Res. 21, 30–46.

Wang, H.L., Morales, M., 2009. Pedunculopontine and laterodorsal tegmental nuclei contain distinct populations of cholinergic, glutamatergic and GABAergic neurons in the rat. Eur. J. Neurosci. 29, 340–358.

Williams, J.A., Reiner, P.B., 1993. Noradrenaline hyperpolarizes identified rat mesopontine cholinergic neurons in vitro. J. Neurosci. 13, 3878–3883.

Williams, J.T., North, R.A., Shefner, S.A., Nishi, S., Egan, T.M., 1984. Membrane properties of rat locus coeruleus neurones. Neuroscience 13, 137–156.

Yang, Q., Michelson, H.B., 2001. Gap junctions synchronize the firing of inhibitory interneurons in guinea pig hippocampus. Brain Res. 907, 139–143.

5

Development and the RAS

Paige Beck, MD, PhD and Edgar Garcia-Rill, PhD

Center for Translational Neuroscience, Department of Neurobiology and
Developmental Sciences, University of Arkansas for Medical Sciences,
Little Rock, AR, USA

DEVELOPMENT OF WAKE–SLEEP STATES

Kleitman described the pattern of wake–sleep cycles from infancy to adulthood in man (Kleitman, 1953). He proposed in the theory of "advanced wakefulness" that the level of consciousness associated with wakefulness shows many gradations and the transition from the primitive nonconscious wakefulness of the neonate to the fully conscious "advanced" wakefulness of the adult does not occur all at once. According to Kleitman, the infant undergoes a greater amount of waking than sleep and a net gain of wakefulness capacity with age, along with the addition of a certain level of consciousness compared to the primitive state. His studies showed that the newborn exhibited periods of 1–2 h of waking with 4–5 h of sleep, but with age, the sleep periods became consolidated into longer and fewer episodes until, by 10 years of age, a single long-uration sleep episode followed by a long waking episode became the norm.

Superimposed on this wake–sleep rhythm, he proposed the basic rest–activity cycle (BRAC), with periods of greater activity during waking as well as during sleep, when rapid eye movement (REM) sleep episodes take place. In other words, Kleitman proposed that the periodic recurrence of the REM state during sleep reflects the operation of the BRAC, the influence of which continued to modulate brain function in wakefulness, that is, an ultradian BRAC of about 90 min cycle duration. His studies showed that the wake–sleep cycle of the human neonate was marked by periods of increased activation every 50–60 min. This periodicity increased gradually with age to the ~90 min cycle of the adult, although it was masked by surges of cortical activity during waking (Figure 5.1).

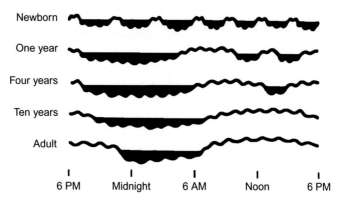

Newborn

One year

Four years

Ten years

Adult

6 PM   Midnight   6 AM   Noon   6 PM

FIGURE 5.1   **Age-related changes in sleep patterns according to Kleitman.** This drawing from Kleitman (1953) shows the alternating wake–sleep periods from birth until adulthood. In the newborn, total sleep time is twice the amount of waking, which is divided into short segments. Superimposed on the wake–sleep cycle is the BRAC with a periodicity of about 50 min. With age, nighttime sleep becomes progressively consolidated into longer episodes until there is a single sleep episode. Superimposed on this cycle is the BRAC that assumes a periodicity of about 90 min. He proposed in the theory of "advanced wakefulness" that the level of consciousness associated with wakefulness shows many gradations, and the transition from the primitive nonconscious wakefulness of the neonate to the fully conscious "advanced" wakefulness of the adult does not occur all at once.

In support of this theory, early studies recorded gross body movements in adults and found that subjects showed discrete periods of spontaneous activity every 90–120 min during both sleep and wakefulness (Sterman and Hoppenbrouwers, 1971). A number of other studies have shown a similar periodicity in multiple measures, including eye movements, muscle tone (electromyographic—EMG) and EEG patterns (Othmer et al., 1969), physiological (especially temperature) and behavioral measures (Kripke and O'Donoghue, 1968), errors in a continuous performance task (Globus et al., 1971), performance of verbal and spatial tasks (Klein and Armitage, 1979), and a number of other physiological and performance measures (Rechstaffen et al., 1963; Friedman and Fisher, 1967; Oswald et al., 1970; Orr and Hoffman, 1974; Lavie and Kripke, 1981; see Monk et al., 1997, for caveats). Additional work has confirmed that ultradian oscillations in delta wave activity in the brain, adrenocorticotropic activity (cortisol secretion), and autonomic activity (heart rate variability) are coupled at 90–110 min (Gronfier et al., 1999). Similar BRACs have been observed in other species such as primates that show REM sleep and EMG cycles around 66 min (Kripke et al., 1976), felines that exhibit REM sleep (Delorme et al., 1964; Sterman et al., 1965; Ursin, 1988) and operant performance (Sterman et al., 1972) cycles around 25 min, and rodents that have REM sleep cycles from 8–17 min (Roldan et al., 1963; Van Twyler, 1969; Timo-Iaria et al., 1970).

There appears to be a linear relationship between the log of the body weight and the log of the period or amplitude of related measures of arousal

(Garcia-Rill, 2002). The sleep state-dependent, midlatency auditory-evoked P50 potential in the human, which has been proposed to be an expression of ascending cholinergic activation of the intralaminar thalamus (reviewed in Garcia-Rill and Skinner, 2002, and described in detail in Chapter 6), shows about a 90 min cycle in the peak amplitude of this "preattentional" measure (Garcia-Rill et al., 1999). The rodent equivalent of the human P50 potential, the P13 potential, also was found to show a periodicity in peak amplitude but at a 13–16 min period, similar to the rodent REM sleep-cycle duration. In addition, human performance on a psychomotor vigilance task, essentially a reaction time task, showed a ~90 min cycle in the lowest number of lapses.

Taken together, these observations suggest that there is a recurrent endogenous activation of ascending, preattentional-related systems, as well as descending, motor, and autonomic resetting-related systems. That is, the role of "early" REM sleep in providing endogenous stimulation and contributing to the maturation of thalamocortical pathways may persist in the form of a "late" REM sleep and vigilance drive in the adult. At face value, these findings suggest that we are more alert every 90 min. Why do we require a periodic shot of vigilance? What is the function of a BRAC that transcends the wake–sleep cycle and is expressed as timed oscillations of CNS activity during waking and as REM sleep periods during sleep? The more obvious possibilities include metabolic variation, pacemaker-like activity, and a fatigue-recovery phenomenon (Kleitman, 1953). Unfortunately, there is little information on the cellular mechanisms behind the BRAC that could shed further light into this process.

EVOLUTIONARY CONSIDERATIONS

For years, it was thought that the resting sleep/active sleep pattern was present only in mammals and birds and that this was linked to homeothermic control. The evolutionary perspective is that reptiles are sluggish in the cold (are ectothermic), but mammals can control their temperature (have endothermic control by using metabolic energy) and thus survive in a greater range of temperatures and environments. However, studies in monotremes (egg-laying mammals) have shown that the platypus has periods of resting sleep with high-amplitude EEG and that REM sleep occurs during moderately high-voltage EEG periods. These results suggest that REM sleep may have been present in the first mammals and evolved from reptilian ancestors (Siegel et al., 1999). The idea that REM sleep (the inactive state in reptiles) is the oldest state due to its homeostatic regulation (low cardiac and respiratory regulation) and that reptilian waking (the active state) is analogous to mammalian slow-wave sleep is gaining ground. This suggests that the newest state to evolve is waking. The term "advanced wakefulness" was originally proposed by Kleitman to describe the increase in waking seen in the human infant with age (Kleitman, 1953): "The ontogenetic evolution of advanced wakefulness manifests itself not only in a total increase in duration but also in a consolidation of the wakefulness phases into three, two, finally one continuous stretch, as the morning and later the afternoon naps are given up (Fig. 36.1)" (pg. 367). The same term was adopted much later as an evolutionary theory of advanced wakefulness (Rial et al., 1993). This theory suggested that a sleeping mammal has its reptilian brain stem either awake or in a reduced activity state (presumably equivalent to mammalian slow-wave and REM sleep), but when it awakens, it reaches a new state of awareness (cortical functionality). This fits in generally with what we know about thermoregulation and sleep, which shifts from waking control by the telencephalon (endothermic) to slow-wave sleep control by the diencephalon (fall in set point) to mesopontine REM sleep control (virtually ectothermic). However, evidence for evolutionary theories is, at best, difficult to muster so that these remain suggestive constructs regarding the evolution of wake–sleep states but basically unproved. The largely discredited theory that "ontogeny recapitulates phylogeny" would seem to describe a parallel between the evolution and the development of wake–sleep patterns. That theory suggested that animals go through stages when developing from the embryo to the adult that resemble stages in evolution. Regardless of whether any of these theories have a basis in fact, the developmental decrease in REM sleep is in favor of additional waking, while there is little change in slow-wave sleep across development.

We will see later in this volume that the process of reaching full consciousness is complicated by the fact that, upon waking, blood flow increases first in the upper brain stem and thalamus and only later increases in the frontal lobes (Balkin et al., 2002). These authors emphasized that the process of awakening entails a rapid reestablishment of consciousness (within a few minutes) followed by a relatively slow (20–30 min) reestablishment of full awareness. Cerebral blood flow was measured using $H_2^{15}O$ positron emission tomography and found to be most rapidly reestablished in the brain stem and thalamus, suggesting that the reactivation of these regions underlies the reestablishment of basic conscious awareness. Over the following 15+ min, further increases in cerebral blood flow were evident primarily in anterior cortical regions, suggesting recovery from "hypofrontality" is a slower process. These results question ideas that insist that the cortex is solely responsible for achieving conscious awareness.

DEVELOPMENTAL DECREASE IN REM SLEEP

The human newborn exhibits an even distribution of waking, REM sleep, and slow-wave sleep, spending about 8 h in each state (Roffwarg et al., 1966). Presumably, prenatal levels of REM sleep are at least that amount, if not more, with the fetus oscillating between slow-wave sleep and REM sleep in utero. After birth, there is a gradual decrease in REM sleep from about 8 h at birth to about 1 h by ~20 years of age, beyond which there is a mild decrease until senescence (Roffwarg et al., 1966) (Figure 5.2). After birth, slow-wave sleep may increase transiently before puberty, then gradually decreases from 8 h/day to 6 h/day in the elderly, but these changes are not significant. It should be noted that the decrease in total sleep time with age observed in this study needs reconsideration. This work did not take into account the possible presence of neurological disorders in the elderly subjects, suggesting that a decrease in total sleep time may not be evident in successful aging (Figure 5.2). Therefore, the gain observed in total waking time, from about 8 h at birth to about 16 h at maturity, is mostly at the expense of REM sleep duration. It is this gradual decrease in REM sleep that is the focus of much attention in the consideration of the development of wake–sleep states. This set of events generates several questions. Why would the brain require a state in which there is elevated high-frequency cortical activity before birth? What circumstances lead to the postnatal decrement in REM sleep duration? What is the role of such a decrease?

It has been suggested that REM sleep has the biological function of directing the course of brain maturation (Marks et al., 1995). This is in keeping with evidence suggesting that activity-dependent development may be a widespread mechanism directing neural connectivity throughout the

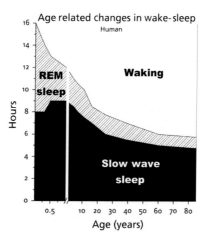

FIGURE 5.2 **Developmental decrease in REM sleep in the human.** This is a linear plot computed from the nonlinear figure in Roffwarg et al. (1966). At birth, total sleep time is 16 hours with one half of that spent in REM sleep. The decrease in REM sleep assumes an adult percentage soon after puberty and remains at 15–20% of total sleep time. Changes in slow-wave sleep are minimal during development, so that the decrease in REM sleep is in the favor of increased waking time. It is not certain that the total sleep time in these data at senescence is descriptive of normal aging without neurological and other brain diseases.

brain (Llinas, 1984; Marks et al., 1995). Under this hypothesis, REM sleep could provide endogenous stimulation at a time when the brain has little or no exogenous input. High-frequency brain stem activation, especially in the form of ponto–geniculo–occipital (PGO) waves, could contribute to the maturation of thalamocortical pathways. If this is the case in the neonate, what is the role of REM sleep in the adult? That is, is REM sleep in the fetus the same as REM sleep in the adult? The fact that there is a rebound in REM sleep after it has been suppressed indicates that it is important for the adult brain. However, REM sleep does not appear to be essential for survival in the adult. When the stressful components of REM sleep deprivation are removed, REM sleep suppression does not lead to death or disease (Jouvet, 1962). In fact, the use of imipramine, a tricyclic antidepressant, was found to lead to the elimination of dreams and other REM sleep signs (assuming this means absence of REM sleep) without major consequences (Jouvet, 1999).

In the rat, the developmental decrease in REM sleep occurs between 10 and 30 days of age, declining from over 75% of total sleep time at birth to about 15% of sleep time by 30 days of age (Figure 5.3) (Jouvet-Mounier et al., 1970). Some studies found that REM sleep rebounds after REM sleep deprivation in the developing rat was absent at 15 days of age, small at 21 days, and larger at 30 days (Feng et al., 2001). That is, the ontogeny of REM sleep rebound was related to the ontogeny of baseline REM sleep.

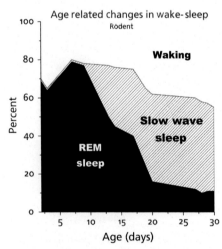

FIGURE 5.3 **Developmental decrease in REM sleep in the rodent.** This is a plot computed from the figure in Jouvet-Mounier et al. (1970). At birth, virtually all sleep is in REM sleep, which decreases between 10 and 30 days of age to the adult level of 10–15% of total sleep time. The decrease in REM sleep is at the favor of slow-wave sleep, so that there is no increase in waking during development in the rodent.

These data suggest that REM sleep rebound is important but mainly in the more developed or adult animal.

The equivalence of REM sleep in the fetus vs. adult has been questioned (Adrien, 1984), suggesting that the initial embryonic REM sleep is replaced by "true" REM sleep when the brain stem structures regulating it are sufficiently mature. If this is the case, is the presence of inhibition of muscle tone necessary or, for that matter, the same during "early" and "late" REM sleep? It appears that REM sleep atonia is a maturing process, leading to increased motor inhibition with age. For example, twitching during REM sleep is more intense in neonates (in both human and rodent), suggesting that motor inhibition may be less effective early compared with later (Siegel, 1999). This would suggest a disassociation between REM sleep itself and the atonia of REM sleep in development.

The "maturation" hypothesis of REM sleep has been extended to suggest that as soon as neurogenesis stops (e.g., around 21 days in the cat and rat), REM sleep appears. Thus, REM sleep has been proposed to replace genetic programs that maintain individuality; in other words, that REM sleep provides a mechanism for repeated reprogramming of individuality (Jouvet, 1999). Even if true, it would seem that such a function would be unnecessary in the fetus, again suggesting a difference between "early" REM sleep and "late" REM sleep. In addition, it would seem that REM sleep deprivation, contrary to observation, would lead to a "loss" of individuality if this theory were correct. Moreover, as will become evident below, the most dramatic changes in this system occur at around 15 days, not 21 days, in the rat.

REM SLEEP INHIBITION

It has been suggested that the direction of the changes in the decrease in REM sleep with development indicates that there is an active REM sleep inhibitory process (RIP). The RIP was proposed to develop during the first two weeks of life in the rat (Vogel et al., 2000) that may or may not be equivalent to the decrease seen in the human across puberty. This hypothesis predicts that (a) one or more inhibitory process becomes progressively stronger during this period (as outlined above, the pedunculopontine nucleus (PPN) receives inhibitory serotonergic, noradrenergic, cholinergic, and GABAergic inputs, all likely candidates) and (b) stimulation or blockade of this process will decrease or increase, respectively, the manifestations of REM sleep (Feng et al., 2001). These studies confirmed that four measures of REM sleep duration (tonic REM sleep, phasic REM sleep, mean REM duration, and number of REM episodes) and two measures of REM sleep delay (REM sleep latency and percent of nonsleep-onset REM sleep periods) all decreased in parallel, prompting the suggestion that an RIP is at play during this stage in development (Vogel et al.,

2000). A number of findings lend support to this idea by indicating that there are a number of changes in (a) the intrinsic properties of and (b) the neurotransmitter inputs to mesopontine cholinergic neurons during this developmental window of time, ~15 days in the rat.

CHANGES IN INTRINSIC PROPERTIES

Morphologically, PPN neurons change markedly from birth to 30 days of age in the rat. PPN neurons in animals aged 0–40 days were labeled histochemically with nicotinamide adenine dinucleotide phosphate (NADPH) diaphorase (a specific marker of mesopontine cholinergic neurons (Vincent et al., 1963)) and their cell areas measured over this developmental period (Skinner et al., 1989). There was a gradual increase in average area from about 200 μm^2 at 0–3 days to about 400 μm^2 at 12–14 days before peaking at about 500 μm^2 by 15–17 days. There was then a gradual decrease in mean cell area to about 300 μm^2 by 30–40 days, which was the mean area in adult animals. The hypertrophy observed was ascribed to developmental hallmarks related to the time of eye and ear opening at around 15 days of age in the rat (Sheets et al., 1988; Kungel et al., 1996). This phenomenon was accompanied by increased labeling attributed to changes in the production of structural proteins and of enzymes involved in metabolic events, axonal growth, and synaptogenesis around 15 days of age. The subsequent decrease in mean cell size did not appear to be due to selective loss of large cells, rather to overall cell shrinkage (Skinner et al., 1989) (Figure 5.4). Another report described an increase in choline acetyltransferase activity during the developmental decrease in REM sleep percent in the rat (Ninomiya et al., 2001), simultaneously with the hypertrophy described previously. How the hypertrophy observed relates to the dramatic decrease in REM sleep percentage remains to be determined. These results were recently confirmed by measurements of cell area in PPN neurons that had been injected intracellularly after recordings between 12 and 21 days in brain stem slices, emphasizing the presence of a maximal cell area around 500 μm^2 at around 15–17 days of age. However, later work showed that noncholinergic PPN neurons did not undergo marked changes in cell size during this developmental window, that is, the hypertrophy observed at ~15 days was evident only in cholinergic PPN neurons (Kobayashi et al., 2004).

Intrinsic membrane properties were measured in PPN neurons in brain stem slices from rats 12–21 days of age, that is, during the most significant decrease in REM sleep time and during a period bridging those studied in other labs (Kobayashi et al., 2002). Recorded neurons were injected with biocytin (processed for Texas Red fluorescent label) and identified as cholinergic using NADPH diaphorase histochemistry. The results suggested that type III (A+LTS) neurons were less evident after 17 days, apparently

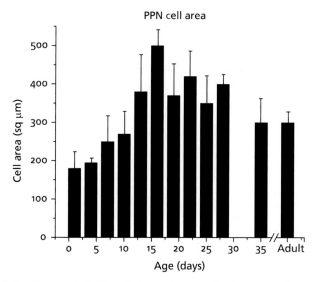

FIGURE 5.4 **Changes in PPN cell size during the developmental decrease in REM sleep in the rat.** Total cell area of cholinergic cells (noncholinergic cells do not undergo change during this period in development) was about $200\,\mu m^2$ at birth, increased to a peak of about $500\,\mu m^2$ at about 15 days and then decreased to the adult size of about $300\,\mu m^2$ by 30 days of age. The hypertrophy observed in PPN cholinergic neurons coincides with eye and ear opening and may be related to the significant increase in sensory afferent information. The hypertrophy is accompanied by increased labeling intensity for nitric oxide synthase, which is present in PPN cholinergic neurons.

differentiating into type I (LTS) neurons (Kobayashi et al., 2002). It is not known if type III (A+LTS) neurons, some of which may be cholinergic, change firing properties across 12–21 days or were simply not sampled. In general, these results suggest that there is a gradual increase across this stage in the number of cholinergic PPN neurons with LTS properties, a mechanism that changes the dynamic pattern of activity in this region (Figure 5.5). In addition, type II cells exhibit higher amplitude and longer duration afterhyperpolarization (AHP) than type I or type III cells and are likely to fire at lower rates. Interestingly, other results show that, after 15 days, type II cells become segregated into two groups based on long- vs. short-duration AHP (Kobayashi et al., 2002). The AHP duration in 12–15 day cells vs. 16–21 day cells was not statistically different, but there were a significant increase in the variability of AHP duration after 16 days and a segregation of AHP duration into two statistically different populations. PPN cells aged 16–21 days had either long AHP duration (~239 ms) or short AHP duration (~86 ms). Regression analysis of these two populations showed that the slopes of the best-fit lines describing the AHP duration had diverging slopes with age (Kobayashi et al., 2002). That is, after day 16, there was a gradually increasing difference in AHP duration

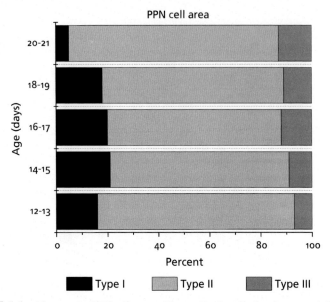

FIGURE 5.5   **Changes in PPN cell type with age.** During the developmental decrease in REM sleep, the proportion of type I cells (LTS, noncholinergic) peaks at about 15 days and decreases to 5–10% by the end of the developmental period. The number of type III cells (A+LTS, 1/3 cholinergic) increases slightly during the same time. The proportion of type II cells (A, 2/3 cholinergic) does not change during this window in development.

between the two populations of type II cells, indicative of a marked change in potential firing rate after 15 days.

Action potential (AP) duration also varied across this time period. The mean AP duration for the different types of PPN neurons was shorter for type II neurons compared with type I and type III cells. Type II neurons also showed significant decreases in AP duration over time (Kobayashi et al., 2002). Previous studies of type II PPN neurons described the presence of two subpopulations of type II cells in >30-day rats in terms of AP duration (Takakusaki and Kitai, 1997). We also found a clear segregation of type II cells in terms of AP duration, suggesting that short and long AP duration neurons are present throughout the 12–21-day epoch. However, our results also showed that there was a decrease in the number of long AP duration neurons with a concomitant increase in short AP duration neurons with age, a shift that occurred around 15 days of age, again indicating a change in intrinsic firing properties at this time (Kobayashi et al., 2002).

The segregation into long and short AP durations was related to the long and short AHP durations described above. There was no correlation between AP duration and AHP duration for cells aged 12–15 days. However, for cells aged 16–21 days, there was a significant correlation between AP duration and AHP duration (Kobayashi et al., 2002).

This relationship may be straightforward since a prolongation in the AP duration may allow additional calcium entry during repolarization, thus, improving the likelihood of activating the long-duration AHP mechanism. Previous studies found that a subpopulation of type II (A) cells can fire faster at around 10 Hz in pacemaker fashion (Leonard and Llinas, 1990). We suspect that these may include more short AHP and AP duration neurons. These properties imply that the period around 15 days of age is a transition time between lower compared with higher firing rates in at least some type II PPN neurons. While changes in membrane properties are evident during the developmental decrease in REM sleep (Figure 5.5), there are also marked changes in the responses to transmitters in the PPN.

CHANGES IN TRANSMITTER RESPONSES

The primary responses to the major inputs to PPN neurons are described in Chapter 3; however, these inputs change during the developmental decrease in REM sleep. Results show that some PPN cells were hyperpolarized by a 5-HT$_2$ agonist, but the effect did not change during the developmental decrease in REM sleep (Figure 5.6, left-pointing triangles). Using the 5-HT$_1$ receptor agonist 5-carboxyamidotryptamine, type I and type III PPN cells aged 12–21 days were depolarized, hyperpolarized, or not affected in about equal numbers. However, 80% of type II PPN cells were hyperpolarized but only 8% were depolarized, and depolarizing responses were evident up to only day 16, not thereafter (Kobayashi et al., 2002). These results suggest that there is a reorganization of serotonergic input to the PPN such that (a) it is both excitatory and inhibitory and then purely inhibitory after approximately 15 days and (b) the 5-HT$_1$ inhibition increases between 12 and 21 days (Figure 5.6, diamonds).

In terms of glutamatergic input, at day 12, most PPN neurons were strongly depolarized by NMDA, while the same neurons were only slightly depolarized by KA (Kobayashi et al., 2004). The NMDA effect gradually decreased, whereas the KA effect gradually increased. By 21 days, PPN neurons were only slightly depolarized by NMDA, while the same neurons were strongly depolarized by KA (Figure 5.6, crossing filled square and filled circle lines). These results indicate a switch in NMDA- vs. KA receptor activation of PPN cells at around 15 days (Kobayashi et al., 2004).

As discussed in Chapter 4, noradrenaline has been reported to hyperpolarize 7- to 15-day (i.e., during the first half of the developmental decrease in REM sleep) cholinergic mesopontine neurons in the laterodorsal tegmental nucleus (LDT) (Williams and Reiner, 1993), although similar studies have never been carried out in the PPN or at later stages (15–30 days) of the developmental decrease in REM sleep. Later results showed that the α_2-adrenergic receptor agonist clonidine hyperpolarized

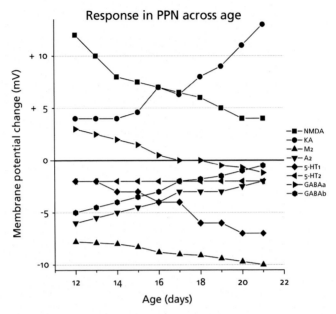

FIGURE 5.6 **Transmitter responses of PPN cells during the developmental decrease in REM sleep.** Both NMDA and KA depolarized PPN neurons; however, during development, the depolarization induced by NMDA decreased, while the depolarization induced by KA increased, as depicted by the crossing lines of filled squares and filled circles. Muscarinic M₂ receptor responses induced hyperpolarization, which increased in amplitude during this period (filled upward-pointing triangles). On the other hand, the hyperpolarization induced by a₂ adrenergic input decreased in amplitude during the developmental decrease in REM sleep (downward-pointing triangles). Serotonergic type 2 input did not change during this period, inducing hyperpolarization of PPN cells (left-pointing triangles), while serotonergic type 1 input increasingly hyperpolarized PPN cells during the same time. Finally, GABAA input was at first excitatory, switching to inhibition after 15 days of age (right-pointing triangles), while GABAB input was initially inhibitory and lessened during the same period (sextagons).

most cholinergic and noncholinergic PPN cells (Bay et al., 2006). This hyperpolarization decreased significantly in amplitude from 12 to 21 days (Figure 5.6, downward-pointing triangles). However, much of these early effects (12–15 days) were indirect and noncholinergic cells that were less hyperpolarized than cholinergic cells. These results suggest that the α₂-adrenergic receptor on cholinergic PPN neurons activated by clonidine may play only a modest role, if any, in the developmental decrease in REM sleep. However, clonidine blocked or reduced the hyperpolarization-activated inward cation conductance I_H, so that its effects on the firing rate of a specific population of PPN neurons could be significant (Bay et al., 2006).

As far as cholinergic input to PPN neurons, results showed that the nicotinic agonist 1,1-dimethyl-4-phenylpiperazinium iodide (DMPP) depolarized PPN neurons early in development and then had dual effects before hyperpolarizing PPN neurons by the end of the period studied (Good et al., 2007). Most of the effects of DMPP persisted following application of the sodium channel blocker tetrodotoxin and in the presence of glutamatergic, serotonergic, noradrenergic, and GABAergic antagonists but were blocked by application of the nicotinic antagonist mecamylamine. These results suggest that PPN neurons exhibit postsynaptic nicotinic receptors, but additional studies need to verify that this is not a tetrodotoxin-resistant effect. The nonspecific cholinergic agonist carbachol hyperpolarized all type II PPN cells and depolarized all types I and III PPN cells, but did not change effects during the developmental decrease in REM sleep. These effects persisted in the presence of tetrodotoxin but were mostly blocked by the muscarinic antagonist atropine, and the remainder by mecamylamine. These results suggest that PPN neurons exhibit postsynaptic muscarinic receptors, as previously shown. While the nicotinic and muscarinic inputs to the PPN may modulate the developmental decrease in REM sleep, the muscarinic inputs appear to modulate different types of cells differentially (Good et al., 2007), increasing its inhibitory effects during the developmental decrease in REM sleep (Figure 5.6, upward-pointing triangles).

Other findings suggest that $GABA_A$ receptor activation leads to depolarization of PPN neurons early in development (12-16 days) but to hyperpolarization later (17-21 days), during the developmental decrease in REM sleep (Figure 5.6, right-pointing triangles) (Bay et al., 2007). These effects persisted in tetrodotoxin and were assumed to be postsynaptic. The majority of depolarized cells early in development were noncholinergic PPN cells, whereas most cholinergic PPN cells were hyperpolarized in both early and later periods. $GABA_B$ receptor activation hyperpolarized both cholinergic and noncholinergic PPN neurons early and to a lesser extent in later periods (Figure 5.6, sextagons). GABA, the primary inhibitory influence in the adult brain, was found to be excitatory during development due to the delayed expression of chloride cotransporters (Cossart et al., 2005). Results in PPN neurons, especially noncholinergic cells, showed depolarization early in development shifting to hyperpolarization later during the developmental decrease in REM sleep. We do not know if the differential responses in cholinergic neurons (i.e., mostly hyperpolarization with few cells depolarizing) are due to earlier maturation of chloride transporters in cholinergic compared with noncholinergic cells or to other factors. This developmental shift occurred only in response to the $GABA_A$ agonist muscimol and not to the $GABA_B$ agonist baclofen. It is unknown if a delay in maturation of chloride transporters in PPN cells is limited to channels activated by $GABA_A$ and not to $GABA_B$ receptors (Bay et al., 2007).

In summary, the pharmacological findings suggest that there is a developmental shift around 15 days of age in the rat, during the most dramatic decrease in REM sleep duration, towards increased 5-HT1 inhibition, decreased NMDA excitation and increased KA activation, decreased noradrenergic inhibition, and increased cholinergic and GABAergic inhibition of PPN neurons (Figure 5.6). It is not clear which of these is responsible for the developmental decrease in REM sleep or whether other mechanisms such as intracellular pathways and metabolism play a role. In Chapter 8, we will consider a number of intracellular pathways and mechanisms that may participate in this phenomenon. Regardless of the driving forces behind the decrease in REM sleep, one important consideration is, what if the developmental decrease in REM sleep does not occur or is reduced? What are the consequences of lifelong increased REM sleep drive? What will be the manifestations of such dysregulation? Regardless whether or not the potential functions of REM sleep described above are correct, a number of severe disorders are marked by a virtually permanent developmental increase in vigilance and REM sleep drive. That is, certain states are characterized by increased level of high-frequency rhythms during both waking, which is diagnosed as hypervigilance, and sleep, which manifests during REM sleep as increased drive resulting in frequent awakenings. These disorders are considered in Chapters 11 and 12, although the disorder of wake–sleep disturbances in obesity is considered below. Nevertheless, it needs to be stressed that the most obvious underlying principle behind the development of the RAS is that there is a dramatic increase in waking as the human develops from birth to adulthood.

PUBERTY AND WAKE–SLEEP

Interestingly, the onset of puberty is induced by the wake–sleep system. The onset of puberty is associated with a rise in gonadotropin-releasing hormone (GnRH) pulsing, which induces a rise in luteinizing hormone (LH) release during sleep (Boyar et al., 1972). The release of LH during sleep appears to occur after episodes of slow-wave sleep (Shaw et al., 2012). Although the cause of the GnRH rise is unknown, there are leptin receptors in the hypothalamus, and this region synthesizes GnRH (Meister and Håkansson, 2001). Individuals who are deficient in leptin do not initiate puberty (Clayton and Trueman, 2000). The levels of leptin increase with the onset of puberty and then decline to adult levels by the end of puberty. However, the leptin control of GnRH induction of LH release during sleep may involve the RAS. Almost every cell in the PPN has leptin receptors, and these have a marked effect on their physiology. Leptin is better known for its role in obesity, and obesity is marked by wake–sleep dysregulation.

Leptin, a hormone that regulates appetite and energy expenditure, is increased in obese individuals, although these individuals often exhibit leptin resistance (Ahima and Flier, 2000). Obesity is characterized by wake–sleep disturbances such as excessive daytime sleepiness in the absence of sleep-disordered breathing, increased REM sleep, increased nighttime arousals, increased total wake time, and decreased percentage of total sleep (Dixon et al., 2007; Vgontzas et al., 1998). Several studies have shown that short sleep duration is highly correlated with decreased leptin levels in both animal and human models (Aldabal and Bahammam, 2011; Spiegel et al., 2005; Taheri et al., 2004). The role of leptin in the adult vs. the postnatal rat is different. While leptin is key in the regulation of energy homeostasis in the adult, it is relatively ineffective in these processes in the postnatal rat (Cottrell et al., 2009). Leptin is essential in the development and maturation of neuronal and glial cells in the fetus and neonate, suggesting a role in brain development (Ahima et al., 1999; Udagawa et al., 2007). In the hypothalamus, there is a postnatal leptin surge in both the rat and mouse, which begins around postnatal day 4, peaks at ~day 7 (~500% increase compared with that of baseline levels), and consistently decreases until puberty (Cottrell et al., 2009). Additionally, there is an increase in leptin receptor ObRb (the long, signaling receptor isoform) mRNA expression in the hypothalamus, which begins at ~day 12, peaks at ~day 15 (~300% increase compared with levels at day 4), and decreases from ~day 16 to puberty (Cottrell et al., 2009), that coincides with the developmental decrease in REM sleep described above.

We described the presence of leptin receptors in both cholinergic and noncholinergic PPN cells (Beck et al., 2013a, 2013b). Figure 5.7 shows that immunocytochemical labeling for the leptin receptor is present in cholinergic as well as noncholinergic PPN cells. We found that leptin affected the intrinsic properties of PPN cells by decreasing AP amplitude (Figure 5.8), suggesting that firing of PPN neurons was downregulated by leptin.

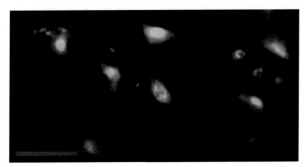

FIGURE 5.7   **Immunocytochemical labeling for the leptin receptor.** Merged view of a fluorescence photomicrograph for both bNOS and leptin receptor immunocytochemical labeling of a section through the PPN of the rat. bNOS-labeled cholinergic PPN cells as solid red cytoplasm (rhodamine filter). Leptin receptor labeling was evident as punctiform green label (FITC filter). This suggests that all or most cholinergic cells in the PPN bear leptin receptors and that many noncholinergic cells also bear leptin receptors. Calibration bar: 100 μm.

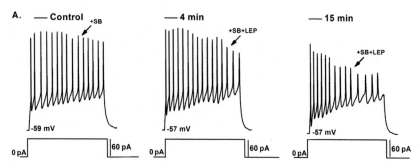

FIGURE 5.8   **Decrease in action potential amplitude of PPN neurons induced by leptin.** Left record, induced train of action potentials (APs) elicited by depolarization (bottom record, 60 pA step) before leptin (100 nM). Study was performed under superfusion of fast synaptic blockers (SB) that eliminated NMDA, KA, glycine, and GABA inputs. After 4 min of leptin exposure, AP amplitude in response to the same amplitude current step began to decrease (middle panel); note lower amplitude at end of step. After 15 min of leptin exposure (right panel), AP amplitude was significantly reduced throughout the current step.

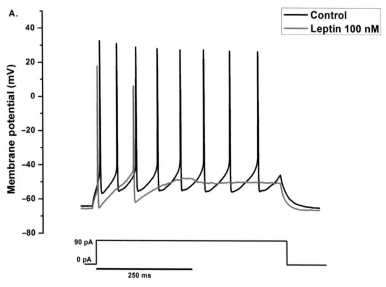

FIGURE 5.9   **Decrease in action potential frequency of PPN neurons induced by leptin.** Records compare responses to depolarizing steps of the same amplitude delivered before (black recordings) and after 15 min exposure to 100 nM leptin (gray record). Note the decreased amplitude and frequency of firing elicited by leptin. These effects are reflective of an effect on sodium currents that generate the AP.

In addition, we determined that AP frequency (Figure 5.9) was also decreased, suggesting that leptin led to decreased firing in these cells. These two effects are indicative of an effect on sodium currents. We therefore studied sodium conductance (Figure 5.10) in PPN cells and found that the

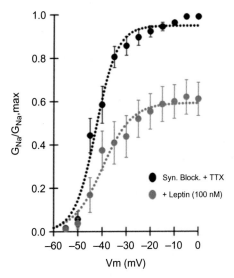

FIGURE 5.10   **Effects of leptin on sodium currents measured in PPN neurons in the presence of synaptic blockers and tetrodotoxin (TTX).** Current steps delivered at varying membrane potentials elicited sodium currents of different amplitudes. In the control condition before leptin exposure (black circles), sodium currents were evident between −45 and −30 mV. After 15 min exposure to 100 nM leptin, the current amplitudes were decreased by about 40% (gray circles). Leptin had a direct effect on sodium currents responsible for downregulating AP amplitude and frequency.

average decrease in sodium conductance by leptin was ~40%. Finally, we observed that leptin decreased the hyperpolarization-activated current (I_H) (Figure 5.11) in PPN neurons. This current is present in 30–40% of PPN cells, as described in Chapter 3, and its decrease would lead to an impairment in recovering from or responding to inhibitory inputs, again abating activity in the PPN. We concluded that leptin normally decreases activity in the PPN by reducing I_H and sodium currents. Our results also suggest that the effects of leptin on the intrinsic properties of PPN neurons are leptin receptor- and G protein-dependent (Beck et al., 2013b). The effects of leptin on PPN neurons were blocked by a leptin antagonist in the presence of tetrodotoxin suggesting that leptin acts directly on leptin receptors on PPN cells. Intracellular GDPβ(a G-protein inhibitor) significantly reduced the effect of leptin on sodium currents (~60% reduction), but not on I_H (~25% reduction). Intracellular GTPγS (a G-protein activator) reduced the effect of leptin on both sodium currents (~80% reduction) and I_H (~90% reduction). In general then, action by leptin would reduce arousal and REM sleep by lowering PPN cell firing, thus favoring slow-wave sleep. Moreover, in states of leptin dysregulation (i.e., leptin resistance seen in obesity), this effect may be blunted at the level of the PPN, therefore causing increased arousal and REM sleep drive and ultimately

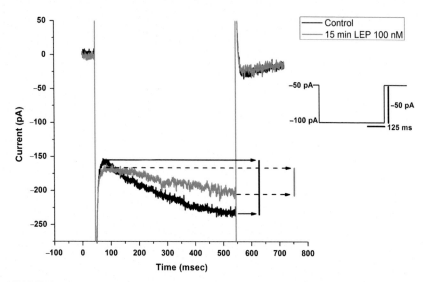

FIGURE 5.11 **Effects of leptin on I_H current measured in PPN neurons during voltage-clamp recordings.** Hyperpolarizing steps in some (~40%) of the PPN neurons reveal the presence of I_H, which tends to speed the return to resting membrane potential and tends to increase firing frequency (black record). After 15 min exposure to 100 nM leptin, I_H current was reduced by about 40% (gray record). The net effect of leptin is to decrease or slow the return of the membrane potential after hyperpolarization, thereby decreasing activity in these cells.

leading to sleep-related obesity disorders. These findings are significant since a recent study found that obesity in children sets in between the ages of 5 and 14 years and perhaps earlier (Cunningham et al., 2014). This is the same period of the developmental decrease in REM sleep, which, if reduced, would lead to more frequent awakenings from sleep and more intense REM sleep.

GONADAL STEROIDS

In general, estrogen and testosterone have an excitatory effect on arousal and attention in females and males, respectively. There is considerable evidence suggesting that exogenous administration of estradiol in women will increase cognitive performance and decrease reaction time (Liu et al., 2003). A similar effect is observed after androgen therapy in men, which also shortens sleep time (Canway et al., 2000). Anabolic steroids lead not only to increased muscle size and therefore strength but also to decreased reaction time and increased attention in athletes (Lustig, 1994). How can gonadal steroids affect brain functions related to gamma band activity such as arousal and attention? If gonadal steroids induce increased electrical coupling (Chapter 3), the effect could resemble that of the stimulant modafinil, that is, increasing arousal and attention by general disinhibition and RAS upregulation through increased expression of gap junctions. Given the advantages in timing conferred by electrical coupling, it would not be surprising if anabolic steroids and other agents that increase coupling would also decrease reaction time, allowing for superior performance in sports. On the other hand, excessive use of anabolic steroids is known to lead to psychotic symptoms, including hypervigilance and increased REM sleep drive (perhaps in the form of intense dreaming and hallucinations), suggesting that overindulgence of anabolic steroids could lead to dysregulation of wake–sleep states, arousal, and attention. These findings also suggest that the reported effects on cognition of gonadal steroid therapy (postmenopause, sexual dysfunction, etc.) could be due to increased arousal in the brain, perhaps promoting longer lasting synchrony, especially of high-frequency activity.

In vitro, gonadal steroids are known to increase gap junction expression in a number of somatic cell lines and cultures. In neurons, they alter neurite outgrowth, neuritic spine development, synaptogenesis, and increase gap junctions in certain brain regions (Matsumoto et al., 1988). Spinal cord neurons in sexually dimorphic nuclei lose gap junctions after castration or ovariectomy that return with exogenous androgen or estrogen treatment, respectively (Coleman and Sengelaub, 2002; Shinohara et al., 2001). Estrogen increases Cx36 mRNA expression in the suprachiasmatic nucleus but not the cortex (Shughrue et al., 1997). These results suggest that gonadal steroids increase neuronal gap junction expression and probably increase electrical coupling, an effect diminished by castration or ovariectomy.

There are two estrogen receptors: α and β. The two estrogen receptors are expressed in different patterns and at different levels throughout the brain (Greco et al., 1999), but the expression in the individual nuclei of the RAS has not been described. The androgen receptor is present in a portion of PPN neurons (Cicirata et al., 2000), but its presence in the SubC and Pf has not been described. There are several ways that estrogen could regulate Cx36 expression. The mouse Cx36 promoter (Klinge, 2001) has an imperfect palindrome with remarkable similarity to the consensus estrogen response element. Although no characterized estrogen response elements are identical to the putative element in the Cx36 promoter, all mismatches are found in other estrogen response elements (O'lone et al., 2004). The putative estrogen response element is conserved in the human Cx36 promoter suggesting that estrogen/estrogen receptors regulate Cx36 expression directly. There are also Sp1-like sites that could synergize with multiple estrogen response element half-sites nearby (Greco et al., 1999). If indeed gonadal steroids modulate gap junction expression and dysregulation of this mechanism occurs, increased Cx36 function may lead to unwanted synchronization promoting a state of excessive vigilance and increased REM sleep drive.

CLINICAL IMPLICATIONS

The evidence presented herein raises exciting possibilities for some of the cellular- and transmitter-related changes that may be responsible for the development of the adult wake–sleep control system. These considerations help explain how one of our most essential, evolutionarily conserved systems develops into its predictable adult form and perhaps how disturbances in its development can explain a number of disorders with devastating circumstances (see developmental onset of various psychiatric disorders in Chapter 11). Future studies will also be essential in determining how such disturbances result in lifelong changes in the function of this system. The development of more effective therapeutic strategies for the disorders mentioned is dependent on detailed information to be gained from investigation of these critical periods.

From a practical point of view, the "advanced wakefulness" seen in development suggests that the newborn may be amenable to entraining in order to promote longer and longer wake–sleep periods. The two stimuli that help entrain waking are light and food. By exposing newborns to daylight, and by providing a meal, early in the day, the first waking episode of the day can be slowly entrained. These stimuli may also help alleviate sleep deprivation in the parents. The type of meal can also modulate wake–sleep rhythms. High-carbohydrate meals, on the one hand, tend to promote transport of precursors of inhibitory transmitters into the brain, promoting somnolence. High-protein meals, on the other hand, tend to preferentially promote transport of precursors of excitatory transmitters, promoting waking.

During puberty, the effects of gonadal steroids can be significant; estrogen and testosterone produce excitatory effects on vigilance, while progesterone tends to be soporific. Exogenous ingestion of anabolic steroids, such as used for boosting athletic performance, not only is illegal in organized sports but also has serious side effects. Such effects are probably exaggerated in young people and may induce long-term deleterious effects. Such steroids are different from those taken, for example, in asthma inhalers, as anti-inflammatories. These are generally low dose and safe, although some patients may develop side effects. However, the first sign that these drugs or any other psychoactive agent is having an undesired side effect will be a marked change in wake–sleep rhythms. Such symptoms will probably be followed by behavioral changes such as mood swings and appetite and sexuality shifts. Upon abstaining from these agents, there should be a return to more normal wake–sleep, appetite, and reproductive behavior rhythms.

Kleine–Levin Syndrome—The synergy between wake–sleep, appetite, and reproductive behavioral control is considerable, so that normalization of one rhythm should lead to correction of the others. Conversely, some disorders effect a change on all homeostatic systems. For example, Kleine–Levin syndrome is a postpubertal onset disorder, mostly in males, that includes marked disturbances in wake–sleep rhythms (periodic hypersomnia, up to 20 h per day), appetite (hyperphagia, especially bingeing on carbohydrates), and hypersexuality (increased sexual urges) (Arnulf et al., 2012). The first episode in many patients follows some sort of infection (virus, mononucleosis, encephalitis, etc.) that can have occurred years earlier, lasts for 1–2 weeks, and recurs for years. Fortunately, the disorder often becomes milder and in many does not recur after the mid-twenties. There is no established effective treatment, but the thalamus and upper brain stem may be involved. This is one example of the comorbidities induced when one homeostatic system is altered (wake–sleep), thereby affecting other systems (appetite).

Obesity—Finally, there has been a worldwide increase in the prevalence of obesity over the last several decades. This increase is paralleled by a trend in reduced sleep duration in both adults and children. Both longitudinal and prospective studies have shown that chronic partial sleep loss is associated with an increase in obesity (Beccuti and Pannain, 2011). As stated above, obesity is characterized by excessive daytime sleepiness (in the absence of sleep-disordered breathing), decreased total sleep time, and decreased total percentage of sleep (Vgontzas et al., 1998). It has also been shown that there is an inverse relationship between body mass index (BMI) and REM sleep duration (Rutters et al., 2012). Therefore, it is clinically imperative that the association between reduced sleep and obesity be revealed. The factors that lead to shortened sleep time are frequent arousals and increased REM sleep drive during sleep. These are precisely the main states that the PPN modulates. In many individuals, obesity may actually be a wake–sleep disorder, one that may be particularly inducible during puberty. This is another example of the meshing of the homeostatic control systems for wake–sleep, appetite, reproduction, and temperature discussed in Chapter 3. At present, it is not known if the leptin resistance of obesity is present in leptin receptors in the PPN, so that this area of research is critical to public health and needs to be pursued aggressively. After all, if leptin resistance can be alleviated in the PPN, wake–sleep dysregulation in obesity could be blunted.

References

Adrien, J., 1984. Ontogenese du sommeil chez le mammifere. In: Benoit, O. (Ed.), Physiologie du Sommeil. Masson, Paris, pp. 19–29.

Ahima, R.S., Flier, J.S., 2000. Leptin. Ann. Rev. Physiol. 62, 413–437.

Ahima, R.S., Bjorbaek, C., Osei, S., Flier, J.S., 1999. Regulation of neuronal and glial proteins by leptin: implications for brain development. Endocrinology 140, 2755–2762.

Aldabal, L., Bahammam, A.S., 2011. Metabolic, endocrine, and immune consequences of sleep deprivation. Open Respir. Med. J. 5, 31–43.

Arnulf, I., Rico, T., Mignot, E., 2012. Diagnosis, disease course, and management of patients with Kleine–Levin syndrome. Lancet Neurol. 11, 918–928.

Balkin, T.J., Braun, A.R., Wesensten, N.J., Jeffries, K., Varga, M., Baldwin, P., Belenky, G., Herscovitch, P., 2002. The process of awakening: a PET study of regional brain activity patterns mediating the re-establishment of alertness and consciousness. Brain 125, 2308–2319.

Bay, K., Mamiya, K., Good, C., Skinner, R.D., Garcia-Rill, E., 2006. Alpha-2 adrenergic regulation of pedunculopontine nucleus (PPN) neurons during development. Neuroscience 141, 769–779.

Bay, K.D., Beck, P., Skinner, R.D., Garcia-Rill, E., 2007. GABAergic modulation of developing pedunculopontine nucleus (PPN). NeuroReport 18, 249–253.

Beccuti, G., Pannain, S., 2011. Sleep and obesity. Curr. Opin. Clin. Nutr. Metab. Care 14, 402–412.

Beck, P., Urbano, F.J., Williams, K.D., Garcia-Rill, E., 2013a. Effects of leptin on pedunculopontine nucleus (PPN) neurons. J. Neural Transm. 120, 1027–1038.

Beck, P., Mahaffey, S., Urbano, F.J., Garcia-Rill, E., 2013b. Role of G-proteins in the effects of leptin on pedunculopontine nucleus (PPN). J. Neurochem. 126, 705–714.

Boyar, R., Finkelstein, J., Roffwarg, H., Kapen, S., Weitzman, E., Hellman, L., 1972. Synchronization of augmented luteinizing hormone secretion with sleep during puberty. New Eng. J. Med. 287, 582–586.

Canway, A.J., Handelsman, D.J., Lording, D.W., Stuckey, B., Zajac, J.D., 2000. Use, misuse and abuse of androgens. The Endocrine Society of Australia consensus guidelines for androgen prescribing. Med. J. Aust. 172, 220–224.

Cicirata, F., Parenti, R., Spinella, F., Giglio, S., Tuorto, F., Zuffardi, O., Gulisano, M., 2000. Genomic organization and chromosomal localization of the mouse Connexin36 (mCx36) gene. Gene 251, 123–130.

Clayton, P.E., Trueman, J.A., 2000. Leptin and puberty. Arch. Dis. Child. 83, 1–4.

Coleman, A.M., Sengelaub, D.R., 2002. Patterns of dye coupling in lumbar motor nuclei of the rat. J. Comp. Neurol. 454, 34–41.

Cossart, R., Bernard, C., Ben-Ari, Y., 2005. Multiple facets of GABAergic neurons and synapses: multiple fates of GABA signaling in epilepsies. Trends Neurosci. 28, 108–115.

Cottrell, E.C., Cripps, R.L., Duncan, J.S., Barrett, P., Mercer, J.G., Herwig, A., Ozanne, S.E., 2009. Developmental changes in hypothalamic leptin receptor: relationship with the postnatal leptin surge and energy balance neuropeptides in the postnatal rat. Am. J. Physiol. Regul. Integr. Comp. Physiol. 296, R631–R639.

Cunningham, S.A., Kramer, M.R., Narayan, K.M.V., 2014. Incidence of childhood obesity in the United States. New Eng. J. Med. 370, 403–411.

Delorme, F., Vimont, P., Jouvet, M., 1964. Etude statistique du cycle veille–sommeils chez le chat. Comp Rend Séances Soc Biol Filial 58, 2128–2130.

Dixon, J.B., Dixon, M.E., Anderson, M.L., Schachte, L., O'Brien, P.E., 2007. Daytime sleepiness in the obese: not as simple as obstructive sleep apnea. Obesity 15, 2504–2511.

Feng, P., Ma, Y., Vogel, G.W., 2001. Ontogeny of REM rebound in postnatal rats. Sleep 24, 645–653.

Friedman, S., Fisher, C., 1967. On the presence of a rhythmic, diurnal, oral instinctual drive cycle in man. Am. Psychoanal. Assoc. 15, 317–343.

Garcia-Rill, E., 2002. Mechanisms of sleep and wakefulness. In: Lee-Chiong, T., Sateia, M.J., Carskadon, M.A. (Eds.), Sleep Medicine. Hanley & Belfus, Philadelphia, PA, pp. 31–39.

Garcia-Rill, E., Skinner, R.D., 2002. The sleep state-dependent P50 midlatency auditory evoked potential. In: Lee-Chiong, T., Sateia, M.J., Carskadon, M.A. (Eds.), Sleep Medicine. Hanley & Belfus, Philadelphia, PA, pp. 697–704.

Garcia-Rill, E., Miyazato, H., Skinner, R.D., Williams, K., 1999. Periodicity in the amplitudes of the sleep state-dependent, midlatency auditory evoked P1 potential in the human and P13 potential in the rat. Neurosci. Abstr. 25, 627.

Globus, G.G., Phoebus, E., Moore, C., 1971. REM "sleep" manifestations during waking. Psychopharmacology 7, 308–312.

Good, C., Bay, K.D., Buchanan, R., Skinner, R.D., Garcia-Rill, E., 2007. Muscarinic and nicotinic responses in the developing pedunculopontine nucleus (PPN). Brain Res. 1129, 147–155.

Greco, B., Edwads, D.A., Michael, R.P., Zumpe, D., Clancy, A.N., 1999. Colocalization of androgen receptors and mating-induced FOS immunoreactivity in neurons that project to the central tegmental field in male rats. J. Comp. Neurol. 408, 220–236.

Gronfier, C., Simon, C., Piquard, F., Ehrhart, J., Brandenberger, G., 1999. Neuroendocrine processes underlying ultradian sleep regulation in man. J. Clin. Endocrinol. Metab. 84, 2686–2690.

Jouvet, M., 1962. Receherches sur les structures nerveuses et mechanismes responsables des differentes phases du sommeil physiologique. Arch. Ital. Biol. 100, 125–206.

Jouvet, M., 1999. The Paradox of Sleep, The Story of Dreaming. The MIT Press, Cambridge, MA.

Jouvet-Mounier, D., Astic, L., Lacote, D., 1970. Ontogenesis of the states of sleep in rat, cat, and guinea pig during the first postnatal month. Dev. Psychobiol. 2, 216–239.

Klein, R., Armitage, R., 1979. Rhythms in human performance: 1 1/2-hour oscillations in cognitive style. Science 204, 1326–1328.

Kleitman, N., 1953. Sleep and Wakefulness. University of Chicago Press, Chicago, IL.

Klinge, C.M., 2001. Estrogen receptor interaction with estrogen response elements. Nucleic Acids Res. 29, 2905–2919.

Kobayashi, T., Homma, Y., Good, C., Skinner, R.D., Garcia-Rill, E., 2002. Developmental changes in the effects of serotonin on neurons in the region of the pedunculopontine nucleus. Dev. Brain Res. 140, 57–66.

Kobayashi, T., Skinner, R.D., Garcia-Rill, E., 2004. Developmental decrease in REM sleep: the shift to kainate receptor regulation. Thalamus Relat. Syst. 2, 315–324.

Kripke, D.F., O'Donoghue, J.P., 1968. Perpetual deprivation, REM sleep and ultradian biological rhythm. Psychopharmacology 5, 231–232.

Kripke, D.F., Halberg, F., Crowley, T.J., Pegram, V.G., 1976. Ultradian spectra in monkeys. Int. J. Chronobiol. 3, 193–204.

Kungel, M., Koch, M., Friauf, E., 1996. Cysteamine impairs the development of acoustic startle response in rats: possible role of somatostatin. Neurosci. Lett. 202, 181–184.

Lavie, P., Kripke, D.F., 1981. Ultradian circa 1 1/2-hour rhythm: a multioscillatory system. Life Sci. 29, 2445–2450.

Leonard, C.S., Llinas, R.R., 1990. Electrophysiology of mammalian pedunculopontine and laterodorsal tegmental neurons in vitro: implications for the control of REM sleep. In: Steriade, M., Biesold, D. (Eds.), Brain Cholinergic Systems. Oxford Science, Oxford, UK, pp. 205–223.

Liu, P.Y., Yee, B., Wishart, S.M., Jimenez, M., Jung, D.G., Grunstein, R.R., Handelsman, D.J., 2003. The short-term effects of high-dose testosterone on sleep, breathing, and function in older men. J. Clin. Endocrinol. Metab. 88, 3605–3613.

Llinas, R.R., 1984. Possible role of tremor in the organization of the nervous system. In: Findley, L.J., Capildeo, R. (Eds.), International Neurological Symposium on Tremor. MacMillan Press, London, UK, pp. 473–478.

Lustig, R.H., 1994. Sex hormone modulation of neural development in vitro. Horm. Behav. 28, 383–395.

Marks, G.A., Shaffery, J.P., Oksenberg, A., Speciale, S.G., Roffwarg, H.P., 1995. A functional role for REM sleep in brain maturation. Behav. Brain Res. 69, 1–11.

Matsumoto, A., Arnold, A.P., Zampighi, G.A., Micevych, P.E., 1988. Androgenic regulation of gap junctions between motoneurons in the spinal cord. J. Neurosci. 8, 4177–4183.

Meister, B., Håkansson, M.L., 2001. Leptin receptors in hypothalamus and circumventricular organs. Clin. Exp. Pharmacol. Physiol. 28, 610–617.

Monk, T.H., Buysse, D.J., Reynolds, C.F., Berga, S.L., Jarrett, D.B., Begley, A.E., Kupfer, D.J., 1997. Circadian rhythms in human performance and mood under constant conditions. J. Sleep Res. 6, 9–18.

Ninomiya, Y., Koyama, Y., Kayama, Y., 2001. Postnatal development of choline acetyltransferase activity in the rat laterodorsal tegmental nucleus. Neurosci. Lett. 308, 138–140.

O'lone, R., Frith, M.C., Karlsson, E.K., Hansen, U., 2004. Genomic targets of nuclear estrogen receptors. Mol. Endocrinol. 18, 1859–1875.

Orr, W.C., Hoffman, H.J., 1974. A 90-min cardiac biorhythm: methodology and data analysis using modified periodograms and complex demodulation. IEEE Trans. Biomed. Eng. 21, 130–143.

Oswald, I., Merrington, J., Lewis, H., 1970. Cyclical "on demand" oral intake by adults. Nature 225, 959–960.

Othmer, E., Hayden, M.P., Segelbaum, R., 1969. Encephalic cycles during sleep and wakefulness: a 24-hour pattern. Science 164, 447–449.

Rechstaffen, A., Wolpert, E., Dement, W., Mitchell, S., Fisher, S., 1963. Nocturnal sleep in narcoleptics. Electroencephalogr. Clin. Neurophysiol. 15, 599–609.

Rial, R., Nicolau, M.C., Lopez-Garcia, J.A., Almirall, H., 1993. On the evolution of waking and sleeping. Comp. Biochem. Physiol. Comp. Physiol. 104, 189–193.

Roffwarg, H.P., Muzio, J.N., Dement, W.C., 1966. Ontogenetic development of the human sleep–dream cycle. Science 152, 604–619.

Roldan, E., Weiss, T., Fifkov, E., 1963. Excitability changes during the sleep cycle of the rat. Electroencephalogr. Clin. Neurophysiol. 15, 775–785.

Rutters, F., Gonnissen, H.K., Hursel, R., Lemmens, S.G., Martens, E.A., Westerterp-Plantenga, M.S., 2012. Distinct associations between energy balance and the sleep characteristics slow wave sleep and rapid eye movement sleep. Int. J. Obes. 36, 1346–1352.

Shaw, N.D., Butler, J.P., McKinney, S.M., Nelson, S.A., Ellenbogen, J.M., Hall, J.E., 2012. Insights into puberty: the relationship between sleep stages and pulsatile LH secretion. J. Clin. Endocrinol. Metab. 97, E2055–E2062.

Sheets, L.P., Dean, K.F., Reiter, L.W., 1988. Ontogeny of the acoustic startle response and sensitization to background noise in the rat. Behav. Neurosci. 102, 706–713.

Shinohara, K., Funabashi, T., Nakamura, T.J., Kimura, F., 2001. Effects of estrogen and progesterone on the expression of connexin-36 mRNA in suprachiasmatic nucleus of female rats. Neurosci. Lett. 309, 37–40.

Shughrue, P.J., Lane, M.V., Merchenthaler, I., 1997. Comparative distribution of estrogen receptor-α and -β mRNA in the rat central nervous system. J. Comp. Neurol. 388, 507–525.

Siegel, J.M., 1999. The evolution of REM sleep. In: Lydic, R., Baghdoyan, H.B. (Eds.), Handbook of Behavioral State Control. Cellular and Molecular Mechanisms. CRC Press, New York, NY, pp. 87–100.

Siegel, J.M., Manger, P.R., Nienhuis, R., Fahringer, H.M., Shalita, T., Pettigrew, J.D., 1999. Sleep in the platypus. Neuroscience 91, 391–400.

Skinner, R.D., Conrad, N., Henderson, V., Gilmore, S., Garcia-Rill, E., 1989. Development of NADPH diaphorase positive pedunculopontine neurons. Exp. Neurol. 104, 15–21.

Spiegel, K., Knutson, K., Leproult, R., Tasali, E., Van Cauter, E., 2005. Sleep loss: a novel risk factor for insulin resistance and Type 2 diabetes. J. Appl. Physiol. 99, 2008–2019.

Sterman, M.B., Hoppenbrouwers, T., 1971. The development of sleep–waking and rest–activity patterns from fetus to adult in man. In: Sterman, M.B., McGinty, D.J., Adinolfi, A.M. (Eds.), Brain Development and Behavior. Academic Press, New York, NY, pp. 203–227.

Sterman, M.B., Knauss, T., Lehmann, D., Clemente, C.D., 1965. Circadian sleep and waking patterns in the laboratory cat. Electroencephalogr. Clin. Neurophysiol. 19, 509–517.

Sterman, M.B., Lucas, E.A., MacDonald, L.R., 1972. Periodicity within sleep and operant performance in the cat. Brain Res. 38, 327–341.

Taheri, S., Lin, L., Austin, D., Young, T., Mignot, E., 2004. Short sleep duration is associated with reduced leptin, elevated ghrelin, and increased body mass index. Public Libr. Sci. Med. 1, e62.

Takakusaki, K., Kitai, S.T., 1997. Ionic mechanisms involved in the spontaneous firing of tegmental pedunculopontine nucleus neurons of the rat. Neuroscience 78, 771–794.

Timo-Iaria, C., Negrao, N., Schmidek, W.R., Hishino, K., De Menezes, E.L., Da Rocha, T.L., 1970. Phases and states of sleep in the rat. Physiol. Behav. 5, 1057–1062.

Udagawa, J., Hatta, T., Hashimoto, R., Otani, H., 2007. Roles of leptin in prenatal and perinatal brain development. Congenit. Anom. 47, 77–83.

Ursin, R., 1988. The two stages of slow wave sleep in the cat and their relation to REM sleep. Brain Res. 11, 347–356.

Van Twyler, H., 1969. Sleep patterns of five rodent species. Physiol. Behav. 4, 901–905.

Vgontzas, A.N., Bixler, E.O., Tan, T.L., Kantner, D., Martin, L.F., Kales, A., 1998. Obesity without sleep apnea is associated with daytime sleepiness. Arch. Intern. Med. 158, 1333–1337.

Vincent, S.R., Satoh, K., Armstrong, D.M., Fibiger, H.C., 1963. NADPH-diaphorase: a selective histochemical marker for the cholinergic neurons in the pontine reticular formation. Neurosci. Lett. 43, 31–36.

Vogel, G.W., Feng, P., Kinney, G.G., 2000. Ontogeny of REM sleep in rats: possible implications for endogenous depression. Physiol. Behav. 68, 453–461.

Williams, J.A., Reiner, P.B., 1993. Noradrenaline hyperpolarizes identified rat mesopontine cholinergic neurons in vitro. J. Neurosci. 13, 3878–3883.

Ascending Projections of the RAS

James Hyde, PhD and Edgar Garcia-Rill, PhD†*

*Institute for Neuroscience, University of Pittsburgh, Pittsburgh, PA, USA
†Department of Neurobiology and Developmental Sciences,
University of Arkansas for Medical Sciences, Little Rock, AR, USA

THE ASCENDING RETICULAR ACTIVATING SYSTEM

Early descriptions of the ascending reticular activating system were made using acetylcholinesterase labeling (Shute and Lewis, 1967). The wedge-shaped nucleus that gave rise to these projections was erroneously labeled as "cuneiform nucleus," the term pedunculopontine nucleus coming into use years later. The "dorsal tegmental pathway" was seen to project to the inferior and superior colliculi, the pretectal region, the medial and lateral geniculate nuclei, and the intralaminar and specific, as well as anterior, thalamic nuclei. The "ventral tegmental pathway" was found to project to the oculomotor nucleus, mammillary bodies, subthalamic nucleus, globus pallidus, posterior and lateral hypothalamus, lateral preoptic region, paraventricular nuclei, and olfactory tubercle (Shute and Lewis, 1967). As described in the previous chapters, ascending noradrenergic and serotonergic RAS outputs travel in parallel to the colliculi, geniculates, thalamus, hypothalamus, basal forebrain, and other regions, while descending outputs travel to the cerebellum and pontine and medullary reticular formation (as we will see in Chapter 7). Thus, there are direct noradrenergic and serotonergic projections to the cortex and the spinal cord, while cholinergic afferents to these areas are relayed. This anatomical heterogeneity allows the RAS to activate a host of brain systems. It is thought that waking is maintained by tonic activity in the RAS reinforced by sensory input.

As we saw in the previous chapter, all of the primary sensory pathways conduct sensation-specific information via their respective lemniscal relays to the "specific" thalamic nuclei and then to the primary cortical

Waking and the Reticular Activating System in Health and Disease

(Pa) areas. These are rapidly conducting pathways with fewer synapses and high synaptic security providing information for sensory discrimination. These projections provide the "content" of conscious experience (Llinas, 2001). However, every sensory system activates the RAS in parallel, setting up a more slowly conducting, multisynaptic, low synaptic security, "nonspecific" input. These afferents travel to the RAS and are relayed to "nonspecific" thalamic nuclei of the intralaminar thalamus before traveling to the cortex. This less sensation-specific information is related to arousal and provides the "context" of conscious experience (Llinas, 2001). That is, the RAS may not only provide the necessary modulation for arousal but also help form the basis for earlier, more basic, "preattentional" processes. Another basic function of the RAS is its participation in fight-or-flight responses so that alerting stimuli simultaneously activate thalamocortical systems (leading to cortical activation) and postural and locomotor systems (leading to the preparation of postural and locomotor systems). Chapter 7 will describe the organization of descending modulation by the RAS, while this chapter summarizes work in relation to the ascending modulation of arousal and preattentional processing by elements of the RAS. Two main regions modulated by the RAS in terms of waking are the intralaminar thalamus and the cortex.

INTRALAMINAR THALAMUS

Membrane properties—The intralaminar thalamic nuclei are a collection of midline cells that have long been considered the "nonspecific," arousal-related portion of the thalamus based on their extensive input from the ascending RAS and their diffuse projections mainly to layers I and II in the cerebral cortex (Jones, 1985). Anatomical studies established the presence of projections from the reticular formation to the rostral intralaminar nuclei, the centrolateral (CL)-paracentral (PC), and to the caudal intralaminar nuclei, the parafascicular (Pf)-centromedial (CM) regions (Pare et al., 1988). The pathway from the RAS to the CL-PC conveys midbrain reticular projections presumably involved in modulating fast rhythms during waking and perhaps REM sleep to areas of the cortex (Steriade and Glenn, 1982). This cortical projection target differentiates the intralaminar thalamic "nonspecific" regions from the "specific" projections of the thalamocortical relay nuclei that project to deeper layers of the cortex, especially layer IV. However, some results indicate that individual intralaminar nuclei have disparate efferent projections, suggesting the existence of more specificity within this system (Bentivoglio et al., 1991; Jones, 2002). For example, many intralaminar neurons also project to subcortical targets like the striatum (Bentivoglio et al., 1991; Jones, 1985, 2002; Lai et al., 2000; Sadikot et al., 1990, 1992; Steriade and Glenn, 1982; Van der Werf et al., 2002). Just as there is functional heterogeneity in the projections of the intralaminar thalamic nuclei, there is intranuclear heterogeneity of morphological, neurochemical, and functionally distinct cell types (Anna et al., 1999; Arai et al., 1994; Celio, 1990; Frassoni et al., 1997; Guillazo-Blanch et al., 1999; Harte et al., 2000; Hermenegildo et al., 2000; Vale-Martinez et al., 1999; Resibois and Rogers, 1992; Van der Werf et al., 2002).

On the one hand, cells in the "specific" thalamic relay nuclei exhibit a fairly homogeneous morphology, regardless of the species or nucleus (Llinas and Steriade, 2006), characterized by radiating primary dendrites with compact bushy dendritic trees. Nearly all of these cells also exhibit a stereotypic bimodal discharge pattern dependent on membrane potential (Jahnsen and Llinas, 1984a). With depolarization, thalamic relay neurons exhibit a "tonic" mode of continuous action potential as occurs during arousal states. However, when hyperpolarized, any depolarizing input leads to the manifestation of low-threshold spike (LTS) "bursting" mode of firing leading, along with I_H current in these cells, to characteristic rhythmic slow oscillations (0.5–4 Hz) as occur during drowsiness and SWS (McCormick and Bal, 1997; Steriade, 1999). Interactions between two voltage-dependent currents present in almost all thalamic relay neurons, the hyperpolarization-activated cation current (I_H) and the low-threshold calcium current (I_T), underlie the "pacemaker" oscillatory "bursting" mode of firing of these cells during SWS (Llinas, 1980). Computational modeling studies indicate that small changes in the amplitude or voltage dependence of these currents renders thalamic neurons incapable of generating the rhythmic LTS spikes that underlie the slow membrane oscillations (Destexhe and Babloyantz, 1993; Destexhe et al., 1993a,b, 1998; McCormick and Huguenard, 1992; Vasilyev and Barish, 2002). Thalamic neurons typically exhibit adultlike firing properties and similar morphology by 12 days postnatally (Warren and Jones, 1997).

On the other hand, anatomical studies showed that many cells in the "nonspecific" intralaminar thalamus have a different morphology compared to thalamic relay neurons (Hazlett et al., 1976; Hazlett and Hazlett, 1977; Parent and Parent, 2005; Pearson et al., 1984). These cells typically contain long unbranching processes in their proximal dendritic trees. Figure 6.1A shows recording sites of Pf neurons on either side (anterior and posterior) of the fasciculus retroflexus (fr), while Figure 6.1B is a high-power view of two recorded cells labeled intracellularly by the recording electrode solution containing neurobiotin (Ye et al., 2009). Note the sparse, long, and unbranching dendrites of Pf cells. Our studies showed that certain membrane properties, specifically a decrease in action potential duration in one of the two types of cells, along with the afterhyperpolarization in these neurons, were still developing until the end of the developmental decrease in REM sleep at 30 days (see Chapter 5). However, the resting membrane potential, input resistance, and time constant values of Pf

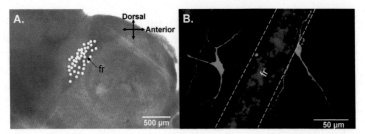

FIGURE 6.1   **Location of the Pf and cell shape.** (A) Sagittal section of the thalamus in the rat showing the *fasciculus retroflexus* at the posterior edge of the thalamus (dark band of fibers). The locations of recorded neurons are denoted by white dots on either side of this fiber tract. (B) Brain slice immunocytochemically labeled for neurobiotin that diffused into the cells during recording. Note the smooth and nonbranching nature of Pf cell dendrites in comparison with the bushy appearance of the thalamic relay neuron dendrites (not shown).

cells were already established at approximately adult levels by postnatal day 12 (Phelan et al., 2005). The most significant difference between these and thalamic relay neurons was that the number of cells expressing the hyperpolarization-activated cation current I_H decreased with age, and in those still expressing the current, there was a decrease in the amplitude of I_H (Phelan et al., 2005). Coupled with this was a relative absence of cells with LTS currents, along with decreased amplitude of LTS currents. The majority of Pf cells in our study exhibited firing properties that lacked the prominent bimodal firing patterns present in "specific" thalamic relay neurons. Even in the minority of Pf cells that did have an LTS-like response, the amplitude of this response was of lower amplitude than in neighboring regions of the thalamus. Changes in the holding potential of Pf cells did little to alter the overall firing pattern of the cells. Not surprisingly, therefore, no Pf cells displayed repetitive slow oscillations ("bursting"), and they lacked the prominent "tonic" firing that characterizes classic thalamic relay neurons (Jahnsen and Llinas, 1984b). This indicates that cells in Pf have different firing properties than cells in neighboring regions that do express rhythmic high-frequency bursting behaviors (Steriade et al., 1993). Thus, the difference between the somatodendritic morphology of "nonspecific" Pf and "specific" thalamic relay neurons extends to differences in their electrophysiological properties.

Push–Pull on "Specific" and Reticular Thalamic Neurons—The ascending RAS projections to the thalamus terminate in the "specific" nuclei, the "nonspecific" nuclei, and the reticular nucleus of the thalamus. The intralaminar projections are described below, but critical to the control of wake–sleep cycles are the projections to the "specific" nuclei and reticular nucleus. The reticular nucleus of the thalamus is located anteriorly to the thalamus, capping its anterodorsal pole. All of the cells in the reticular nucleus are GABAergic and are electrically coupled. As such, they represent a massive synchronizing inhibitory pulse delivered to thalamic relay nuclei. When the reticular nucleus is active, it delivers a recurrent inhibitory oscillation to thalamic relay cells. Thalamic relay nuclei in turn send excitatory projections to the reticular nucleus. This reciprocal interaction ensures positive feedback for repetitive inhibitory oscillatory activity, ensuring that thalamic relay neurons maintain a "bursting" mode of firing. Rhythmic bursting is induced by the I_H and LTS "pacemaker" currents present in thalamic relay neurons described above, thus repetitive slow oscillations are guaranteed (Llinas and Steriade, 2006). This is the mechanism that produces slow oscillations during SWS and occurs in the absence of RAS drive to these cells. However, sensory input or a change in state activates RAS ascending projections, especially from the PPN, some of which project to thalamic relay nuclei and to GABAergic cells of the reticular nucleus. The cholinergic input to reticular thalamic GABAergic cells is inhibitory, rapidly stopping the "bursting" slow inhibitory drive to thalamic relay neurons that produce slow waves. At the same time, the cholinergic input to thalamic relay neurons is excitatory, driving these cells into a depolarized "tonic" mode of firing. This helps maintain the tonic activity of waking and REM sleep states (Steriade, 1994). The "push" towards tonic depolarization of thalamic relay neurons and the "pull" away from inhibition by reticular nucleus GABAergic cells represent the shift from SWS to the activated, high-frequency states of waking and REM sleep (Figure 6.2).

Inputs from PPN to "Nonspecific" Thalamus—How then does the RAS modulate the intralaminar thalamus? The caudal intralaminar thalamic nuclei, including Pf and CL, are the main thalamic targets of the PPN (Capozzo et al., 2003; Erro et al., 1999; Grunwerg et al., 1992; Hallanger and Wainer 1988; Oakman et al., 1999; Scarnati et al., 1987; Sofroniew et al., 1985;

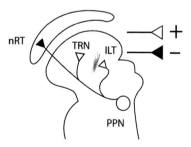

FIGURE 6.2 **Ascending PPN projections to the thalamus.** The PPN sends projections to the intralaminar thalamus (ILT) in the region of the *fasciculus retroflexus* and also to thalamic relay neurons (TRN), as well as the thalamic reticular nucleus (nRT). Projections to the ILT and TRN are excitatory, but projections to nRT are inhibitory. This promotes depolarization of ILT and TRN neurons while blocking the recurring inhibition from the nRT that promotes slow waves. That is, the PPN exercises a "push" towards waking and a "pull" away from sleep.

Steriade and Glenn, 1982; Sugimoto and Hattori 1984). These two nuclei project to the cerebral cortex and basal ganglia structures (Bentivoglio et al., 1991; Erro et al., 1999; Jones, 1985, 2002; Lai et al., 2000; Otake and Nakamura 1998; Rudkin and Sadikot 1999; Sadikot et al., 1990, 1992; Steriade and Glenn, 1982; van der Werf et al., 2002). As such, ascending PPN inputs to these cells are likely to play an important role in regulating both arousal and motor activity. In this regard, we demonstrated that stimulation of the PPN elicits a prolonged response in cells in the intralaminar thalamus (Kobayashi et al., 2004). However, differences in the frequency dependence of firing induced by different trains of PPN stimuli (10 Hz vs 60 Hz vs 90 Hz) in these nuclei indicate that they differ in the manner in which they relay PPN information to the cortex and basal ganglia (Kobayashi et al., 2004). Figure 6.3 summarizes these results. Basically, we found that stimulation of the PPN induced the highest firing frequency in intralaminar cells when stimulating at 60 Hz compared with 10 Hz, 30 Hz, or 90 Hz. The amplitude of the membrane depolarization in intralaminar cells induced by PPN stimulation was also highest after 60 Hz, compared with 10 Hz, 30 Hz, or 90 Hz stimulation. That is, responses were maximal when stimulation was carried out at gamma frequencies.

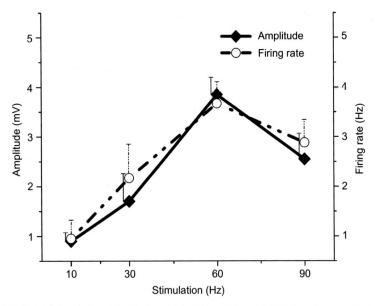

FIGURE 6.3   **Amplitude and frequency of Pf responses following the stimulation of the PPN at different frequencies.** Stimulation at 10 versus 30 versus 60 versus 90 Hz induced increasing firing rates that peaked at 60 Hz and then decreased at 90 Hz. The amplitude of the membrane responses observed was highest at 60 Hz compared to lower (10 and 30 Hz) or higher (90 Hz) frequencies. That is, the optimal responses induced in the intralaminar thalamus by PPN stimulation were elicited by gamma frequency stimulation.

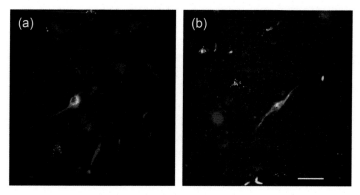

FIGURE 6.4 **Retrograde labeling and recording of PPN cells projecting to the Pf.** (a) Injections of fluorescent retrobeads into the Pf 3 days before recordings in retrogradely labeled PPN cells appear as green dots. Processing for bNOS immunocytochemistry labeled all cholinergic PPN neurons (red cytoplasm) and processing for neurobiotin showed the recorded cell (purple). The presence of retrobeads along with neurobiotin make the beads appear white. In this case, the recorded cell was found to project to the Pf and to be cholinergic. (b) In another slice, the recorded cell was positive for neurobiotin and retrobeads, confirming a thalamic-projecting neuron, but the cytoplasm was not labeled by bNOS, showing that the cell was not cholinergic. Calibration bar: 50 μm.

We then studied the responses of thalamic-projecting cells in the PPN. Figure 6.4(a) shows the labeling protocol in which injection of dye was made into the Pf two days before harvesting brain slices. Recordings were made in the PPN of these slices ((Figure 6.4(b)), but only from retrogradely labeled cells observed under fluorescence microscopy ((Figure 6.4(c)). That is, only PPN cells that project to the Pf were studied. Responses to cholinergic agonists in these cells showed that 65% of Pf cells are hyperpolarized and 35% are depolarized by the activation of muscarinic and nicotinic receptors (Ye et al., 2009).

This input to the Pf differs slightly from the responses of PPN neurons to cholinergic input, in which 73% of thalamic-projecting PPN neurons are hyperpolarized but only 13% of thalamic-projecting PPN cells are depolarized (Ye et al., 2010). The mechanisms behind the primary hyperpolarization to cholinergic input are discussed in Chapter 9, answering the question how does tonic cholinergic inhibition of the PPN lead to high-frequency oscillations?

We also found that decreases in the frequency of miniature EPSCs, and amplitude of electrical stimulation-evoked EPSCs, were blocked by a muscarinic M_2 antagonist, suggesting the presence of M_2 muscarinic receptors at terminals of presynaptic glutamatergic neurons. That is, muscarinic agonists induced multiple types of postsynaptic responses, enhancing both inhibitory and excitatory fast transmissions to PPN thalamic-projecting neurons through muscarinic receptors. At present, it is not known if PPN glutamatergic neurons also project to the intralaminar thalamus, but we do know that not all the projections from the PPN to the Pf were cholinergic.

Intralaminar thalamic neurons send diffuse projections to the cortex and to the basal ganglia, as described above. In addition, recordings of intralaminar thalamic neurons in vivo have demonstrated that cells projecting to the cortex fire high-frequency action potentials in the 20–40 Hz range in relation to EEG desynchronization (Steriade et al., 1993). These authors proposed that intralaminar thalamic cells are "well suited for the distribution of fast rhythms during arousal and rapid eye movement sleep over the cerebral cortex" (Steriade et al., 1993). Such a function suggests that spontaneous activity in the cortex as reflected in the background EEG is modulated by the intralaminar thalamus to generate high-frequency rhythms. How then is the activity of individual arousing stimuli arriving in the RAS and transmitted to the intralaminar thalamus manifested at the level of the cortex? One technique for measuring responses to a single sensory stimulus is the evoked potential, in which a stimulus is presented and numbers of trials are averaged in the filtered EEG over the desired cortical region in order to extract the response that recurs following each stimulus.

CORTEX

The Midlatency Auditory P50 Potential—The human P50 potential is a midlatency auditory evoked response induced by a click stimulus and recorded at the vertex. The P50 potential peaks at a latency of 40–70 ms. The P50 potential was also referred to as the P1 potential because it is the first positive wave following the early brain stem auditory evoked response (BAER) that occurs at a 5–10 ms latency and the primary auditory response, called Pa, at the superior temporal gyrus, that occurs at a 25 ms latency. The Pa response is directly related to the auditory lemniscal nuclei, the "specific" thalamic nuclei, and the auditory cortex, while the P50 potential is related to the nonlemniscal, reticular, intralaminar, or "nonspecific" thalamic nuclei ascending pathways. The human P50 potential has three main characteristics: (1) It is sleep state-dependent, that is, it is present during waking and REM sleep, but is absent during deep SWS (Erwin and Buchwald, 1986a), so that it is present when the cortex is activated and the EEG shows high-frequency activity and, of course, when the PPN is active, (2) it is blocked by muscarinic cholinergic antagonists, such as scopolamine, so that it may be mediated, at least in part, by cholinergic neurons (Buchwald et al., 1991), (3) it undergoes rapid habituation at stimulation rates greater than 2 Hz. For example, the primary auditory cortex Pa potential can follow stimulus frequencies close to 20 Hz. However, the P50 potential cannot follow such high frequencies of stimulation, implying that it is not generated by a primary afferent pathway, but by multisynaptic, low-security synaptic elements of the RAS (Erwin and Buchwald, 1986a, b). Figure 6.5 shows the

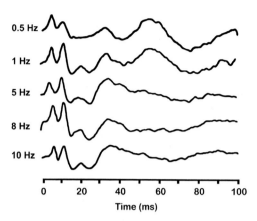

FIGURE 6.5 **Frequency following of auditory evoked responses recorded at the vertex.**
The peaks evident at 10 ms latency or less are part of the BAER, while the peak evident at
~25 ms latency is the Pa response, and the peak at ~50 ms latency is the P50 potential. Note
that, as stimuli were delivered at faster frequencies from 0.5 Hz to 10 Hz, the BAER and Pa responses persisted, but the P50 potential decreased at 1 Hz and disappeared at 5 Hz frequency
(from Erwin and Buchwald (1986a, 1986b)).

frequency of the BAER, Pa, and P50 responses when stimuli are applied
at increasing frequencies.

The P50 potential decreases and disappears with progressively deeper
stages of SWS and then reappears during REM sleep at full amplitude
(Kevanishvili and von Specht, 1979; Erwin and Buchwald, 1986a). None
of the earlier latency auditory evoked potentials (BAER or Pa primary auditory cortex potentials) possess this characteristic. This suggests that the
P50 potential is functionally related to the states of arousal. In addition,
the P50 potential is most likely generated, at least in part, by a cholinergic
mesopontine cell group such as the PPN, since its cells are preferentially
active during waking and REM sleep, but inactive during SWS (Garcia-
Rill, 1997; Garcia-Rill and Skinner, 2001; Reese et al., 1995a). We described
the methodology essential for recording the P50 potential reliably and re-
producibly (Garcia-Rill and Skinner, 2001). Essentially, while the threshold
for auditory stimuli at the level of the primary auditory cortex Pa potential
is in the order of 15 dB, the threshold for auditory P50 potential responses
is in the order of 70 dB. This is in keeping with the higher thresholds necessary to elicit arousal-related responses.

Our studies on the normal P50 potential were carried out comparing
postpubertal adolescents (12–19 years of age), young adults (24–39 years of
age), middle-aged adults (40–55 years of age), and older adults (55–78 years
of age) (Rasco et al., 2000). There were no statistically significant differences
in the mean amplitude or mean peak latency of the P50 potential between
males and females or between age groups. Using a paired-stimulus

paradigm, the degree of habituation, which subserves the process of sensory gating, was tested at three different interstimulus intervals (ISIs): 250, 500, and 1000 ms. That is, the response to the first stimulus of a pair represents the initial responsiveness of the system, which sets up descending modulation on which is superimposed the response to the second stimulus of a pair. The amplitude of the second stimulus is more inhibited or "gated" at early (250 ms) compared with later (1000 ms) intervals. That is, a recovery curve can be calculated from the three ISIs. There were no statistically significant differences in habituation between males and females; however, there was a decrease in habituation in the youngest group at the 250 ms ISI. These results suggested that (a) the recovery cycle of the P50 potential using an auditory click stimulus was greater than 1 s, that is, a full amplitude P50 potential is induced only if the second stimulus is applied at about a 0.5 Hz frequency, and (b) there is a sensory gating deficit in adolescents compared with adults (Rasco et al., 2000). These results suggest a mild anxiety-like responsiveness in teenagers, of which most parents are well aware.

Source of the P50 potential—Early work in the human, based on topographical studies of auditory evoked potentials, concluded that the source of the N100-P200 potential complex was the primary auditory cortex (Vaughan and Ritter, 1970). Their sink/source calculations indicated a polarity reversal of this waveform across the Sylvian fissure. They paid little attention to the earlier latency P50 potential, probably because the frequency of stimulation was too high (more than 1 Hz) to allow its full expression. A later study showed that stimulation faster than 0.5 Hz leads to a significant decrement in the amplitude of the P50 potential (Erwin and Buchwald, 1986b). However, another group failed to confirm the reversal of early auditory components across the Sylvian fissure (Kooi et al., 1971). While a subsequent study found that auditory evoked potentials in the 20–60 ms latency range were generated in the posterior temporal lobe (Wood and Wolpaw, 1982), other workers found the P50 potential still to be present after bilateral lesions of the temporal lobe (Woods et al., 1984). Others concluded that the source for the P50 potential was indeed the temporal lobe, but could not exclude the contribution of other sources (Reite et al., 1988). Perhaps the most telling study involved surface and depth recordings in human subjects (Goff et al., 1980). Depth-recording electrodes passing vertically and medially to the primary auditory cortex failed to find a reversal across the Sylvian fissure, as did electrodes passing along the long axis of the temporal lobe from the occipital cortex to the tip of the temporal lobe. These authors concluded that their data "failed to select from the two theories" on the origin of this potential (Goff et al., 1980), the temporal auditory cortex (Vaughan and Ritter, 1970), versus a modality nonspecific response (Williamson et al., 1970), that is, a volume-conducted brain stem potential. The studies by Buchwald cited above (Buchwald et al., 1991; Erwin and Buchwald, 1986a,b) demonstrated the labile nature of the P50 potential (i.e., its sleep-state dependence, rapid habituation, and blockade by nonsoporific doses of scopolamine), in contrast to the robust nature of the primary auditory cortex Pa potential, which occurs at a 25 ms latency, is generated by the primary auditory cortex, is present during all wake-sleep states, failed to habituate at high frequencies of stimulation, and was unaffected by scopolamine. While the source of the Pa potential has been localized to the primary auditory cortex in the posterior temporal lobe, the nature of the P50 potential is more consistent with waveforms generated by the RAS.

The P50 potential is not generated by the hippocampus. A comprehensive study using depth recordings in epilepsy patients with electrodes implanted in the hippocampus and parahippocampal gyrus found "no P50-like activity within the hippocampus in any of our subjects" (Grunwald et al., 2003). Using subdural electrodes for cortical recordings, these authors observed responses at ~50 ms latency that habituated to repetitive stimulation in only 6/24 patients, and these were confined to the prefrontal cortex and temporoparietal regions near, but not in, the primary auditory cortex (Grunwald et al., 2003). As such, the P50 potential appears to represent, rather than a primary auditory response, a "preattentional" process useful for the study of a number of conditions that manifest arousal/preattentional deficits.

Other authors have questioned the relationship between the P50 potential recorded at the vertex and the auditory cortical response recorded over the superior temporal gyrus. The differences stem from a lack of understanding of the role of the two potentials. The primary auditory cortex response or Pa is a response in relation to sensory discrimination or the content of sensory experience, while the P50 potential is a response to arousal and the context of sensory experience. The primary auditory cortex has a lower threshold and can be elicited by stimuli at less than 65 dB, in keeping with auditory discrimination. The threshold for the P50 potential is higher at ~70 dB, and a louder stimulus is required to elicit the full response, in keeping with its alerting function. In addition, the auditory cortex is activated by tones as easily as clicks, but the P50 potential, being an arousal-related response, requires the use of a wide spectrum of click stimuli. In addition, some investigators use very long-duration stimuli in the range of 100–200 ms. That is, the stimuli outlast the latency of the response. This introduces another variable in that very brief stimuli in the order of 100 µs will elicit a response, so that the long-duration activation of the system introduces unknown variables. Finally, a common mistake is to stimulate too rapidly. The primary auditory response can follow up to 10 or even 20 Hz stimulation, but the P50 potential will habituate at rates above 0.5 Hz. Because of the rapid habituation, the choice of the ISI to assess habituation is also critical. Most decrements in habituation are better observed at a 250 ms ISI, but many authors opt to use a 500 ms ISI, which is less likely to show a significant effect since it lies on the ascending phase of the recovery curve, not at its peak. For a detailed discussion of the technology behind recording the P50 potential, see Garcia-Rill et al. (2008).

The M50 Response—In terms of the P50 EEG potential versus the M50 magnetoencephalography (MEG) response, one study failed to localize a specific auditory M50 response using MEG (Yvert et al., 2005), while another reported clear differences between the P50 and M50 potentials in response to cholinergic agents (Pekkonen et al., 2005). These reports suggest that, based only on latency (without regard for habituation), the originally described characteristics of the P50 potential remain unexplained and suggest a contribution from a subcortical cholinergic source related to arousal level or state. One study interpreted both EEG (P50 potential) and MEG (M50 response) data based on habituation to repetitive stimuli (at a single intertrial interval), concluding that the two measures may be unrelated and that additional generators other than the primary auditory cortex may contribute to the P50 potential (Edgar et al., 2003). That study, however, assumed a superior temporal gyrus source for the M50 response and used click stimuli only 35 dB above threshold, which may generate little activation of the RAS. Another MEG study used unilateral stimulation but at 2.7 and 3.3 Hz, which would induce rapid habituation of RAS responses. These authors described small sources at ~50 ms in the *planum temporale*, but not in Heschl's gyrus (Inui et al., 2006). This 38-channel instrument is unlikely to detect sources in the frontal regions, especially in the region of the vertex where the P50 potential is at its highest amplitude.

We performed a study using MEG, and rather than assuming that M50 sources would be localized in and around the primary auditory cortex on the superior temporal gyrus, we determined localization from (a) the greatest activation at ~50 ms latency that would (b) habituate to repetitive stimulation (see Figure 6.6). Our results showed that each subject exhibited localization of a source on inflated maps for human primary auditory evoked responses in the region of the auditory cortex at a 20–30 ms latency. However, responses at a 40–70 ms latency that also decreased following the second stimulus of a pair, detected as intertrial coherence (ITC) activation after each stimulus, were generally not localizable to the auditory cortex, rather showing multiple sources, usually including the frontal lobes, especially in the region of the vertex (Garcia-Rill et al., 2008). This study used independent component analysis to localize sources without averaging, did not assume a superior temporal gyrus source, and used three ISIs to generate a recovery curve for the same response. Our findings suggest that M50 response sources were present in all subjects and showed diffuse localization usually including frontal lobes, especially in the region of the vertex. The consistency of the density of activation following the first stimulus of a pair lends further credence to the fact that it was the same phenomenon that was measured across ISIs in each subject.

In addition, only the ITC showed a decrement with repetitive stimulation at the same latency. Moreover, only the ITC revealed a recovery curve for this effect at increasing ISIs. This suggests that the activation induced

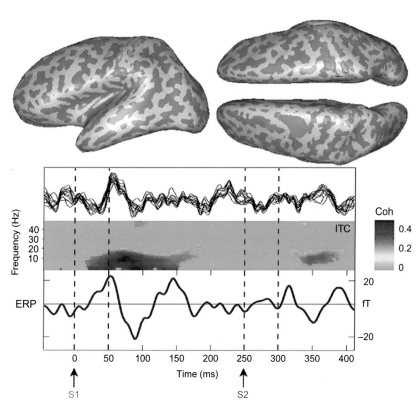

FIGURE 6.6 **M50 response following paired stimuli delivered at the 250 ms ISI.** Top: Localization of the response at a 50 ms latency that also decreased following the second stimulus of a pair. The vertex and regions of the frontoparietal cortex were evident on maps of inflated brains. Bottom: Responses following the two stimuli delivered at time 0 ms (S1) and 250 ms (S2) later. Note the ITC around 50 ms (full response) and 300 ms (habituated response) latency reflected in the event-related potential (ERP). These results suggest that the M50 response is maximal in the region of the vertex and differed considerably from the Pa response that was localized over the auditory cortex in the temporal lobe.

essentially led to phase resetting, in keeping with an oscillatory model of function. The M50 response may thus represent a process leading to an increase in phase concentration of oscillations. We concluded that the M50 response represents a phase resetting induced by the stimulus on ongoing oscillations (Garcia-Rill et al., 2008). This is in keeping with a model of sensory function based on thalamocortical oscillations, with the primary afferent ("specific" thalamic) pathways representing the content of sensory experience and the intralaminar diffuse ("nonspecific" thalamic) pathways representing the context of sensory experience (Llinas, 2001). The first may be reflected in the Pa response, while the latter may be represented by intralaminar diffuse (P50/M50) activation. This is in keeping with the proposed origin of the P50 potential as a preattentional/arousal-related, nonauditory response (Erwin and Buchwald, 1986a,b). Such activation would be characterized by rapid habituation in the absence of significant changes in the initial response, as observed in the present studies. That is, while Pa activation is consistent in modulating ongoing oscillations, habituation of the M50 response would provide a change in phase resetting with repetitive stimulation.

Study of the M50 response is complicated by the use of inappropriate paradigms for studying an arousal-related response, for example, using low-amplitude stimuli below the threshold for the M50 response, using very long-duration stimuli, and using tones that activate a limited auditory population (Golubic et al., 2014). The lack of understanding of the widely different functions of the two waveforms introduces unnecessary controversy.

The localization of the M50 response at the level of the vertex is virtually identical to the localization of changes in blood flow following PPN deep brain stimulation (DBS). We will discuss the use of PPN DBS in detail in Chapter 12; however, results have shown that PPN DBS leads to the most significant changes in cortical blood flow in the region of the vertex (Ballanger et al., 2009). This further supports the idea that sensory afferent information triggering the RAS leads to the activation of the intralaminar thalamus, which in turn induces the activation of the region of the vertex, along with additional frontal lobe sites.

Feline and rodent equivalents of the P50 potential—As discussed above, following an auditory stimulus, the evoked responses evident in the human are, within the first 10 ms, the BAER; then a positive wave, the Pa potential at around 25 ms latency; then the P50 potential at ~50 ms latency; and the P2 or P200 potential at about 200 ms latency. In the feline, the sequence of waveforms following the BAER is "wave 7" at about 15 ms latency, followed by "wave A" at about 20–25 ms latency, and then "wave C" at about 50 ms latency (Erwin and Buchwald, 1987). The feline equivalent to the human P50 potential, which has the same characteristics of sleep-state dependence, rapid habituation, and blockade by scopolamine, is "wave A" (Erwin and Buchwald, 1987). The Pa response (equivalent to the human Pa potential) in the cat appears to be "wave 7" and, like the Pa potential in humans, is present during all wake–sleep states, does not habituate readily, and is unresponsive to scopolamine (Erwin and Buchwald, 1987). Lesions of the cat PPN reduced "wave A" in parallel with cholinergic cell loss (Harrison et al., 1990), lending further evidence to its subcortical source. In physiological studies, we found in decerebrate cats that a large proportion of single units recorded in the PPN responded to auditory stimuli at 10–20 ms latencies and habituated rapidly (Reese et al., 1995b,c). We suggested that "wave A" was not generated by the responses of PPN neurons to auditory stimulation, rather that it was a manifestation of PPN outputs to various subcortical regions (Reese et al., 1995b,c). In keeping with this suggestion, depth-recorded evoked potentials at a latency of

20–25 ms were present in the intralaminar thalamus and other regions to which PPN projects (Hinman et al. 1983; Reese et al., 1995b). Interestingly, depth recordings in and around the cholinergic partner of the PPN, the laterodorsal tegmental (LDT) nucleus, failed to reveal auditory evoked responses in the cat or rat (Reese et al., 1995b), suggesting that this portion of ascending mesopontine cholinergic projections does not participate in the manifestation of these auditory evoked arousal responses. This lends further evidence that the LDT is not at all equivalent to the PPN in its functional role.

In the rodent, the waveforms following an auditory stimulus are the BAER, the P7 potential at about 7 ms latency, followed by the P13 potential at ~13 ms latency, and then by the P25 potential at ~25 ms latency. The midlatency auditory evoked P13 potential is to be the rodent equivalent of the human P50 potential and the feline "wave A." It is the only midlatency auditory evoked response that shares the same characteristics as the human P50 potential and cat "wave A" as described above (Miyazato et al., 1995, 1996, 1998). Since the P13 potential can be elicited after decerebration (Simpson and Knight, 1991), or after ablation of the primary auditory cortex bilaterally (Reese et al., 1995b), it appears to have a subcortical origin. Convincing evidence for the subcortical origin of the vertex-recorded P13 potential is that injections of various neuroactive agents known to inhibit PPN neurons, when injected into the region of the PPN, will reduce or block the vertex-recorded P13 potential while not affecting the primary auditory P7 potential. Noradrenergic, GABAergic, serotonergic, and cholinergic agents are all know to inhibit PPN neurons (Miyazato et al., 1999b; Reese et al., 1995a), and injection of each of these agents in the region of the PPN reduced or blocked the P13 potential in a dose-dependent manner (Miyazato et al., 1999b, 2000; Teneud et al., 2000). Moreover, interventions that modulate arousal such as various anesthetics, head injury, and ethanol all selectively reduced or blocked the P13 potential in a dose-dependent manner (Miyazato et al., 1999a). There is little doubt that the P13 potential is state-dependent, reticular in origin, and modulated in the same manner as the PPN and is therefore the rodent equivalent of the human P50 potential.

A longer latency auditory evoked response, the N40 potential, despite the fact that it has the opposite polarity, has been proposed to be the rodent equivalent of the human P50 potential (Luntz-Leybman et al., 1992). The N40 rodent potential does share with the P50 potential the characteristic of habituation to repeated stimulation (sensory gating properties), and it can be modulated by cholinergic and noradrenergic agents (Stevens et al., 1991, 1993). However, rather than being affected by the muscarinic cholinergic antagonist scopolamine, the N40 potential is affected by nicotinic cholinergic antagonists (Freedman et al., 1994; Luntz-Leybman et al., 1992). In addition, the N40 potential is present during waking and SWS, but absent during REM sleep, which differs from the sleep-state dependence of the human P50 potential and the rodent P13 potential. Moreover, the N40 potential can be recorded in anesthetized animals (Bickford-Wymer et al., 1990; Luntz-Leybman et al., 1992), which makes it unlikely to have arousal-related and sleep-state dependence. In addition, the origin of the N40 potential has been localized to the hippocampus (Bickford-Wymer et al., 1990; Luntz-Leybman et al., 1992); therefore, not only is it not a measure of RAS function, but also we know that the P50 potential is not localizable to the hippocampus (Grunwald et al., 2003). Taken together, these characteristics suggest that the N40 potential is not the rodent equivalent of the human P50 potential, feline "wave A," or rodent P13 potential, despite its obvious usefulness in the study of hippocampal sensory gating.

Figure 6.7 above shows recordings from the vertex in the human and the rat. The general shapes of these waveforms are remarkably similar despite the differences in latency, which are probably due to brain or body size. Figure 6.8 is a graph of the log of the latency versus the log of the body weight for midlatency auditory evoked potentials in three species, rodent, feline, and human. Linear regression lines show over 99% correlation between species weight and the latencies of the primary auditory potentials (bottom line: P7, "wave 7," and Pa), the RAS-mediated responses (middle line: P13, "wave A," and P50), and the long latency potentials (top line: P25, "wave C," and P2).

In summary, the data suggest that, when an auditory input is received, the primary auditory pathway conveys the characteristics of the sound rapidly and effectively (i.e., with high synaptic security) to the "specific" thalamic nuclei and all the way to the primary auditory cortex. However,

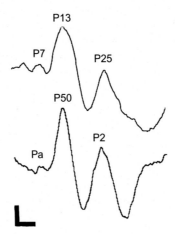

FIGURE 6.7   **Waveforms of the P13 potential in the rat (top) and the P50 potential in the human (bottom).** Note the similarity of the waveforms in manifesting the earlier primary cortex response (P7 in the rat and Pa in the human), followed by the arousal-related responses that are maximal at the vertex (P13 response in the rat and P50 response in the human). The two recordings differ in the time base, with the horizontal calibration bar for the top waveforms being 5 ms and for the bottom waveforms being 20 ms. The vertical calibration bar for the top waveforms is 10 μV and 1 μV for the bottom.

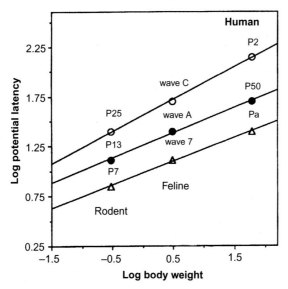

FIGURE 6.8 **Log plot of latency versus body weight.** The rodent auditory responses occur at 7 ms (P7), 13 ms (P13), and 25 ms (P25), while the feline responses have 12 ms (wave 7), 26 ms (wave A), and 50 ms (wave C) latencies. The human auditory responses occur at 25 ms (Pa), 50 ms (P50), and 200 ms (P2). The latency of each of these potentials is linearly correlated with the species' body weight.

there is a parallel activation of the RAS as long as the input is of sufficient amplitude, which is more slowly conducted and with low synaptic security that is relayed by the intralaminar thalamus to the cortex at a later latency. It is this "reticular" and "nonspecific" pathway that mediates the role of generalized arousal. This phenomenon would be expected to lead to cortical synchronization of fast frequencies (i.e., sleep-state dependence) but at longer latency and with lower synaptic security (i.e., rapidly habituating). It is these characteristics that make the human P50 potential, the feline "wave A," and the rodent P13 response such unique measures of RAS function.

CLINICAL IMPLICATIONS

The P50 potential has a number of potential uses in exploring the functions of the ascending modulation of the RAS. Using a paired-stimulus paradigm, the amplitude of the response following the first stimulus is a measure of the initial activation of the system, while the ratio of the amplitude of the second response in relation to the first is a measure of habituation to repetitive stimulation, which subserves the process of sensory gating or distractibility. The P50 potential has excellent potential for studies on the effects of, for example, sleep deprivation and fatigue on distractibility and attention. Because the P50 potential is a measure of ascending modulation by the RAS, it is a more relevant marker of "preattentional" mechanisms underlying cognitive performance than, for example, the startle response, which is, after all, a measure of descending RAS modulation. Sensory gating of the P50 potential also can be achieved in one-fourth of the time as the startle response sensory gating, due to the marked habituation of the startle response and the need to significantly slow the frequency of stimulation (intertrial intervals for the startle response need to be 30 s or greater). What needs to be determined is the degree to which decreased sensory gating of the P50 potential will influence cognitive performance. It can be assumed that processes requiring basic preattentional mechanisms will affect P50 potential measures to a greater extent than those that do not. Now that the rodent equivalent of the P50 potential has been identified and validated (Miyazato et al., 1995, 1996, 1998, 1999a,b), the P13 potential can be used to explore animal models of a host of neurological and psychiatric disorders. The ability to manipulate developmental factors allows the determination of long-term effects of early insults on subsequent arousal, attentional activity, and distractibility. For example, developmental insults, such as mother deprivation, pain, or early drug exposure, can be explored for their effects on P13 potential amplitude and habituation. In Chapters 11 and 12, we will explore the role of the RAS in a number of neurological and psychiatric diseases. In addition to examining the changes in EEG measures in each disorder, we will also describe the manifestation of the P50 potential in the same disorders. Therefore, while the EEG provides a background or spontaneous measure of wake–sleep rhythms, the P50 potential provides a measure of the responsiveness of the system and its habituation to repetitive stimulation. The two measures provide a telling spectrum of quantitative indicators of RAS function.

References

Anna, V.M., Gemma, G.B., Laura, A.V., Pilar, S.T., Margarita, M.N., 1999. Intracranial self-stimulation in the parafascicular nucleus of the rat. Brain Res. Bull. 48, 401–406.

Arai, R., Jacobowitz, D.M., Deura, S., 1994. Distribution of calretinin, calbindin-D28 k, and parvalbumin in the rat thalamus. Brain Res. Bull. 33, 595–614.

Ballanger, B., Lozano, A.M., Moro, E., van Elmeren, T., Hamani, C., Chen, R., Cilia, R., Houle, S., Poon, Y.Y., Lang, A.E., Strafella, A.P., 2009. Cerebral blood flow changes induced by pedunculopontine nucleus stimulation in patients with advanced Parkinson's disease: A [(^{15})O] H$_2$O PET study. Hum. Brain Mapp. 30, 3901–3909.

Bentivoglio, M., Balercia, G., Kruger, L., 1991. The specificity of the nonspecific thalamus: The midline nuclei. Brain Res. 87, 53–80.

Bickford-Wymer, P.C., Nagamoto, H., Johnson, R., Adler, L.E., Egan, M., Rose, G.M., Freedman, R., 1990. Auditory sensory gating in hippocampal neurons: A model system in the rat. Biol Psychiatry 27, 183–192.

Buchwald, J.S., Rubinstein, E.H., Schwafel, J., Strandburg, R.J., 1991. Midlatency auditory evoked responses: Differential effects of a cholinergic agonist and antagonist. Electroencephalogr. Clin. Neurophysiol. 80, 303–309.

Capozzo, A., Florio, T., Cellini, R., Moriconi, U., Scamati, E., 2003. The pedunculopontine nucleus projection to the parafascicular nucleus of the thalamus: an electrophysiological investigation in the rat. J. Neural Transm. 110, 733–747.

Celio, M.R., 1990. Calbindin D-28 k and parvalbumin in the rat nervous system. Neuroscience 35, 375–475.

Destexhe, A., Babloyantz, A., 1993. A model of the inward current I_h and its possible role in thalamocortical oscillations. Neuroreport 4, 223–226.

Destexhe, A., Babloyantz, A., Sejnowski, T.J., 1993a. Ionic mechanisms for intrinsic slow oscillations in thalamic relay neurons. Biophys. J. 65, 1538–1552.

Destexhe, A., McCormick, D.A., Sejnowski, T.J., 1993b. A model for 8–10 Hz spindling in interconnected thalamic relay and reticularis neurons. Biophys. J. 65, 2473–2477.

Destexhe, A., Contreras, D., Steriade, M., 1998. Mechanisms underlying the synchronizing action of corticothalamic feedback through inhibition of thalamic relay cells. J. Neurophysiol. 79, 999–1016.

Edgar, J.C., Huang, M.X., Weisend, M.P., Sherwood, A., Miller, G.A., Adler, L.E., Canive, J.M., 2003. Interpreting abnormality: An EEG and MEG study of P50 and the auditory paired-stimulus paradigm. Biol. Psychol. 65, 1–20.

Erro, E., Lanciego, J.L., Gimenez-Amaya, J.M., 1999. Relationships between thalamostriatal neurons and pedunculopontine projections to the thalamus: a neuroanatomical tract-tracing study in the rat. Exp. Brain Res. 129, 159–170.

Erwin, R.J., Buchwald, J.S., 1986a. Midlatency auditory evoked responses: Differential effects of sleep in the human. Electroencephalogr. Clin. Neurophysiol. 65, 383–392.

Erwin, R.J., Buchwald, J.S., 1986b. Midlatency auditory evoked responses: Differential recovery cycle characteristics. Electroencephalogr. Clin. Neurophysiol. 64, 417–423.

Erwin, R.J., Buchwald, J.S., 1987. Midlatency auditory evoked responses in the human and in the cat model. In: Johnson, R., Rohrbaugh, J.W., Parasweraman, R. (Eds.), Current Trends in Event-Related Potential Research. Elsevier Press, Amsterdam, pp. 461–467.

Frassoni, C., Spreafico, R., Bentivoglio, M., 1997. Glutamate, aspartate and co-localization with calbindin in the medial thalamus an immunohistochemical study in the rat. Exp. Brain Res. 115, 95–104.

Freedman, R., Adler, L.E., Bickford, P., Byerley, W., Coon, H., Cullum, C.M., Griffith, J.M., Harris, J.G., Leonard, S., Miller, C., 1994. Schizophrenia and nicotinic receptors. Harv. Rev. Psychiatry 2, 179–192.

Garcia-Rill, E., 1997. Disorders of the reticular activating system. Med. Hypotheses 49, 379–387.

Garcia-Rill, E., Skinner, R.D., 2001. The sleep state-dependent P50 midlatency auditory evoked potential. In: Lee-Chiong, T., Carskadon, M.A., Sateia, M.J. (Eds.), Sleep Medicine. Hanley & Belfus, Philadelphia, PA, pp. 697–704.

Garcia-Rill, E., Moran, K., Garcia, J., Findley, W.M., Walton, K., Strotman, B., Llinas, R.R., 2008. Magnetic sources of the M50 response are localized to frontal cortex. Clin. Neurophysiol. 119, 388–398.

Goff, W.R., Williamson, P.D., VanGilder, J.C., Allison, T., Fisher, T.C., 1980. Neural origins of long latency evoked potentials recorded from the depth and from the cortical surface of the brain in man. Prog. Clin. Neurophysiol. 7, 126–145.

Golubic, S.J., Aine, C.J., Stephen, J.M., Adair, J.C., Knoefel, J.E., Supek, S., 2014. Modulatory role of prefrontal generator within the auditory M50 network. Neuroimage 92, 120–131.

Grunwald, T., Boutros, N.N., Pezer, N., van Oertzen, J., Fernandez, G., Elger, C.E., 2003. Neuronal substrates of sensory gating within the human brain. Biol. Psychiatry 53, 511–519.

Grunwerg, B.S., Krein, H., Krauthamer, G.M., 1992. Somatosensory input and thalamic projection of pedunculopontine tegmental neurons. Neuroreport 3, 673–675.

Guillazo-Blanch, G., Vale-Martinez, A.M., Marti-Nicolovius, M., Morgado-Bernal, M., Cool-Andreu, I., 1999. The parafascicular nucleus and two-way active avoidance: Effects of electrical stimulation and electrode implantation. Exp. Brain Res. 129, 605–614.

Hallanger, A.E., Wainer, B.H., 1988. Ascending projections from the pedunculopontine tegmental nucleus and the adjacent mesopontine tegmentum in the rat. J. Comparative Neurol. 274, 483–515.

Harrison, J.B., Woolf, N.J., Buchwald, J.S., 1990. Cholinergic neurons of the feline pontomesencephalon. I. Essential role in "wave A" generation. Brain Res. 520, 43–54.

Harte, S.E., Lagman, A.L., Borszcz, G.S., 2000. Antinociceptive effects of morphine injected into the nucleus parafascicularis thalami of the rat. Brain Res. 874, 78–86.

Hazlett, J.C., Hazlett, L.D., 1977. Long axon neurons in the parafascicular and parafascicular posterolateral nuclei of the opossum: A Golgi study. Brain Res. 136, 543–546.

Hazlett, J.C., Dutta, C.R., Fox, C.A., 1976. The neurons in the centromedian–parafascicular complex of the monkey (Macaca mulatta): A Golgi study. J. Comp. Neurol. 168, 41–73.

Hermenegildo, S.H., Anna, V.M., Gemma, G.B., Margarita, M.N., Roser, N.A., Ignacio, M.B., 2000. Differential effects of parafascicular electrical stimulation on active avoidance depending on the retention time, in rats. Brain Res. Bull. 52, 419–426.

Hinman, C.L., Buchwald, J.S., 1983. Depth evoked potential and single unit correlates of vertex midlatency auditory evoked responses. Brain Res. 264, 57–67.

Inui, K., Okamoto, H., Miki, K., Gunji, A., Kakigi, R., 2006. Serial and parallel processing in the human auditory cortex: A magnetoencephalographic study. Cereb. Cortex 16, 18–30.

Jahnsen, H., Llinas, R.R., 1984a. Voltage-dependent burst-to-tonic switching of thalamic cell activity: An in vitro study. Arch. Ital. Biol. 122, 73–82.

Jahnsen, H., Llinas, R.R., 1984b. Electrophysiological properties of guinea-pig thalamic neurones: An in vitro study. J. Physiol. 349, 205–226.

Jones, E.G., 1985. The Thalamus. Plenum Press, New York.

Jones, E.G., 2002. Thalamic organization and function after Cajal. Brain Res. 136, 333–357.

Kevanishvili, Z., von Specht, H., 1979. Human auditory evoked potentials during natural and drug-induced sleep. Electroencephalogr. Clin. Neurophysiol. 47, 280–288.

Kobayashi, T., Good, C., Biedermann, J., Barnes, C., Skinner, R.D., Garcia-Rill, E., 2004. Developmental changes in pedunculopontine nucleus (PPN) neurons. J. Neurophysiol. 91, 1470–1481.

Kooi, R.A., Tipton, A.C., Marshall, R.E., 1971. Polarities and field configurations of the vertex components of the human auditory evoked response: A reinterpretation. Electroencephalogr. Clin. Neurophysiol. 31, 166–169.

Lai, H., Tsumori, T., Shiroyama, T., Yokota, S., Nakano, K., Yasui, Y., 2000. Morphological evidence for a vestibulo–thalamo–striatal pathway via the parafascicular nucleus in the rat. Brain Res. 872, 208–214.

Llinas, R.R., 1980. The intrinsic electrophysiological properties of mammalian neurons: Insights into central nervous system function. Science 242, 1654–1664.

Llinas, R.R., 2001. I of the Vortex, From Neurons to Self. MIT Press, Cambridge.

Llinas, R.R., Steriade, M., 2006. Bursting of thalamic neurons and states of vigilance. J. Neurophysiol. 95, 3297–3308.

Luntz-Leybman, V., Bickford, P.C., Freedman, R., 1992. Cholinergic gating of response to auditory stimuli in rat hippocampus. Brain Res. 587, 130–136.

McCormick, D.A., Bal, T., 1997. Sleep and arousal: Thalamocortical mechanisms. Annu. Rev. Neurosci. 20, 185–215.

McCormick, D.A., Huguenard, J.R., 1992. A model of the electrophysiological properties of thalamocortical relay neurons. J. Neurophysiol. 8, 1384–1400.

Miyazato, H., Skinner, R.D., Reese, N.B., Boop, F.A., Garcia-Rill, E., 1995. A middle-latency auditory-evoked potential in the rat. Brain Res. Bull. 37, 247–255.

Miyazato, H., Skinner, R.D., Reese, N.B., Mukawa, J., Garcia-Rill, E., 1996. Midlatency auditory evoked potentials and the startle response in the rat. Neuroscience 75, 289–300.

Miyazato, H., Skinner, R.D., Garcia-Rill, E., 1998. Sensory gating of the P13 midlatency auditory evoked potential and the startle response in the rat. Brain Res. 822, 60–71.

Miyazato, H., Skinner, R.D., Cobb, M., Andersen, B., Garcia-Rill, E., 1999a. Midlatency auditory evoked potentials in the rat—effects of interventions which modulate arousal. Brain Res. Bull. 48, 545–553.

Miyazato, H., Skinner, R.D., Garcia-Rill, E., 1999b. Neurochemical modulation of the P13 midlatency auditory evoked potential in the rat. Neuroscience 92, 911–920.

Miyazato, H., Skinner, R.D., Crews, T., Williams, K., Garcia-Rill, E., 2000. Serotonergic modulation of the P13 midlatency auditory evoked potential in the rat. Brain Res. Bull. 51, 387–391.

Oakman, S.A., Faris, P.L., Cozzari, C., Hartman, B.K., 1999. Characterization of the extent of pontomesencephalic cholinergic neurons projecting to the thalamus: comparison with projections to midbrain dopaminergic groups. Neurosci. 94, 529–547.

Otake, K., Nakamura, T., 1998. Single midline thalamic neurons projecting to both the ventral striatum and the prefrontal cortex in the rat. Neurosci. 86, 635–649.

Pare, D., Smith, Y., Parent, A., Steriade, M., 1988. Projections of brainstem core cholinergic and non-cholinergic neurons of cat to intralaminar and reticular thalamic nuclei. Neuroscience 25, 69–86.

Parent, M., Parent, A., 2005. Single-axon tracing and three-dimensional reconstruction of centre median-parafascicular thalamic neurons in primates. J. Comp. Neurol. 48, 127–144.

Pearson, J.C., Norris, J.R., Phelps, C.H., 1984. The cytoarchitecture and some efferent projections of the centromedian-parafascicular complex in the lesser bushbaby (Galago senegalensis). J. Comp. Neurol. 225, 554–569.

Pekkonen, E., Jaaskelainen, I.P., Kaakkola, S., Ahveninen, J., 2005. Cholinergic modulation of preattentive auditory processing in aging. Neuroimage 27, 387–392.

Phelan, K.D., Mahler, H.R., Deere, T., Cross, C.B., Good, C., Garcia-Rill, E., 2005. Postnatal maturational properties of rat parafascicular thalamic neurons recorded in vitro. Thalamus Relat. Syst. 3, 89–113.

Rasco, L.M., Skinner, R.D., Garcia-Rill, E., 2000. Effect of age on sensory gating of the sleep state-dependent P1/P50 midlatency auditory evoked potential. Sleep Res. Online 3, 97–105.

Reese, N.B., Garcia-Rill, E., Skinner, R.D., 1995a. The pedunculopontine nucleus—auditory input, arousal and pathophysiology. Prog. Neurobiol. 47, 102–133.

Reese, N.B., Garcia-Rill, E., Skinner, R.D., 1995b. Auditory input to the pedunculopontine nucleus: I. Evoked potentials. Brain Res. Bull. 37, 257–264.

Reese, N.B., Garcia-Rill, E., Skinner, R.D., 1995c. Auditory input to the pedunculopontine nucleus: II. Unit responses. Brain Res. Bull. 37, 265–273.

Reite, M., Teale, P., Zimmerman, J., Davis, K., Whalen, J., 1988. Source location of a 50 msec latency auditory evoked field potential. Electroencephalogr. Clin. Neurophysiol. 70, 490–498.

Resibois, A., Rogers, J.H., 1992. Calretinin in rat brain: An immunohistochemical study. Neuroscience 46, 101–134.

Rudkin, T.M., Sadikot, A.F., 1999. Thalamic input to parvalbumin-immunoreactive gabaergic interneurons: organization in normal striatum and effect of neonatal decortication. Neurosci. 88, 1165–1175.

Sadikot, A.F., Parent, A., Francois, C., 1990. The centre median and parafascicular thalamic nuclei project respectively to the sensorimotor and associative-limbic striatal territories in the squirrel monkey. Brain Res. 510, 161–165.

Sadikot, A.F., Parent, A., Francois, C., 1992. Efferent connections of the centromedian and parafascicular thalamic nuclei in the squirrel monkey: A PHA-L study of subcortical projections. J. Comp. Neurol. 315, 137–159.

Safroniew, M.V., Priestley, J.V., Consolazione, A., Eckenstein, F., Cuello, A.C., 1985. Cholinergic projections from the midbrain and pons to the thalamus in the rat, identified by combined retrograde tracing and choline acetyltransferase immunohistochemistry. Brain Res. 329, 213–223.

Scarnati, E., Gasbarri, A., Campana, E., Pacitti, C., 1987. The organization of nucleus tegmenti pedunculopontinus neurons projecting to basal ganglia and thalamus: a retrograde fluorescent double labeling study in the rat. Neurosci. Lett. 79, 11–16.

Shute, C.C.D., Lewis, P.R., 1967. The ascending cholinergic reticular system: Neocortical, olfactory and subcortical projections. Brain 90, 497–520.

Simpson, G.V., Knight, R., 1991. Multiple brain systems generating the rat auditory evoked potential. II. Dissociation of auditory cortex and non-lemniscal generator systems. Brain Res. 602, 251–263.

Steriade, M., 1994. Sleep oscillations and their blockage by activating systems. J. Psychiatry Neurosci. 19, 354–358.

Steriade, M., 1999. Cellular substrates of oscillations in corticothalamic systems during states of vigilance. In: Lydic, R., Baghdoyan, H.A. (Eds.), Handbook of Behavioral State Control. Cellular and Molecular Mechanisms. CRC Press, New York, NY, pp. 327–347.

Steriade, M., Glenn, L.L., 1982. Neocortical and caudate projections of intralaminar thalamic neurons and their synaptic excitation from midbrain reticular core. J. Neurophysiol. 48, 352–371.

Steriade, M., Curro Dossi, R., Contreras, D., 1993. Electrophysiological properties of intralaminar thalamocortical cells discharging rhythmic (approximately 40 Hz) spike-bursts at approximately 1000 HZ during waking and rapid eye movement sleep. Neuroscience 56, 1–9.

Stevens, K.E., Fuller, L.L., Rose, G.M., 1991. Dopaminergic and noradrenergic modulation of amphetamine-induced changes in auditory gating. Brain Res. 555, 91–98.

Stevens, K.E., Meltzer, J., Rose, G.M., 1993. Disruption of sensory gating by the alpha-2 selective noradrenergic antagonist yohimbine. Biol. Psychiatry 33, 130–132.

Sugimoto, T., Hattori, T., 1984. Organization and efferent projections of nucleus tegmenti pedunculopontinus pars compacta with special reference to its cholinergic aspects. Neurosci. 11, 931–946.

Teneud, L., Miyazato, H., Skinner, R.D., Garcia-Rill, E., 2000. Cholinergic modulation of the sleep state-dependent P13 midlatency auditory evoked potential in the rat. Brain Res. 884, 196–200.

Vale-Martinez, A., Guillazo-Blanch, G., Aldavert-Vera, L., Segura-Tores, P., Marti-Nicolovius, M., 1999. Intracranial self-stimulation in the parafascicular nucleus of the rat. Brain Res. Bull. 48, 401–406.

Van der Werf, Y.D., Witter, M.P., Groenewegen, I.I.J., 2002. The intralaminar and midline nuclei of the thalamus. Anatomical and functional evidence for participation in processes of arousal and awareness. Brain Res. Rev. 39, 107–140.

Vasilyev, D.V., Barish, M.E., 2002. Postnatal development of the hyperpolarization-activated excitatory current I_h in mouse hippocampal pyramidal neurons. J. Neurosci. 22, 8992–9004.

Vaughan, H.G., Ritter, W., 1970. The sources of auditory evoked responses recorded from the human scalp. Electroencephalogr. Clin. Neurophysiol. 28, 360–367.

Warren, R.A., Jones, E.G., 1997. Maturation of neuronal form and function in a mouse thalamo-cortical circuit. J. Neurosci. 17, 277–295.

Williamson, P.D., Goff, W.R., Allison, T., 1970. Somato-sensory evoked responses in patients with unilateral cerebral lesions. Electroencephalogr. Clin. Neurophysiol. 28, 566–575.

Wood, C.D., Wolpaw, J.R., 1982. Scalp distribution of human auditory evoked potentials. II. Evidence for overlapping sources and involvement of auditory cortex. Electroencephalogr. Clin. Neurophysiol. 54, 25–38.

Woods, D.L., Knight, R.T., Neville, H.J., 1984. Bitemporal lesions dissociate auditory evoked potentials and perception. Electroencephalogr. Clin. Neurophysiol. 57, 208–220.

Ye, M., Hayar, A., Garcia-Rill, E., 2009. Cholinergic responses and intrinsic membrane properties of developing thalamic parafascicular neurons. J. Neurophysiol. 102, 774–785.

Ye, M., Hayar, A., Strotman, B., Garcia-Rill, E., 2010. Cholinergic modulation of fast inhibitory and excitatory transmission to pedunculopontine thalamic projecting neurons. J. Neurophysiol. 103, 2417–2432.

Yvert, B., Fischer, C., Bertrand, O., Pernier, J., 2005. Localization of human supratemporal auditory areas from intracerebral auditory evoked potentials using distributed source methods. Neuroimage 28, 140–153.

Descending Projections of the RAS

Charlotte Yates, PT, PhD and*
Edgar Garcia-Rill, PhD†

*Center for Translational Neuroscience, Department of Physical Therapy,
University of Central Arkansas, Conway, AR, USA
†Department of Neurobiology and Developmental Sciences,
University of Arkansas for Medical Sciences, Little Rock, AR, USA

DESCENDING TARGETS

The main regions affected by descending projections of the pedunculo-pontine nucleus (PPN) are the subcoeruleus (SubC) nucleus, the pontine and medullary reticular formation, and, to a lesser extent, the spinal cord. Briefly, projections to the (a) SubC modulate its control of ponto-geniculo-occipital (PGO) waves, atonia, and outputs to the hippocampus that are involved in rapid eye movement (REM) sleep, (b) pontine and medullary reticular formation (nucleus magnocellularis (NMC) and nucleus gigan-tocellularis (NGC)) modulate reticulospinal projections that affect loco-motion and postural control, and (c) spinal cord are sparse and the exact terminations have not been determined (Figure 7.1). As such, the PPN represents a critical center for the control of postural and locomotor events related to waking as well as to REM sleep.

THE SUBCOERULEUS NUCLEUS

The PPN sends widespread projections throughout the pontomedul-lary reticular formation (Reese et al., 1995), including the anterior pontine region (Datta et al., 1998; Jones, 1990; Mitani et al., 1988; Shiromani et al., 1988). Injections of the nonspecific cholinergic agonist carbachol (CAR) into this region induced a REM sleep-like state, including atonia and PGO

http://dx.doi.org/10.1016/B978-0-12-801385-4.00007-0

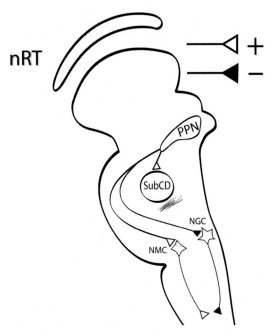

FIGURE 7.1   **Descending projections of the PPN.** Sagittal view of the pontomedullary region showing inputs to the SubCD (a) have a net excitatory effect on the nucleus, which is located immediately anterior to the VII cranial nerve (fiber bundle), (b) have a net excitatory effect on magnocellular neurons that will drive locomotion, and (c) have a net inhibitory effect on gigantocellular neurons that are involved in extensor inhibition and the startle response.

waves (Baghdoyan et al., 1984; Datta et al., 1998; Marks et al., 1980; Mitler and Dement, 1974; Mouret et al., 1967; Morrison, 1988; Vanni-Mercier and Debilly, 1989; Yamamoto et al., 1990). Stimulation of the PPN leads to acetylcholine release in this region (Lydic and Baghdoyan, 1993). Lesion of the pontine area termed the subcoeruleus dorsalis (SubCD) can produce REM sleep without atonia (Mouret et al., 1967; Morrison, 1988; Jouvet and Delorme, 1965; Karlsson et al., 2005; Sanford et al., 1994) or REM sleep without P-waves, the rodent equivalent of PGO waves (Mavanji et al., 2004). Neurons in this region are depolarized or hyperpolarized by muscarinic agonists (Gerber et al., 1991; Greene et al., 1989; Imon et al., 1996; Stevens et al., 1993). The use of *c-Fos* labeling has shown that the SubCD contains a population of "REM-on" neurons (Boissard et al., 2002). The anatomical and lesion studies described above suggest that the SubCD plays a major role in the production of REM sleep atonia via descending projections to the medulla and spinal cord. Additionally, there is evidence that the SubCD may also affect REM sleep processes via ascending projections to forebrain structures like the hippocampus (Bandyopadhya et al., 2006; Datta et al., 2003).

An in vitro study of SubCD cells reported the presence of populations of neurons that were either excited or inhibited by CAR, some of which

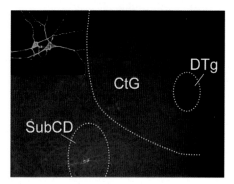

FIGURE 7.2 **Location of a pair of recorded SubCD neurons.** Coronal view of the dorsal pontine region showing at the top left a magnified view of two neurons that were recorded individually, labeled using neurobiotin, and found to be electrically coupled. The methods for such determinations were described (Heister et al., 2007). The location of the SubCD is lateroventrally to the central gray (CtG), at the level of the dorsal tegmental nucleus (DTg).

had low-threshold spikes (LTS) (Brown et al., 2006). However, this study included recordings from regions bearing cholinergic and noradrenergic neurons, suggesting that sampling went well beyond the boundaries of the SubCD. Histological analyses have shown this nucleus to consist of a mixture of mostly GABAergic and glutamatergic neurons, but not cholinergic or noradrenergic cells (Datta, 2006; Datta et al., 1999; Datta and MacLean, 2007). We described the presence of electrical coupling and spikelets in the SubCD and provided evidence for the presence of electrically coupled neurons in the SubCD (Garcia-Rill et al., 2007; Heister et al., 2007) (Figure 7.2). Moreover, we found that CAR acted directly on one-half of SubCD neurons by inducing an inward current, via both nicotinic and muscarinic M_1 receptors. CAR induced a potassium-mediated outward current via activation of M_2 muscarinic receptors in almost one-half of SubCD cells. Evoked stimulation established the presence of NMDA, AMPA, GABA, and glycinergic postsynaptic potentials in the SubCD. CAR also was found to decrease the amplitude of evoked excitatory postsynaptic currents (EPSCs) in most SubCD cells but to decrease the amplitude of evoked inhibitory postsynaptic currents (IPSCs) in a few SubCD cells. Spontaneous EPSCs were decreased by CAR in over one-half of cells recorded, while spontaneous IPSCs were increased in one-quarter of SubCD cells. These findings indicate that cholinergic input to the SubCD, mainly from the PPN, exerts a predominantly inhibitory role on fast synaptic glutamatergic activity and a predominantly excitatory role on fast synaptic GABAergic/glycinergic activity in the SubCD (Heister et al., 2009) (Figure 7.3). We hypothesized that, during REM sleep, cholinergic (ACh) "REM-on" neurons from the PPN that project to the SubCD induce excitation of inhibitory interneurons (GABA cells, some of which are electrically coupled), which leads to coherent inhibition of excitatory

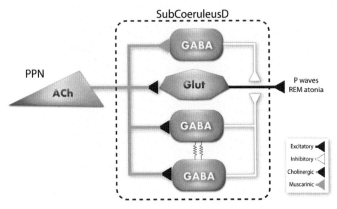

FIGURE 7.3 **Wiring diagram of PPN–SubCD interactions.** We hypothesized that, during REM sleep, cholinergic (ACh) "REM-on" neurons from the PPN (triangle left side) that project to the SubCD induce excitation of inhibitory interneurons (GABA rectangles, some of which are electrically coupled-resistor lines), which leads to coherent inhibition of excitatory (glutamatergic—Glut neurons in polygon) SubCD cells, leading to the production of coordinated activity in SubCD projection neurons. The net inhibition induced would hyperpolarize (open triangles) SubCD cells, and those with LTS would generate bursting activity. The coordination of these projection neurons, of the modulation of the SubCD by the PPN, may be essential for the production of REM sleep signs such as PGO waves by SubCD output neurons.

(glutamatergic) SubCD cells, leading to the production of coordinated activity in SubCD projection neurons. The net inhibition induced would hyperpolarize some SubCD cells, and those with LTS would generate bursting activity. The coordination of these projection neurons, of the modulation of the SubCD by the PPN, may be essential for the production of REM sleep signs such as PGO waves by SubCD output neurons. As we will see below, the PPN also exercises push-pull modulation of reticulospinal systems involved in posture and locomotion.

THE PPN AND THE MLR

As will be discussed in Chapter 12, the PPN has become a target for deep brain stimulation (DBS) for the treatment of Parkinson's disease (PD) in human subjects. We predicted such use long ago based on animal studies using stimulation in the region of the PPN, which at the time was thought to be in the mesencephalic locomotor region (MLR) (Garcia-Rill, 1986, 1991, 1997). The MLR was originally described as a region that required stimulation at increasing current amplitude using long-duration (0.5–1.0 ms) pulses at 40–60 Hz in the precollicular-postmamillary transected cat to induce controlled locomotion on a treadmill (Shik et al., 1966). The lowest-threshold sites were localized to the lateral cuneiform nucleus, dorsal to the lateral portion of the superior cerebellar peduncle. The same year, the PPN was proposed as a descending target of basal ganglia projections (Nauta and Mehler, 1966). We were studying the electrophysiology of striatal neurons (Garcia-Rill et al., 1979) and became interested in the PPN (Garcia-Rill et al., 1981, 1983). We undertook a series of experiments to determine if the PPN and the MLR were overlapping structures. We used neuroactive agents injected into the region of the PPN to induce locomotion (Garcia-Rill et al., 1985) and recorded from locomotion-related PPN neurons, with some cells being active in relation to limb alternation, while others were related to the duration of the locomotor episode (Garcia-Rill and Skinner, 1988). We then addressed why, in order to induce locomotion, (a) lateral, but not medial, cuneiform nucleus stimulation was required, (b) ramping up of the current was required, (c) stimulation at 40–60 Hz, but not higher or lower, was required, and (d) long-duration pulses were required. These requirements have finally been satisfactorily resolved (see below).

Figure 7.4 presents examples of both electrically induced locomotion and chemically induced locomotion. In Figure 7.4(a), electrical stimulation within the region of NADPH-diaphorase-labeled cells (gray region—cholinergic neurons in the PPN are selectively labeled by this histochemical method) induced locomotion on a treadmill in the precollicular-postmamillary transected cat (T). Alternation between the left (LAE) and right (LAE) ankle extensors, the right knee extensor (RKE), and the right ankle flexor (RAF) is evident in the electromyograms. In Figure 7.4(b), instead of electrical stimulation, 10 mM NMDA was injected into the region of cholinergic cell labeling in a similarly transected cat. Alternation between the left forelimb extensor (LFL) and the right forelimb extensor (RFL), as well as the left hind limb extensor (LHL) and the right hind limb extensor (RHL), is observed. We found similar responses in the cat and the rat (Garcia-Rill, 1986, 1991; Garcia-Rill et al., 1981, 1985, 1986, 1987, 1991, 1996; Garcia-Rill and Skinner, 1988, 1991; Skinner and Garcia-Rill, 1990, 1994; Skinner et al., 1990a). Interestingly, locomotion was elicited after ~1–2 seconds of stimulation, a latency similar to that shown to induce low-amplitude, high-frequency EEG in the cortex following stimulation of the region of the PPN (Moruzzi and Magoun, 1949).

Because we were recording or injecting agents into a region that we had first physiologically identified to induce locomotion, we used the term "MLR" in most of our studies to describe the findings. However, we became convinced that the lowest-threshold sites for inducing locomotion were located within the PPN (Garcia-Rill et al., 1987). We questioned the interpretation of the MLR as a locomotion-specific area, showing instead that the region stimulated was a "rhythmogenic" region that was part of the reticular activating system (RAS) (Garcia-Rill, 1991, 1997; Garcia-Rill and Skinner, 1988, 1991; Garcia-Rill et al., 1996; Skinner and

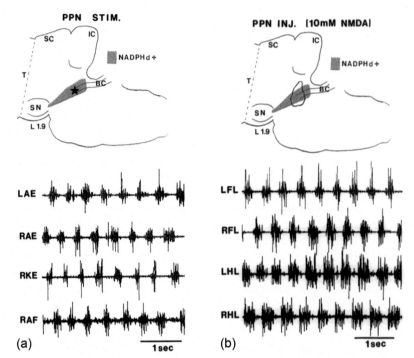

FIGURE 7.4   **Electrically and chemically induced locomotion in the decerebrate cat.**
(a) Electrical stimulation site (star) within the region of NADPH-diaphorase-labeled cells
(gray region—cholinergic neurons in the PPN are selectively labeled by this histochemical
method) induced locomotion on a treadmill in the precollicular-postmamillary transected
cat (T). Alternation between the left (LAE) and right (LAE) ankle extensors, the right knee extensor (RKE), and the right ankle flexor (RAF) is evident in the electromyograms (Garcia-Rill
et al., 1981, 1983). (b) Instead of electrical stimulation, 10 mM NMDA (and other agents) was
injected into the region of cholinergic cell labeling in a similarly transected cat. Alternation
between the left forelimb extensor (LFL) and the right forelimb extensor (RFL), as well as the
left hind limb extensor (LHL) and the right hind limb extensor (RHL), is observed (Garcia-
Rill et al., 1985, 1988). Later studies described similar effects in the rat (Garcia-Rill et al., 1986,
1987; Skinner et al., 1990b).

Garcia-Rill, 1990, 1994; Reese et al., 1995). However, this distinction is still
blurred by those not familiar with the complex literature, therefore requiring some explanation.

   Localization—Early descriptions of cholinergic pathways in the midbrain
localized the dorsal and ventral tegmental cholinergic pathways, based on
acetylcholinesterase labeling, as arising from the cuneiform nucleus in the
posterior midbrain (Shute and Lewis, 1967). In their Figure C, the wedge-
shaped nucleus that gave rise to these cholinergic projections was labeled
as "cuneiform nucleus" instead of "pedunculopontine nucleus." The PPN
was later identified as a cholinergic cell group in the posterior midbrain

(Mesulam et al., 1983), but recent studies using in situ hybridization established that this nucleus is composed of mostly nonoverlapping populations of cholinergic, glutamatergic, and GABAergic neurons (Wang and Morales, 2009). Sagittal sections are ideal for visualizing the PPN, which reveal its wedge shape extending from the dorsolateral to the ventrolateral mesopontine region (Figure 7.5). The *pars compacta* of the PPN is located posteriorly and dorsal to the lateral portion of the superior cerebellar peduncle (SCP), in the optimal site for inducing locomotion on a treadmill at low thresholds (Garcia-Rill et al., 1987). As the nucleus ranges anteroventrally, it intermixes with the superior cerebellar peduncle and terminates at the posterior end of the substantia nigra. The lowest-threshold sites for inducing locomotion were located within the PPN *pars compacta*, which is embedded in the lateral part of the cuneiform nucleus (Garcia-Rill et al., 1987, 1996). Others reported stimulation sites in the cuneiform nucleus dorsal to the superior cerebellar peduncle, but did not label for PPN cells in the stimulated animals (Takakusaki et al., 2003). This led to the erroneous conclusion that the MLR was located dorsal to the superior cerebellar peduncle and that the PPN lay more ventrally only within, but not dorsal to, the superior cerebellar peduncle. However, cholinergic PPN neurons are evident well dorsal to the superior cerebellar peduncle, in the region we reported as low-threshold locomotion-inducing sites (Figure 7.5) (Garcia-Rill et al., 1987; Garcia-Rill and Skinner, 1988). Stimulation at more

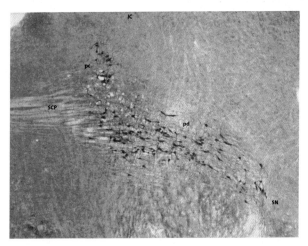

FIGURE 7.5   **Sagittal view of the PPN.** Sagittal sections are ideal for visualizing the PPN, which reveal its wedge shape ventrally to the inferior colliculus (IC), extending from the dorsolateral (top left) to the ventrolateral midbrain region (bottom right). The *pars compacta* (pc) of the PPN is located posteriorly and dorsal to the lateral portion of the superior cerebellar peduncle (SCP), in the optimal site for inducing locomotion on a treadmill at low thresholds (Garcia-Rill et al., 1987). The *pars dissipata* (pd) is located anteriorly and abuts the posterior end of the substantia nigra (SN).

ventral sites did not induce stepping, but rather changes in muscle tone (Takakusaki et al., 2003), an effect that may also have been due to suddenly switching on the stimulus (see below). Stimulation of more medial regions such as the laterodorsal tegmental nucleus (LDT), the medial partner of the PPN, or anteroventrally in the region of the substantia nigra did not induce locomotion (Garcia-Rill, 1991; Garcia-Rill and Skinner, 1991; Skinner and Garcia-Rill, 1990, 1994; Reese et al., 1995). This explains why stimulation of only lateral, but not medial, cuneiform, in which the PPN is embedded, produces reliable stepping on a treadmill.

Figure 7.6 is a three-dimensional reconstruction of sagittal sections containing four neuronal groups in the posterior midbrain of the rat. Dopaminergic substantia nigra neurons are represented by green spheres. PPN cells are represented by red spheres. Locus coeruleus neurons are represented by purple spheres. LDT cells are represented by yellow

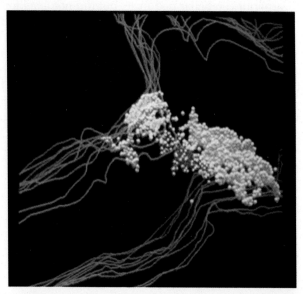

FIGURE 7.6 **Three-dimensional reconstruction of sagittal sections containing four neuronal groups in the posterior midbrain of the rat.** Dopaminergic substantia nigra neurons (green spheres), PPN cells (red spheres), locus coeruleus neurons (purple spheres), and LDT cells (yellow spheres) are represented. This view is from posterolateral to anteromedial, with the PPN laterally and the LDT most medially along the midline and embedded in the central gray. Note that locus coeruleus neurons are interspersed with PPN cells laterally and LDT cells medially. There is a transition from the PPN laterally to the LDT medially that contains scattered cholinergic neurons, and these scattered cells are within the anterior pole of the locus coeruleus. The overlap between cholinergic and noradrenergic neurons is easily evident in double-labeled sections (Garcia-Rill and Skinner, 1991). The wedge shape of the PPN is evident stretching from the peribrachial area posteriorly, moving ventrally to the posterior edge of the substantia nigra. The PPN does not range very far caudally beyond the level of the inferior colliculus.

spheres. This view is from lateral to medial, with the PPN lateral and the LDT most medially along the midline and embedded in the central gray. Note that locus coeruleus neurons are interspersed with PPN cells laterally and LDT cells medially. There is a transition from the PPN laterally to the LDT medially that contains scattered cholinergic neurons, and these scattered cells are within the anterior pole of the locus coeruleus. The overlap between cholinergic and noradrenergic neurons is easily evident in double-labeled sections (Garcia-Rill and Skinner, 1991). The wedge shape of the PPN is evident stretching from the peribrachial area posteriorly and moving ventrally to the posterior edge of the substantia nigra. The PPN does not range very far caudally beyond the level of the inferior colliculus. Stimulation of sites caudally to the PPN would engage the peribrachial region that has a number of cell groups related to pain and micturition (the so-called pontine micturition center is located caudally to PPN in the peribrachial region) (Garcia-Rill et al., 1991; Garcia-Rill and Skinner, 1988; Reese et al., 1995).

Stimulation at 40–60 Hz—Figure 7.7(a) shows the effects of delivering current steps in single patch-clamped PPN neurons (Simon et al., 2010). We found that virtually all PPN cells increased firing frequency and then

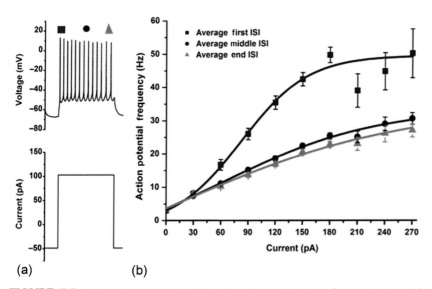

(a)   **(b)**

FIGURE 7.7 **Firing frequency of PPN cells after application of current steps.** (a) Example of the response of a PPN neuron (top record) at the beginning (blue square), middle (red sphere), and end (green triangle) of a current step; in this case 100 pA. (b) Graph of firing frequency in Hz during the first interspike interval (ISI) (blue squares), middle (red spheres), and end (green triangles) of current steps between 30 and 270 pA. Note that firing frequency at all ISIs increased gradually and then plateaued between 25 and 50 Hz for this cell. All PPN cells, regardless of cell type, showed maximal plateaus at 40–60 Hz (Simon et al., 2010).

plateaued at 40–60 Hz as current levels were increased (Figure 7.7(b)), regardless of neurotransmitter or electrophysiological type. That is, these neurons could not be induced to fire any faster than 40–60 Hz, regardless of the level of current delivered intracellularly. This property was further investigated to determine the mechanism responsible for such a unique characteristic by recording patch-clamped neurons in the presence of synaptic blockers (to prevent afferent signals) and tetrodotoxin (to prevent action potential generation). In our investigation of the intrinsic properties of these cells, we applied current steps to attempt to drive high-threshold channels, but the steps were unable to depolarize the membrane sufficiently to maintain depolarization to reveal membrane oscillations (Figure 7.8A), probably due to the activation of potassium channels by the sudden depolarization (Kezunovic et al., 2011). However, when we applied ramps to gradually depolarize the membrane to activate high-threshold channels, we were able to elicit membrane oscillations in the beta and gamma band range (Figure 7.8B). We found that all PPN neurons, regardless of transmitter or electrophysiological type, manifested voltage-dependent, high-threshold P-/Q- and N-type calcium channels that mediated the plateau in maximal membrane oscillatory activity at 40–60 Hz (Kezunovic et al., 2011). That is, the presence of these calcium channels determines the maximal oscillatory frequency displayed by ALL of these cells. We used specific calcium channel blockers (especially ω-agatoxin for P-/Q-type calcium channels) to identify these as the main channels responsible for beta/gamma band oscillations. We should note that cells outside of the PPN do not show similar properties, making the cellular boundaries of the nucleus specific to this kind of activity. The

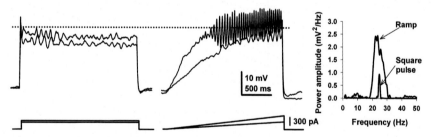

FIGURE 7.8   **Intrinsic membrane oscillations in PPN neurons.** Left panel: We applied current steps to attempt to drive high-threshold channels, but the steps were unable to depolarize the membrane sufficiently to maintain depolarization to reveal membrane oscillations, probably due to the activation of potassium channels by the sudden depolarization. Middle panel: However, when we applied ramps to gradually depolarize the membrane to activate high-threshold channels, we were able to elicit membrane oscillations. Right panel: The power spectrum shows that the oscillations elicited by ramps were in the beta and gamma band range (Kezunovic et al., 2011).

presence of a frequency plateau in PPN cells explains the requirement to stimulate this region at 40–60 Hz to optimally induce locomotion. The use of high-frequency trains applied to this region (e.g., Lai and Siegel, 1990) would be expected to lead to depolarization block and decrease in PPN output, which may induce a decrement in muscle tone.

Ramping up current—Early studies revealed that suddenly switching on the current at previously determined threshold levels elicited only decrements in muscle tone instead of stepping (Garcia-Rill et al., 1986; Takakusaki et al., 2003). Pulses delivered to the PPN had to be increased gradually until locomotion was induced, in many cases then backing off the current, and only then was continuous stepping evident (Garcia-Rill et al., 1986, 1991; Reese et al., 1995). We referred to this characteristic as "recruiting" locomotion. Single cells in the PPN also need to be "recruited" to fire at gamma band frequencies. Figure 7.8 shows the effects of using sudden current steps (Figure 7.8A) vs. current *ramps* (Figure 7.8B) to induce membrane oscillations. The requirement to use ramps to elicit oscillations helps explain why current has to be gradually increased to depolarize PPN neurons slowly, that is, to avoid activating potassium channels, and thus induce stepping. Moreover, current ramps can depolarize membrane potential with a similar time course, as previously observed during temporal summation of EPSPs during gamma band stimulation (Pedroarena and Llinas, 2001). Sudden application of high-amplitude currents would activate potassium channels and fail to sufficiently depolarize neurons. This would instead lead to inactivation of PPN neurons, explaining the decrement in muscle tone evident with sudden onset stimulation. This explains the need for ramping up current, instead of suddenly switching it on, in order to "recruit" stepping. In order to maintain such depolarization, the current needs to be reduced slightly; otherwise, the depolarization will exceed the window for high-threshold calcium channels. This explains the characteristic need to back off the current once locomotion is induced.

Long-duration pulses—We hypothesize that the need to use long-duration pulses is related to the high-threshold calcium channels present in PPN neurons. We used calcium imaging to visualize ramp-activated calcium channels in PPN cells (Hyde et al., 2013). Figure 7.9 shows that voltage-dependent high-threshold P-/Q-type calcium channels are located in the dendrites of PPN neurons. This explains the need to apply high levels of depolarization of the cell body in order to ultimately depolarize the higher-resistance dendrites sufficiently to activate the high-threshold calcium channels and promote gamma band oscillations. Long-duration pulses, along with ramping up current levels, would be effective methods for activating dendritic calcium channels that would in turn allow PPN neurons to fire maximally at gamma band frequencies.

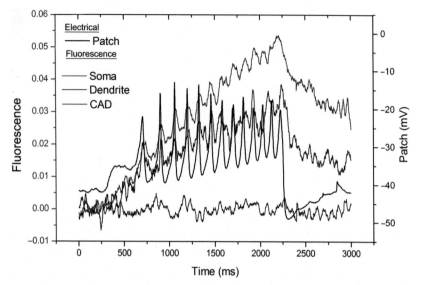

FIGURE 7.9   **Fast calcium imaging of membrane oscillations in PPN cells.** Electrical recordings (black record) exhibited membrane oscillations using current ramps as previously described. Calcium imaging using a ratiometric method (described in Hyde et al., 2013) revealed oscillations in the calcium signal from the cell body (soma—red record) as well as from a proximal dendrite (blue record). Note that the peaks of the calcium signal in the dendrite (blue record) corresponded only with some of the peaks in the electrical signal (black record); however, the peaks in the soma (red record) corresponded with each peak in the electrical signal. This suggests that the oscillations observed in the cell body summate from the signals in the different dendrites. The electrical and calcium signal were abolished after adding the calcium channel blocker cadmium (CAD) (purple record).

POSTURE AND LOCOMOTION

The fact that the PPN is involved in wake–sleep control and arousal led us to conclude that the "MLR" was not a locomotion-specific region, but rather a phenomenon elicited when activating a region with very specific "rhythmogenic" characteristics. But why does the RAS modulate postural and locomotor pathways? The RAS is a phylogenetically conserved system that modulates fight-or-flight responses. During waking, man's ability to detect predator or prey is essential for survival. Under these circumstances, it is not surprising that the RAS can modulate muscle tone and locomotion. This system is thus intrinsically linked to the control of the motor system in order to optimize attack or escape. During REM sleep, the atonia keeps us from acting out our dreams. In fact, only our diaphragm and eye muscles appear to be acting out dream content. Therefore, during both waking and REM sleep, two states modulated by the PPN, the RAS can influence muscle tone and locomotion via the same reticulospinal systems. For example, in a standing individual, there is tonic activation of antigravity, mainly extensor, muscles (the same ones inhibited in the atonia of REM sleep). Before the first step can be taken, there must be flexion of the leg; therefore, there must be a release, or extensor inhibition, from this postural extensor bias. It should be noted that the first sign of stepping from a standing position will always be extensor inhibition to unlock the knees, only then followed by flexion. Extensor inhibition is thus the first action modulated by descending PPN outputs. The question is whether or not the extensor inhibition is prolonged to induce postural collapse or whether it does lead to flexion–extension alternation and locomotion.

Outputs from the PPN and/or SubCD activate reticulospinal systems that lead to profound hyperpolarization of motoneurons, which is the mechanism responsible for the atonia of REM sleep (Chase and Morales, 1994). Cholinergic projections from the PPN to the medioventral medulla elicit locomotion (Kinjo et al., 1990). Outputs from this medullary region in turn activate reticulospinal systems that lead to the triggering of spinal pattern generators to induce stepping (Garcia-Rill et al., 1991, 2004a,b; Reese et al., 1995). In general, electrical stimulation of the pontine and medullary reticular formation is known to induce decreased muscle tone at some sites while producing stepping movements at other sites. This suggests the presence of a heterogeneous, distributed system of reticulospinal motor control. The required parameters of stimulation for eliciting these differing effects are important such that instantaneous, high-frequency trains (similar to high-frequency bursting activity in the range of PGO burst neurons that may drive the atonia of REM sleep) trigger pathways that lead to decreased muscle tone, while lower-frequency tonic stimulation lead gradually to the "recruitment" of locomotor movements (Reese et al., 1995). Therefore, given the extensive evidence, it is to be expected that the PPN should modulate both posture and locomotion.

PUSH–PULL ON RETICULOSPINAL CELLS

The PPN sends diffuse mostly cholinergic projections throughout the pontine and medullary reticular formation (Garcia-Rill, 1991; Garcia-Rill and Skinner, 1987; Garcia-Rill et al., 1986; Jones, 1990; Mitani et al., 1988; Rye et al., 1988; Semba et al., 1990; Shiromani et al., 1988). While carbachol (CAR) injected into the medioventral medulla induces stepping (Skinner et al., 1990a), electrical stimulation of the medioventral medulla locomotion-inducing sites has different characteristics from stimulation of the PPN. Medioventral medulla requires stimulation at 10–20 Hz, while the PPN requires stimulation at 40–60 Hz, in order to induce locomotion (Kinjo et al., 1990; Skinner et al., 1990a). Stimulation of the PPN at 40–60 Hz led to prolonged depolarization of posterior pontine and medioventral medulla neurons, which manifested maximal firing rates at ~10–20 Hz (Mamiya et al., 2005). This may be the natural frequency of firing for this region and why stimulation of this region requires stimulation at 10–20 Hz to induce stepping. Stimulation of the PPN at higher frequencies (~100 Hz) led to cessation of firing in these cells, emphasizing the frequency dependence of descending PPN output, which elicits excitation at 40–60 Hz but inhibition at higher frequencies. The nonspecific cholinergic agonist CAR induced prolonged depolarization in most reticular formation neurons tested, and the muscarinic cholinergic antagonist scopolamine reduced or blocked CAR-induced and PPN stimulation-induced prolonged depolarization in these cells. These findings suggest that PPN (electrical) or CAR (pharmacological) stimulation-induced prolonged depolarization is due to activation of muscarinic receptor-sensitive channels, allowing medioventral medulla neurons to respond to a transient, frequency-dependent depolarization with long-lasting stable firing rates (Mamiya et al., 2005). Interestingly, those neurons that responded with prolonged depolarization were not reticulospinal, but interneurons, suggesting that the signal descending to the spinal cord is mediated by reticulospinal neurons that do not receive direct projections from the PPN (Garcia-Rill et al., 2001).

However, the organization of descending PPN projections is far more specific. Interneurons in the caudal pontine and medioventral medulla that undergo prolonged depolarization following PPN stimulation are small to medium cells (Homma et al., 2002). The mechanism proposed involved the closure of potassium channels by CAR, leading to depolarization and higher firing frequencies. As mentioned above, PPN stimulation at 40–60 Hz was optimal for inducing these cells to fire maximally at 10–20 Hz (Homma et al., 2002), the preferred frequency of stimulation for inducing locomotion following medioventral medulla stimulation (Kinjo et al., 1990). The requirement for lower-frequency stimulation of this region suggests that the mechanism driving this effect does not involve high-threshold calcium channels as in the PPN. However, a low number (~1%) of these neurons were instead hyperpolarized and were found to be large in cell size. That is, descending PPN projections tend to depolarize small to medium cells for prolonged periods ("push" towards locomotion) and to hyperpolarize large cells ("pull" away from decreased muscle tone) to facilitate stepping (Figure 7.10). Some of the large cells that are inhibited by descending PPN projections appear to be those involved in the startle response.

THE STARTLE RESPONSE

The startle response is a short latency motor response to a supramaximal stimulus. In response to a loud auditory stimulus, the startle response pathway travels from the cochlea, to the cochlear nucleus, to the nucleus of

Intracellular recordings in PnC

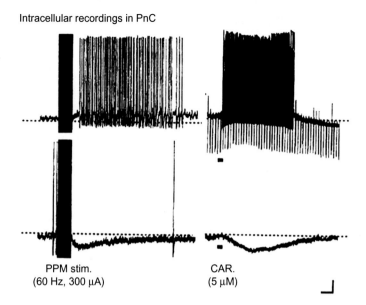

PPM stim.
(60 Hz, 300 μA)

CAR.
(5 μM)

Locomotion after PPN stimulation

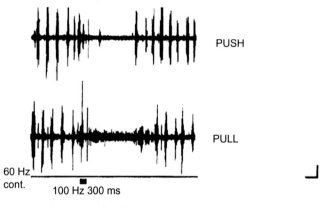

PUSH

PULL

60 Hz
cont.

100 Hz 300 ms

Muscle tone

FIGURE 7.10 **Inputs from the PPN to pontomedullary neurons.** Top records: Intracellular recordings in a magnocellular neuron following stimulation of the PPN for 1s using 300μA pulses delivered at 60Hz. Note the long-lasting depolarization induced, thought to be due to a sodium-dependent plateau potential (left side). The same cell was also depolarized by a 20s application of carbachol (CAR) (right side). On the other hand, intracellular recording from a gigantocellular neuron was found to manifest a long-lasting hyperpolarization following electrical stimulation (left side) as well as CAR application (right side). The magnocellular neurons are thought to participate in driving locomotion, while the gigantocellular neurons (representing only ~1% of the population) are thought to participate in the inhibition of extension underlying the startle response. Bottom records: Electromyographic recordings from left and right hind limb flexors during continuous stimulation of the PPN (60Hz Cont.) as described above, showing induction of alternating stepping. A short-duration train of 300ms at 100Hz was superimposed, which immediately led to the inhibition of stepping. These results suggest that the PPN exercises a "push" towards stepping and a "pull" away from extensor, antigravity tone necessary for posture (Mamiya et al., 2005).

the lateral lemniscus and then activates giant cells in the posterior pontine reticular formation (Davis, 1984). The giant cells of the posterior pontine reticular formation project to the spinal cord to induce a short latency inhibition of extensor muscles. These cells make up about 1% of the population (Koch, 1999), and, when lesioned, the startle response is reduced (Koch et al., 1992). PPN lesions were shown to reduce prepulse inhibition of the startle response (Swerdlow and Geyer 1993), and auditory-responsive neurons in the posterior pontine reticular formation were inhibited by CAR (Koch et al., 1993). These results suggest that descending PPN projections to these large cells participate in modulation of startle response sensory gating. Such connectivity allows the PPN to modulate fight-or-flight responses.

Interestingly, the startle response is basically a flexor response, placing the body in a "ready" position. This response shifts the standing individual from extensor to flexor bias, as if going from a standing position (extended) to a "ready" position (flexed). The startle response is composed of a short latency activation of muscle activity (the "ready" condition) that occurs too fast for RAS modulation, followed by a brief inhibition (the "reset" state) and a long latency activation (the "go" condition). The brief, intermediate latency inhibition is thought to be part of the modulation of the startle response by the PPN and may represent a "resetting" of motor programs that allow the subsequent selection of response strategies, the triggering of attack or escape movements (Garcia-Rill et al., 1996). This intermediate extensor inhibition response coincides with the P13 potential described in Chapter 6, which is an ascending manifestation of PPN output. The complexity of these electromyographic responses unfortunately is missing when the startle response is measured using whole-body or segmental/reflex movement (Figure 7.11).

The initial inhibition induced by PPN outputs is followed by prolonged activation to maintain muscle contraction and postural stability. This phenomenon is related to the mechanisms at play that are described in Chapter 9. Briefly, initial cholinergic release acts on M_2 muscarinic receptors to initially hyperpolarize the membrane; however, continued release induces G protein activity that permits prolonged excitation. This intracellular mechanism that elicits an initial inhibition followed by a prolonged excitation may participate in the activation of gamma band maintenance within the PPN, in the modulation of the startle response, and in the induction of locomotion through reticulospinal systems.

SPINAL CORD

Direct projections from the PPN to the spinal cord are sparse and they originate from noncholinergic PPN neurons (Skinner et al., 1990b), but it is not clear on which cells they terminate or their role in direct activation

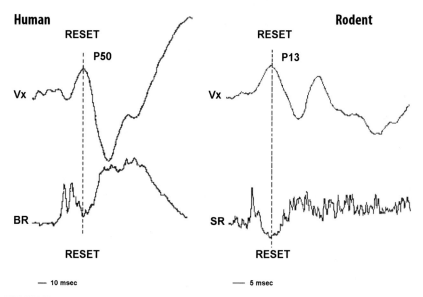

FIGURE 7.11   **Auditory midlatency potentials and neck EMG responses in the human and the rodent.** Left side: The top record is the human vertex-recorded (Vx) midlatency auditory P50 potential and the bottom record is the face muscle EMG recorded blink reflex (BR). Note the early response in the BR followed by an inhibitory or lack of activity that coincides with the peak of the Vx P50 potential, which is then followed by a long latency EMG response (Garcia-Rill et al., 1991, 1996). Right side: The top record is the rodent vertex-recorded (Vx) P13 potential and the bottom record is the neck muscle EMG recorded startle response (SR). Note the early response in the SR followed by an inhibitory or lack of activity that coincides with the peak of the Vx P13 potential, which is then followed by a long latency EMG response. Note also the differences in time base calibrations (Garcia-Rill et al., 1996; Skinner et al., 1990a).

of the spinal cord. At present, we assume that the main effects of PPN outputs are mediated by interneurons in the pontine and medullary reticular formation and reticulospinal projections, as discussed above. The spinal cord has locomotion pattern generators referred to as central pattern generators (CPGs) (Figure 7.12). These contain the flexion–extension sequences necessary for locomotion. Sherrington (1910) believed that locomotion was an entirely reflexive activity and showed that he could transect the spinal cord and induce stepping by stimulating the cut end of the cord or by the reflexive input generated by leg swing when the animal was suspended over a treadmill. However, Graham Brown (1911) showed that he could transect the spinal cord and also induce locomotor movements by stimulating the cord even after cutting the dorsal roots, thus concluding that there are CPGs that contain the preprogrammed sequence of movements required for locomotion. These CPGs are presumably under both central (descending pathways) regulation and peripheral (afferent input) regulation. Central inputs serve to trigger the locomotor episode,

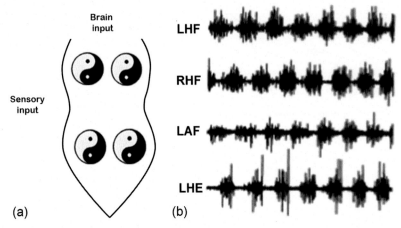

FIGURE 7.12   **Spinal pattern generator.** (a) The locomotor pattern is available in the cervical and lumbar enlargements of the spinal cord. There are half-centers (yin yang diagrams) in the left and right ventral/intermediate zones of the spinal that control the flexor–extensor sequences for each limb needed for stepping. (b) EMG recordings in limb muscles following stimulation of the MLR or spinal cord reveal an adultlike stepping pattern composed of alternation between agonists in different limbs such as left and right hind limb flexors (LHL and RHF), a proximodistal delay between proximal leg and ankle flexors (LHF and LAF), and alternation between agonist and antagonist (LHF and LHE) in the same limb such as proximal leg flexor and extensor (Atsuta et al., 1988, 1991a, 1991b; Iwahara et al., 1991).

which is then maintained by feedback generated by leg swing. Forssberg et al. (1980) demonstrated that it is possible to induce locomotor movements in the chronic spinal cat and expanded our conceptual knowledge of CPGs. Stimulation of the PPN, as described above, provided a method for "recruiting" locomotion and triggering CPGs using brain stem stimulation. Our results showing that the PPN is involved in such modulation and that it is part of basal ganglia circuitry led to the use of PPN DBS for PD described in Chapter 12.

Based on the observation that festinating gait in Parkinson's disease (PD) is similar to the digitigrade step cycle of the newborn, it was hypothesized that the locomotor deficits in PD represented a *regression to a neonatal state* (Forssberg et al., 1984). We reasoned that the spinal cord below the level of the lesion in spinal cord injury (SCI) is deafferented from central, descending control and reverts to an earlier, "neonatal" state (Garcia-Rill and Skinner, 1991). Therefore, we needed to know what parameters of stimulation would be required to drive stepping in the neonatal spinal cord. Thus, we developed the in vitro brain stem–spinal cord preparation in the 0–4-day-old rat (Atsuta et al., 1988). We were the first to show that chemical stimulation of the brain stem or spinal cord bath could be used to drive locomotion (Atsuta et al., 1991a, 1991b). We also determined that stimulation of the spinal cord in the neonate rat required the use of long-duration (0.5–1 ms) pulses delivered at low frequencies (0.5–20 Hz) (Atsuta et al., 1988, 1991a, 1991b; Iwahara et al., 1991). Stimulation of the spinal cord in the neonatal in vitro brain stem–spinal cord preparation induced not only alternating hind limb movements along with electromyographic (EMG) alternation of agonists on the left and right sides but also other characteristics of an advanced step cycle in eliciting alternation of antagonists in the same limb and a proximodistal delay in EMG patterning (Atsuta et al., 1988, 1991a, b; Iwahara et al., 1991). We then undertook to use the same parameters of stimulation but applied epidurally over the lumbar enlargement in the adult, acutely transected cat (Iwahara et al., 1992). We were the first to determine that epidural stimulation of the preenlargement and enlargement segments of the lumbar cord, using the same parameters of stimulation used in the neonatal in vitro brain stem–spinal cord preparation, induced locomotion on a treadmill within a few hours after midthoracic transection of the spinal cord in the adult cat (Iwahara et al., 1992). That is, the same parameters that induced stepping in a neonatal spinal cord were effective in eliciting locomotion in the adult transected spinal cord, suggesting that the spinal cord after SCI was responding like a neonatal spinal cord.

We explored the idea that epidural spinal cord stimulation could be used in SCI patients to induce stepping. Epidural stimulation of the spinal cord in humans had been used for many years for the treatment of chronic pain. Such stimulation is still being used for pain treatment and involves the implantation of epidural stimulating electrodes over the cervical spinal cord in patients with chronic pain. This method employs short-duration (~0.1 ms), high-frequency (>100 Hz) pulses to disrupt pain signals. We proposed the use of similar units for implantation epidurally in patients with an SCI over the cervical or lumbar preenlargement and enlargement segments, but using long-duration, low-frequency pulses, and secured a patent for the method and device (Garcia-Rill et al., 1991). Figure 7.13 shows the patented site of implantation for the electrodes, which are driven by a subcutaneous receiver that is activated by an external transmitter. A similar method was later used by others to induce stepping in a patient implanted with a similar device that used similar parameters of stimulation (Herman et al., 2002). The patient was classified using the American Spinal Injury Association (ASIA) scale as ASIA C, having some motor function below the level of the injury. The results showed that the pattern and duration of locomotion were improved by epidural stimulation. Recently, our use of epidural stimulation (using shorter-duration and higher-frequency stimulation than we patented) was replicated to drive standing by stimulating more caudally in the lumbar enlargement, and also some locomotor movements were induced, in a patient classified as ASIA B (Harkema et al., 2011). Different parameters of stimulation are required to induce changes in posture (standing) than

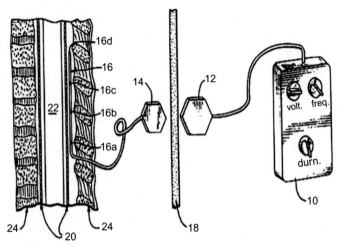

FIGURE 7.13   **Patented method for inducing locomotion following epidural stimulation of the spinal cord.** Patented site of epidural implantation of the electrodes (16a–16d), which are driven by a subcutaneous receiver (14) that is activated by an external transmitter (12). Cross sections of the spinal cord (20) and vertebrae (24) are also shown. A controller (10) induces long-duration (100–200 ms) stimuli at 5–20 Hz (Garcia-Rill et al., USPTO #5,002,053, 1991).

in stepping, in keeping with work showing that standing and treadmill training in spinal animals are possible but animals must be trained in each activity separately (Frigon and Rossignol, 2006). Nevertheless, the technology of epidural stimulation is receiving renewed attention because it shows promise for compensating some of the deficits induced by SCI.

   We reasoned further that the effects of epidural spinal cord stimulation would be improved if we could also compensate for other deficits, especially muscle atrophy, hyperreflexia, and spasticity, all of which impede progress in regaining locomotor function. To that end, subsequent research involved the use of passive exercise to normalize hyperactive reflexes after SCI in the rat (Skinner et al., 1996; Reese et al., 2006; Yates et al., 2008a). We found that daily passive exercise of the hind limbs in a rat that had undergone a midthoracic spinal transection would normalize hyperactive reflexes within 30 days. These studies determined that passive exercise required less training if undertaken early after injury but could still have beneficial effects in alleviating excessive reflexes if undertaken long after injury (Yates et al., 2008a). We went on to develop a motorized bicycle exercise trainer (MBET) for patients with SCI (Garcia-Rill et al., 2004a,b) and determined that daily use for 8–10 weeks normalized the excessive reflexes induced by SCI (Kiser et al., 2005). The machine has wings that keep

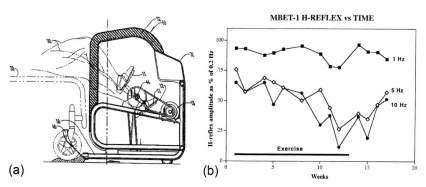

(a)

(b)

FIGURE 7.14 **Patented motorized bicycle exercise trainer (MBET) and effects on H-reflex.** (a) Patented MBET showing attachment of the wheelchair wheel (46), and ankles (11, 12) onto pedals driven by a motor (13). The wings (22) maintain the hip. Knee, and ankle joints aligned, so that the leg (36) is passively cycled. (b) Amplitude of the H-reflex when tested at 1 Hz (block squares), 5 Hz (open diamonds), and 10 Hz (black circles) over a 17 week course of daily MBET. Note the decrease in the amplitude of the H-reflex towards normal amplitudes after 8–10 weeks and a return of excessive reflexes within 2–3 weeks of stopping MBET (Garcia-Rill et al., USPTO #6,755,768, 2004).

the hip, knee, and ankle joints aligned, while a motor passively cycles the feet strapped in stirrups. That is, passive MBET gradually modified intraspinal circuits to alleviate hyperreflexia in patients with long-term SCI (Figure 7.14).

We went on to show that hyperreflexia in the rat is delayed after transection, as it is in the human, suggesting that there is a period of reorganization that takes place after SCI but before reflexes become abnormal and that the excessive reflexes are not merely due to direct effects of the transection in terms of eliminating transmitter inputs (Yates et al., 2008b). We then identified one of the mechanisms involved in hyperreflexia as a dysregulation of electrical coupling (Yates et al., 2009), which could be treated with modafinil for its ability to increase electrical coupling and help normalize excessive reflexes, even without passive exercise (Yates et al., 2009). Detailed descriptions of these avenues, however, are beyond the scope of the present chapter.

In summary, descending projections from the PPN modulate postural muscle tone and locomotion, the atonia of REM sleep and the startle response, and the onset and duration of stepping. Because of the presence of spinal pattern generators (CPGs), the task of activating stepping is greatly simplified. Descending PPN projections need only initiate ("recruit") cycling by the CPG, and the feedback generated by leg swing, and minimal brain stem output, can maintain stepping. This facilitates the role of the RAS in fight-or-flight responses.

CLINICAL IMPLICATIONS

There are a host of clinical conditions that are related to disruption of descending PPN modulation, and they occur during waking and sleep. Armed with the information presented above, the symptoms and manifestations evident in these disorders can be better understood and treated. However, only a few of these will be discussed here since there are a number of texts amply describing such disorders as sleep paralysis, sleep walking, bruxism, and sleep terrors.

Narcolepsy—This disorder is perhaps the best known, but quite rare, wake–sleep disease. Briefly, narcolepsy is marked by excessive daytime sleepiness, cataplexy (loss of postural muscle tone), and rapid onset of REM sleep upon falling asleep, along with hypnagogic hallucinations soon after falling asleep and occasional sleep paralysis. The origin of the disease is related to various factors, including human leukocyte antigen (HLA) disturbance, which suggests a genetic predisposition to autoimmune disorder. In many patients, there is degeneration of hypothalamic orexin neurons (Mignot, 2001). The role of these cells was discussed in Chapter 3. We should stress that the role of orexin neurons may not be frank arousal but that these cells are activated by goal-directed behavior and help maintain survival and survival-related behaviors (Chase, 2013; Torterolo et al., 2011). How are the symptoms of this disorder related to PPN function? The excessive daytime sleepiness is obviously a lack of maintenance of arousal and vigilance by the RAS (see Chapter 9). The lack of driving of RAS cells by descending orexin projections can help explain this symptom. That is, decreased excitation from orexin input will downregulate PPN output. The symptom of cataplexy occurs upon affective incitement such as laughter, anger, startle, or similar sudden activation of the PPN and RAS. Instead of producing the normal startle response (mild extensor inhibition) and assumption of the "ready" (flexed) position to fight or flee, the extensor or postural inhibition is profound and lasting, leading to postural collapse. This set of events suggests that descending PPN projections to the SubCD and pontomedullary reticular formation elicit excessive postural inhibition when activated by a sudden arousing stimulus in narcolepsy. The prolonged activation normally present, for example, in the later phases of the startle response, is missing. The symptom of fast onset REM sleep upon falling asleep suggests that there is increased REM sleep drive, and the hypnagogic hallucinations suggest that REM sleep-related dreaming episodes are stronger and more vivid. The sleep paralysis serves to emphasize the presence of increased REM sleep drive in which the atonia of REM sleep is particularly deep and prolonged. That is, the patients awaken but are still suffering from the prolonged atonia of REM sleep. This causes some anxiety since motoneurons are inhibited so that movement is prevented and respiration is slowed but the patient is fully awake. The spectrum of symptoms suggest that, in general, there is increased REM sleep drive (accounting for early onset REM sleep, hallucinations, etc.) but decreased waking drive (accounting for daytime sleepiness) in narcolepsy. That is, there is a differentiation of waking vs. REM sleep control, which is a concept addressed in Chapter 9. Interestingly, the onset of this disease is postpubertal in most patients, pointing to this critical period as discussed in Chapter 5. Fortunately, the stimulant modafinil is now one treatment of choice. Its mechanism of action was described in Chapter 4, basically increasing electrical coupling to improve coherence. The increased coherence at high frequencies promotes waking during the day and helps normalize sleep architecture at night through increased coherence of low frequencies. Narcolepsy is also treated with gamma hydroxybutyrate (GHB), especially for those patients manifesting cataplexy. GHB is a GABA agonist with depressant properties that paradoxically inhibits cataplexy in narcolepsy. Great care is needed when using this agent chronically due to its abuse potential, potentiation of the depressant actions of alcohol, and other side effects.

REM sleep behavior disorder (RBD)—This disorder is characterized by abnormal behavior during REM sleep, in which the patient acts out their dreams. Basically, the atonia of REM sleep is absent. The usual inhibition of motoneurons during REM sleep is missing. The disorder is rare, occurs mainly in males and has a late onset (in the 60s), and is marked by an HLA disturbance similar to that in narcolepsy (Schenck et al., 1996a). The degeneration in this disease, instead of attacking the orexin system, appears to attack the noradrenergic cells of the locus coeruleus (Schenck et al., 1996b), although degeneration in the olfactory bulb and dorsal nucleus of the vagus is common to those seen in PD (see Chapter 12). Almost one-half of patients go on to develop PD 7–10 years after developing RBD, and many PD patients develop RBD years after the onset of PD (Boeve et al., 2003). This suggests that the degeneration of catecholaminergic cells in PD begins in the substantia nigra and progresses to the locus coeruleus, while in RBD, the degeneration begins in the locus coeruleus and progresses to the substantia nigra. It is not clear if all PD and RBD patients exhibit olfactory bulb and dorsal vagal nucleus degeneration or if arteriosclerosis and other etiologies may be present in these clinical entities. Treatment is quite successful, with most patients responding well to clonazepam. This suggests that the disorder, assuming that locus coeruleus degeneration is a major cause, is one in which REM sleep atonia is disinhibited. That is, the lack of inhibition to PPN or SubCD neurons modulating REM sleep atonia is presumably restored by the clonazepam, a classic benzodiazepine. As described in Chapter 4, the locus coeruleus normally inhibits PPN cells, and decreased inhibitory input to output neurons mediating

REM sleep atonia could account for these manifestations. However, other possibilities exist such as decreased locus coeruleus input to SubCD and/or pontomedullary neurons also controlling REM sleep atonia.

Parkinson's disease (PD)—As will be seen in Chapter 12, PPN deep brain stimulation (DBS) is useful in ameliorating the postural and locomotor deficits in PD and may in the future be useful in modulating other movement and postural muscle tone disorders. Briefly, the technique involves proper localization with minimal damage and bleeding, testing of leads, and assessment of changes in motor manifestations. The DBS is applied for 20/24 hours and appears to have beneficial effects. The details of the benefits of this therapy are described in Chapter 12.

There are a number of other clinical conditions that involve descending PPN pathways, including a number of psychiatric (Chapter 11) and neurological (Chapter 12) disorders that are characterized by hypervigilance (increased ascending tone) and increased REM sleep drive and exaggerated startle response and other reflexes such as blink reflex (increased descending tone). These will be considered below.

References

Atsuta, Y., Garcia-Rill, E., Skinner, R.D., 1988. Electrically induced locomotion in the in vitro brainstem–spinal cord preparation. Brain Res. 470, 309–312.

Atsuta, Y., Abraham, P., Iwahara, T., Garcia-Rill, E., Skinner, R.D., 1991a. Control of locomotion in vitro: II. Chemical stimulation. Somatosen. Motor Res. 8, 55–63.

Atsuta, Y., Garcia-Rill, E., Skinner, R.D., 1991b. Control of locomotion in vitro: I. Deafferentation. Somatosen. Motor Res. 8, 45–53.

Baghdoyan, H.A., Rodrigo-Angulo, M.L., McCarley, R.W., Hobson, J.A., 1984. Site-specific enhancement and suppression of desynchronized sleep signs following cholinergic stimulation of three brainstem regions. Brain Res. 306, 39–52.

Bandyopadhya, R.S., Datta, S., Saha, S., 2006. Activation of pedunculopontine tegmental protein kinase A: a mechanism for rapid eye movement sleep generation in the freely moving rat. J. Neurosci. 26, 8931–8942.

Boeve, B.F., Silber, M.H., Parisi, J.E., Dickson, D.W., Ferman, T.J., Benarroch, E.E., Schmeichel, A.M., Smith, G.E., Petersen, R.C., Ahlskog, J.E., Matsumoto, J.Y., Knopman, D.S., Schenck, C.H., Mahowald, M.W., 2003. Synucleinopathy pathology and REM sleep behavior disorder plus dementia or parkinsonism. Neurology 61, 40–45.

Boissard, R., Gervasoni, D., Schmidt, M.H., Barbagli, B., Fort, P., Luppi, P.H., 2002. The rat ponto-medullary network responsible for paradoxical sleep onset and maintenance: a combined microinjection and functional neuroanatomical study. Eur. J. Neurosci. 16, 1959–1973.

Brown, R.E., Winston, S., Basheer, R., Thakkar, M.M., McCarley, R.W., 2006. Electrophysiological characterization of neurons in the dorsolateral pontine rapid-eye-movement sleep induction zone of the rat: intrinsic membrane properties and responses to carbachol and orexins. Neurosci. 143, 739–755.

Chase, M.H., 2013. A unified survival theory of the functioning of the hypocretinergic system. J. Appl. Physiol. 115, 954–971.

Chase, M.H., Morales, F.R., 1994. The control of motoneurons during sleep. In: Kryger, M.H., Roth, T., Dement, W.C. (Eds.), Principles and Practice of Sleep Medicine. WB Saunders, London, pp. 163–176.

Datta, S., 2006. Activation of phasic pontine-wave generator: A mechanism for sleep-dependent memory processing. Sleep Biol. Rhyth. 4, 16–26.

Datta, S., MacLean, R.R., 2007. Neurobiological mechanisms for the regulation of mammalian sleep–wake behavior: reinterpretation of historical evidence and inclusion of contemporary cellular and molecular evidence. Neurosci. Behav. Rev. 31, 775–824.

Datta, S., Siwek, D.F., Patterson, E.H., Cipolloni, P.B., 1998. Localization of pontine PGO wave generation sites and their anatomical projections in the rat. Synapse 30, 409–423.

Datta, S., Patterson, E.H., Siwek, D.F., 1999. Brainstem afferents of the cholinoceptive pontine wave generation sites in the rat. Sleep Res. Online 2, 79–82.

Datta, S., Mavanji, V., Patterson, E.H., Ulloor, J., 2003. Regulation of rapid eye movement sleep in the freely moving rat: local microinjection of serotonin, norepinephrine, and adenosine into the brainstem. Sleep 26, 513–520.

Davis, M., 1984. The mammalian startle response. In: Eaton, R.C. (Ed.), Neural Mechanisms of Startle Behavior. Plenum Press, New York, pp. 287–351.

Forssberg, H., Grillner, S., Halbertsma, J., 1980. The locomotion of the low spinal cat. I. Coordination within a hindlimb. Acta Physiol. Scand. 108 (3), 269–281.

Forssberg, H., Johnels, B., Steg, G., 1984. Is parkinsonian gait caused by a regression to an immature walking pattern? Adv. Neurol. 40, 375–379.

Frigon, A., Rossignol, S., 2006. Functional plasticity following spinal cord lesions. Prog. Brain Res. 157, 231–260.

Garcia-Rill, E., 1986. The basal ganglia and the locomotor regions. Brain Res. Rev. 11, 47–63.

Garcia-Rill, E., 1991. The pedunculopontine nucleus. Prog. Neurobiol. 36, 363–389.

Garcia-Rill, E., 1997. Disorders of the reticular activating system. Med. Hypoth. 49, 379–387.

Garcia-Rill, E., Skinner, R.D., 1988. Modulation of rhythmic function in the posterior midbrain. Neuroscience 17, 639–654.

Garcia-Rill, E., Skinner, R.D., 1991. Modulation of rhythmic functions by the brainstem. In: Shimamura, M., Grillner, S., Edgerton, V.R. (Eds.), Neurobiology of Human Locomotion. Japan Scientific Society Press, Tokyo, pp. 137–158.

Garcia-Rill, E., Hull, C.D., Levine, M.S., Buchwald, N.A., 1979. The spontaneous firing patterns of forebrain neurons. IV. Effects of bilateral and unilateral frontal cortical ablations on firing of caudate, globus pallidus and thalamic neurons. Brain Res. 165, 23–36.

Garcia-Rill, E., Skinner, R.D., Gilmore, S.A., 1981. Pallidal projections to the mesencephalic locomotor region (MLR) in the cat. Amer. J. Anat. 161, 311–322.

Garcia-Rill, E., Skinner, R.D., Fitzgerald, J.A., 1983. Activity in the mesencephalic locomotor region (MLR) during locomotion. Exp. Neurol. 82, 609–622.

Garcia-Rill, E., Skinner, R.D., Fitzgerald, J.A., 1985. Chemical activation of the mesencephalic locomotor region. Brain Res. 330, 43–54.

Garcia-Rill, E., Skinner, R.D., Conrad, C., Mosley, D., Campbell, C., 1986. Projections of the mesencephalic locomotor region in the rat. Brain Res. Bull. 17, 33–40.

Garcia-Rill, E., Houser, C.R., Skinner, R.D., Smith, W., Woodward, D.J., 1987. Locomotion-inducing sites in the vicinity of the pedunculopontine nucleus. Brain Res. Bull. 18, 731–738.

Garcia-Rill, E., Skinner, R.D., Atsuta, Y., 1991. Method and device for inducing locomotion by electrical stimulation of the spinal cord, US Patent Trade Off, 5,002,053.

Garcia-Rill, E., Reese, N.B., Skinner, R.D., 1996. Arousal and locomotion: from schizophrenia to narcolepsy. In: Holstege, G., Saper, C.D. (Eds.), The Emotional Motor System. In: Progress in Brain Research, 107. pp. 417–434.

Garcia-Rill, E., Skinner, R.D., Miyazato, H., Homma, Y., 2001. Pedunculopontine stimulation induces prolonged activation of pontine reticular neurons. Neuroscience 104, 455–465.

Garcia-Rill, E., Homma, Y., Skinner, R.D., 2004a. Arousal mechanisms related to posture and movement. I. Descending modulation. In: Mori, S., Stuart, D.G., Wiesendanger, M. (Eds.), Brain Mechanisms for the Integration of Posture and Movement. Progress in Brain Research, 143, pp. 283–290.

Garcia-Rill, E., Skinner, R.D., Reese, N.B., 2004b. Motorized Bicycle Exercise Trainer, US Patent Trade Office, 6,755,768.

Garcia-Rill, E., Heister, D.S., Ye, M., Charlesworth, A., Hayar, A., 2007. Electrical coupling: novel mechanism for sleep-wake control. Sleep 30, 1405–1414.

Gerber, U., Stevens, D.R., McCarley, R.W., Greene, R.W., 1991. Muscarinic agonists activate an inwardly rectifying potassium conductance in medial pontine reticular formation neurons of the rat in vitro. J. Neurosci. 11, 3861–3867.

Graham Brown, T., 1911. The intrinsic factors in the act of progression in the mammal. Proc. Royal Soc. London 84, 309–319.

Greene, R.W., Gerber, U., McCarley, R.W., 1989. Cholinergic activation of medial pontine reticular formation neurons in vitro. Brain Res. 476, 154–159.

Harkema, S., Gerasimenko, Y., Hodes, J., Burdick, J., Angeli, C., Chen, Y., Ferreira, C., Willhite, A., Rejc, E., Grossman, R.G., Edgerton, V.R., 2011. Effect of epidural stimulation of the lumbosacral spinal cord on voluntary movement, standing, and assisted stepping after motor complete paraplegia: a case study. Lancet 377, 1938–1947.

Heister, D.S., Hayar, A., Charlesworth, A., Yates, C., Zhou, Y.H., Garcia-Rill, E., 2007. Evidence for Electrical Coupling in the SubCoeruleus (SubC) Nucleus. J. Neurophysiol. 97, 3142–3147.

Heister, D.S., Hayar, A., Garcia-Rill, E., 2009. Cholinergic modulation of GABAergic and glutamatergic transmission in the dorsal Subcoeruleus: mechanisms for REM sleep control. Sleep 32, 1135–1147.

Herman, R., He, J., D'Luzansky, S., Willis, W., Dilli, S., 2002. Spinal cord stimulation facilitates functional walking in a chronic, incomplete spinal cord injured. Spinal Cord 40, 65–68.

Homma, Y., Skinner, R.D., Garcia-Rill, E., 2002. Effects of pedunculopontine nucleus (PPN) stimulation on caudal pontine reticular formation (PnC) neurons in vitro. J. Neurophysiol. 87, 3033–3047.

Hyde, J., Kezunovic, N., Urbano, F.J., Garcia-Rill, E., 2013. Spatiotemporal properties of high speed calcium oscillations in the pedunculopontine nucleus. J. Appl. Physiol. 115, 1402–1414.

Imon, H., Ito, K., Dauphin, L., McCarley, R.W., 1996. Electrical stimulation of the cholinergic laterodorsal tegmental nucleus elicits scopolamine-sensitive excitatory postsynaptic potentials in medial pontine reticular formation neurons. Neuroscience 74, 393–401.

Iwahara, T., Atsuta, Y., Garcia-Rill, E., Skinner, R.D., 1991. Locomotion induced by spinal cord stimulation in the neonate rat in vitro. Somatosen. Motor Res. 8, 281–287.

Iwahara, T., Atsuta, Y., Garcia-Rill, E., Skinner, R.D., 1992. Spinal cord stimulation-induced locomotion in the adult cat. Brain Res. Bull. 28, 99–105.

Jones, B.E., 1990. Immunohistochemical study of choline acetyltransferase-immunoreactive processes and cells innervating the pontomedullary reticular formation in the rat. J. Comput. Neurol. 295, 485–514.

Jouvet, M., Delorme, F., 1965. Locus coeruleus et sommeil paradoxal. Comput. Rend. Soc. Biol. 159, 895–899.

Karlsson, K.A., Gall, A.J., Mohns, E.J., Seelke, A.M., Blumberg, M.S., 2005. The neural substrates of infant sleep in rats. Public Lib. Sci. Biol. 3, e143.

Kezunovic, N., Urbano, F.J., Simon, C., Hyde, J., Smith, K., Garcia-Rill, E., 2011. Mechanism behind gamma band activity in the pedunculopontine nucleus (PPN). Eur. J. Neurosci. 34, 404–415.

Kinjo, N., Atsuta, Y., Webber, M., Kyle, R., Skinner, R.D., Garcia-Rill, E., 1990. Medioventral medulla-induced locomotion. Brain Res. Bull. 24, 509–516.

Kiser, T.S., Reese, N.B., Maresh, T., Hearn, S., Yates, C., Skinner, R.D., Garcia-Rill, E., 2005. Use of a motorized bicycle exercise trainer to normalize frequency-dependent habituation of the H-reflex in spinal cord injury. J. Spinal Cord Med. 28, 241–245.

Koch, M., 1999. The neurobiology of startle. Prog. Neurobiol. 59, 107–128.

Koch, M., Lingelhohl, K., Pilz, P.K.D., 1992. Loss of the acoustic startle response following neurotoxic lesions of the caudal pontine reticular formation: possible role of giant neurons. Neuroscience 49, 617–625.

Koch, M., Kungel, M., Herbert, H., 1993. Cholinergic neurons in the pedunculopontine tegmental nucleus are involved in the mediation of prepulse inhibition of the acoustic startle response in the rat. Exp. Brain Res. 97, 71–82.

Lai, M., Siegel, J., 1990. Muscle tone suppression and stepping produced by stimulation of midbrain and rostral pontine reticular formation. J. Neurosci. 10, 2727–2730.

Lydic, R., Baghdoyan, H., 1993. Pedunculopontine stimulation alters respiration and increases Ach release in the pontine reticular formation. Amer. J. Physiol. 264, R544–R554.

Mamiya, N., Bay, K., Skinner, R.D., Garcia-Rill, E., 2005. Induction of long-lasting depolarization in medioventral medulla (MED) neurons by cholinergic input from the pedunculopontine nucleus (PPN). J. Appl. Physiol. 99, 1127–1137.

Marks, G.A., Farber, J., Roffwarg, H.P., 1980. Metencephalic localization of ponto–geniculo–occipital waves in the albino rat. Exp. Neurol. 69, 667–677.

Mavanji, V., Ulloor, J., Saha, S., Datta, S., 2004. Neurotoxic lesions of phasic pontine-wave generator cells impair retention of 2-way active avoidance memory. Sleep 27, 1282–1292.

Mesulam, M.M., Mufson, E.J., Wainer, B.H., Levey, A.I., 1983. Central cholinergic pathways in the rat: an overview based on an alternative nomenclature (Ch1–Ch6). Neuroscience 10, 1185–1201.

Mignot, E., 2001. A commentary on the neurobiology of the hypocretin/orexin system. Neuropharmacology 25, S5–S13.

Mitani, A., Ito, K., Hallanger, A.E., Wainer, B.H., Kataoka, K., McCarley, R.W., 1988. Cholinergic projections from the laterodorsal and pedunculopontine tegmental field to the pontine gigantocellular tegmental field in the cat. Brain Res. 451, 397–402.

Mitler, M.M., Dement, W.C., 1974. Cataplectic-like behavior in cats after micro-injections of carbachol in pontine reticular formation. Brain Res. 68, 335–343.

Morrison, A.R., 1988. Paradoxical sleep without atonia. Arch. Ital. Biol. 126, 275–289.

Moruzzi, G., Magoun, H.W., 1949. Brain stem reticular formation and activation of the EEG. Electroenceph. Clin. Neurophysiol. 1, 455–473.

Mouret, J., Delorme, F., Jouvet, M., 1967. Lesions of the pontine tegmentum and sleep in rats. Comput. Rend. Séances Soc. Biol. Filial. 161, 1603–1606.

Nauta, W.J.H., Mehler, W.R., 1966. Projections of the lentiform nucleus in the monkey. Brain Res. 1, 3–42.

Pedroarena, C., Llinas, R.R., 2001. Interactions of synaptic and intrinsic electroresponsiveness determine corticothalamic activation dynamics. Thalamus Related Syst. 1, 3–14.

Reese, N.B., Garcia-Rill, E., Skinner, R.D., 1995. The pedunculopontine nucleus–auditory input, arousal and pathophysiology. Prog. Neurobiol. 47, 105–133.

Reese, N.B., Skinner, R.D., Mitchell, D., Yates, C., Barnes, C., Kiser, T., Garcia-Rill, E., 2006. Restoration of frequency-dependent depression of the H-reflex by passive exercise in spinal rats. Spinal Cord 44, 28–34.

Sanford, L.D., Morrison, A.R., Graziella, L.M., Harris, J.S., Yoo, L., Ross, R.J., 1994. Sleep patterning and behavior in cats with pontine lesions creating REM without atonia. J. Sleep Res. 3, 233–240.

Schenck, C., Garcia-Rill, E., Segall, M., Noreen, H., Mahowald, M.W., 1996a. HLA class II genes associated with REM sleep behavior disorder. Ann. Neurol. 39, 261–263.

Schenck, C., Garcia-Rill, E., Skinner, R.D., Anderson, M., Mahowald, M.W., 1996b. A case of REM sleep behavior disorder with autopsy-confirmed Alzheimer's Disease: post mortem brainstem histochemical analyses. Biol. Psychiat. 40, 422–425.

Sherrington, C.S., 1910. Flexion-reflex of the limb, crossed extension-reflex, and reflex stepping and standing. J. Physiol. 40, 28–121.

Shik, M.L., Severin, F.V., Orlovskii, G.N., 1966. Control of walking and running by means of electric stimulation of the midbrain. Biofizika 11, 659–666.

Shiromani, P., Armstrong, D.M., Gillin, J.C., 1988. Cholinergic neurons from the dorsolateral pons project to the medial pons: a WGA-HRP and choline acetyltransferase immunohistochemical study. Neurosci. Lett. 95, 12–23.

Shute, C.C.D., Lewis, P.R., 1967. The ascending cholinergic reticular system: neocortical, olfactory and subcortical projections. Brain 90, 497–520.

Simon, C., Kezunovic, N., Ye, M., Hyde, J., Hayar, A., Williams, D.K., Garcia-Rill, E., 2010. Gamma band unit and population responses in the pedunculopontine nucleus. J. Neurophysiol. 104, 463–474.

Skinner, R.D., Garcia-Rill, E., 1990. Brainstem modulation of rhythmic functions and behaviors. In: Klemm, W.R., Vertes, R.P. (Eds.), Brainstem Mechanisms of Behavior. John Wiley & Sons, New York, pp. 419–445.

Skinner, R.D., Garcia-Rill, E., 1994. Mesolimbic interactions with mesopontine modulation of locomotion. In: Kalivas, P., Barnes, C. (Eds.), Limbic Motor Circuits and Neuropsychiatry. CRC Press, New York, NY, pp. 155–191.

Skinner, R.D., Kinjo, N., Ishikawa, Y., Biedermann, J.A., Garcia-Rill, E., 1990a. Locomotor projections from the pedunculopontine nucleus to the medioventral medulla. NeuroReport 1, 207–210.

Skinner, R.D., Kinjo, N., Henderson, V., Garcia-Rill, E., 1990b. Locomotor projections from the pedunculopontine nucleus to the spinal cord. NeuroReport 1, 183–186.

Skinner, R.D., Houle, J.D., Reese, N.B., Berry, C.L., Garcia-Rill, E., 1996. Effects of exercise and fetal spinal cord implants on the H-reflex in chronically spinalized adult rats. Brain Res. 729, 127–131.

Stevens, D.R., Birnstiel, S., Gerber, U., McCarley, R.W., Greene, R.W., 1993. Nicotinic depolarizations of rat medial pontine reticular neurons studied in vitro. Neuroscience 57, 419–424.

Takakusaki, K., Habaguchi, T., Ohtinata-Sugimoto, J., Saitoh, K., Sakamoto, T., 2003. Basal ganglia efferents to the brainstem centers controlling postural muscle tone and locomotion: a new concept for understanding motor disorders in basal ganglia dysfunction. Neuroscience 119, 293–308.

Torterolo, P., Ramos, O.V., Sampogna, S., Chase, M.H., 2011. Hypocretinergic neurons are activated in conjunction with goal-oriented survival-related motor behaviors. Physiol. Behav. 104, 823–830.

Vanni-Mercier, G., Debilly, G., 1989. A key role for the caudoventral pontine tegmentum in the simultaneous generation of eye saccades in bursts and associated ponto–geniculo-occipital waves during paradoxical sleep in the cat. Neuroscience 86, 571–585.

Wang, H.L., Morales, M., 2009. Pedunculopontine and laterodorsal tegmental nuclei contain distinct populations of cholinergic, glutamatergic and GABAergic neurons in the rat. Eur. J. Neurosci. 29, 340–358.

Yamamoto, K., Mamelak, A.N., Quattrochi, J.J., Hobson, J.A., 1990. A cholinoceptive desynchronized sleep induction zone in the anterolateral pontine tegmentum: locus of the sensitive region. Neuroscience 39, 279–293.

Yates, C., Charlesworth, A., Reese, N.B., Skinner, R.D., Garcia-Rill, E., 2008a. The effects of passive exercise therapy initiated prior to or after the development of hyperreflexia following spinal transection. Exp. Neurol. 213, 405–409.

Yates, C., Charlesworth, A., Allen, S., Reese, N.B., Skinner, R.D., Garcia-Rill, E., 2008b. The onset of hyperreflexia in the rat following complete spinal cord transection. Spinal Cord 46, 798–803.

Yates, C., Charlesworth, A., Reese, N.B., Ishida, K., Skinner, R.D., Garcia-Rill, E., 2009. Modafinil normalized hyperreflexia after spinal transection in adult rats. Spinal Cord 47, 481–485.

The 10 Hz Fulcrum

Edgar Garcia-Rill, PhD

Center for Translational Neuroscience, Department
of Neurobiology and Developmental Sciences, University of Arkansas for
Medical Sciences, Little Rock, AR, USA

ALPHA FREQUENCY (~10 HZ) ACTIVITY

What is the relationship between activity at beta/gamma band (20–60 Hz) and activity at alpha band (10 Hz) in the pedunculopontine nucleus (PPN)? Studies on humans have shown that alpha frequency oscillations in the PPN correlate with gait performance (Thevathasan et al., 2012). PPN neurons show frequency of firing at alpha and beta bands in relation to voluntary movements in patients performing self-paced wrist and ankle movements (Tsang et al., 2010). While passive movement increased alpha activity, imagined movement tended to decrease it, uncoupling alpha phase locking (Tattersall et al., 2014). In animal studies, PPN neurons fire at ~10 Hz during cortical slow oscillations but still support nested gamma oscillations (Mena-Segovia et al., 2008). Our in vitro studies have shown that, while PPN and Pf cells have the capacity to fire maximally at gamma band frequencies, in the absence of ramp stimuli, the "resting" firing frequency is in the 8–12 Hz range (Kezunovic et al., 2010, 2011, 2012, 2013; Simon et al., 2010). These findings are interpreted to mean that the PPN manifests higher-frequency activity (beta/gamma) when receiving sensory input and initiating, triggering, or recruiting movement. However, once the sequence is started, activity decreases in frequency (to alpha) in order to maintain the activity or return to rest. This is very much like stepping on the accelerator to begin moving an automobile, then coasting at idle speed to maintain progress. The relegation of maintenance of movement to a lower frequency allows the system to respond to additional sensory inputs with higher frequencies in order to alter the course. That is, PPN activity in relation to movement is of higher frequency when planning and recruiting movement, becoming lower in frequency once inertia is overcome. Why is alpha an "automatic" frequency? Why is alpha frequency the "idle" or "resting" speed? Why are there so many 10 Hz rhythms?

Waking and the Reticular Activating System in Health and Disease
http://dx.doi.org/10.1016/B978-0-12-801385-4.00008-2

ALPHA RHYTHM

The "alpha rhythm" is the best-known 10 Hz rhythm with a history stemming from the earliest days of the electroencephalogram (EEG). Alpha waves were originally described by Hans Berger, which resulted when the subjects being EEG recorded closed their eyes. When the subjects opened their eyes, the EEG frequency increased to beta and gamma, and alpha waves were "blocked." Alpha waves are also reduced or "suppressed" when the subjects become drowsy or fall asleep, and the EEG shifts to theta and lower-frequency waves. The frequency range for alpha was originally set at 7.5–12.5 Hz, and theta waves were said to be 4–7 Hz. It is not clear how correlated, if at all, cortical EEG theta waves are to hippocampal theta waves, which are in the 6-10 Hz range. However, in human magnetoencephalography (MEG) recordings, the peak of the cortical theta wave is closer to 8 Hz, while the peak of the alpha wave is closer to 10 Hz with eyes closed (Llinas et al., 2001).

In sleep studies, subjects who are awake manifest beta/gamma activity, then shift to alpha upon closing the eyes, and later drift into sleep that exhibits mainly delta and slow waves in the 0.5–4 Hz range. Slower frequencies below 10 Hz signal drowsiness and the transition towards frank sleep, while faster frequencies above 10 Hz signal more activation and awakening. Alpha waves themselves are most evident when the eyes are closed. Why? Alpha waves are thought to signal relaxation or calmness, and they are the easiest waves to entrain when using biofeedback. Why?

Our hypothesis is that occipital alpha waves are manifested when we eliminate one of our most important sensory systems, the visual system. When we eliminate the constant flow of visual input flowing into our brains by closing the eyes, alpha waves are expressed because it is the basic idling speed of the awake brain. If we become drowsy, we shift into theta and delta waves towards sleep, but if we open our eyes, the EEG immediately shifts to beta and then gamma when we focus attention. That is, the 10 Hz rhythm that in the EEG is referred to as occipital "alpha" waves is actually the speed of the awake but unchallenged brain at its "resting state." Any slower and we shift into drowsiness; any faster and we begin to pay attention and cogitate.

As we will discuss in Chapter 9, the EEG amplifier has inappropriate band-pass filters for detecting events as fast as action potentials, which occur in the 1–2 ms range. EEG amplifiers more faithfully reflect the activity of slow potentials such as dendritic postsynaptic potentials and intrinsic membrane oscillations, which occur in the 10–20 ms range. That is, the EEG is a measure of dendritic potentials and membrane oscillations, not of action potential frequencies (Murakami and Okada, 2006). That means that EEG recordings exhibit large-amplitude, low-frequency waveforms when large ensembles of cells are "beating together," rather than separately, such as when low-amplitude, high-frequency activity is evident. The most simple way to think about high-amplitude, low-frequency waves is that large groups of cortical columns are firing in unison, which means they have little, if any, lateral inhibition separating the activity of individual columns. On the other hand, low-amplitude, high-frequency EEG recordings (above 10 Hz) reflect the firing of many different columns acting independently. This suggests that the lack of sensory perception seen during slower rhythms (below 10 Hz) is a reflection of a lack of individuality of columnar organization and thus a lack of sensory perception. As independent frequencies increase, more localized activity generates lower-amplitude waves, and the higher EEG frequencies suggest disparate firing patterns across these smaller groups of cells.

Therefore, the 10 Hz rhythm is a frequency fulcrum between the large-ensemble (high-amplitude) firing of slower EEG frequencies and the small-ensemble (low-amplitude) firing of faster EEG frequencies. As more and more aggregation occurs, the rhythms become slower and of higher amplitude, eliminating differential activity across cortical columns, thus leading to the absence of sensory perception. As activity becomes more and more localized, the rhythms become faster and of lower amplitude, signaling more independent activity in cortical columns and detection of sensory events. This is probably why sensation and perception are eliminated with slow EEG states. In the extreme case of epilepsy, very large ensembles are synchronized, and thus perception, memory, and learning are not possible. That is why seizures are marked by an absence of memory of the event. On the other hand, as the eyes are opened and sensory flow is reestablished, high-frequency activity in small ensembles now begins the work of perception and more complex activities like attention, selective attention, learning, and memory formation. We should note that a variety of afferent stimuli, not only visual stimuli but also auditory, olfactory, etc., "block" or "suppress" alpha waves and induce beta/gamma activity (see, e.g., Green and Arduini, 1954). That is, activation of any sensory afferent input results in activating the brain beyond the 10 Hz "idle" speed. The idea that the alpha rhythm or other resting rhythms are "blocked" by sensory or motor events is a recurrent concept; however, consideration must given to the idea that this rhythm is not being "blocked," but merely replaced by a higher-frequency rhythm, for example, beta or gamma, necessary for sensory discrimination and/or for movement.

There is less evidence for a 10 Hz rhythm localized to the superior temporal lobe that has been termed the *tau* rhythm. MEG recordings in humans described this rhythm in the region of the primary auditory cortex that is "suppressed" by auditory stimulation (Lehtela et al., 1997). To continue the metaphor of driving an automobile, 10 Hz is the idle speed of the engine of the brain, and lower frequencies represent the engine sputtering and almost dying altogether. When we receive sensory inputs or, for that matter, when we perform movements, it is similar to pressing the accelerator to increase engine speed to begin motion towards a target. When we are "at rest," the idle speed is 10 Hz, which is a rhythm to entrain if we wish to relax, such as when using biofeedback. We relax by decreasing high-frequency activity from the beta and gamma ranges to the "resting state" alpha range. In Chapter 9, we will see that waking is a metabolically expensive state. But, if alpha waves reflect the "idle" speed of the brain, there should be other indicators of 10 Hz "resting state."

THE MU RHYTHM

The *mu* rhythm was first described by Gastaut (Gastaut et al., 1954) as a 8–12 Hz wave present over the vertex and bilaterally across the precentral motor cortex, basically at the EEG C3, Cz, and C4 electrode placements. The *mu* rhythm is present when the body is at rest, but the rhythm is "suppressed" or "blocked" when the person performs a motor action, or, after practice, when the person views another or visualizes a motor action. This "suppression" is also referred to as "desynchronization," which is a term that has been questioned because the EEG does not "desynchronize," but merely increases synchronization of slower, or synchronization of faster, frequencies (Steriade et al., 1996). That is, it is more likely that, just as with the alpha rhythm, the *mu* rhythm is merely *replaced* by faster activity when the function of that region is called for. As discussed above, the alpha rhythm localized over the occipital cortex is shifted, not "blocked," from idle speed to higher frequencies when the eyes are opened. Similarly, it is likely that the *mu* rhythm over the vertex and precentral cortex is shifted, not "suppressed," from idle speed to higher frequencies when motor events are called for.

The *mu* rhythm is also suppressed during tactile stimulation (Cheyne et al., 2003), as is the alpha rhythm (Green and Arduini, 1954). Both the occipital alpha rhythm and the precentral *mu* rhythm have been referred to as "idle rhythms" or "resting state" activity (Pfurtscheller, 1992). The *mu* rhythm has been associated with somatosensory information, while a faster beta rhythm in the precentral region has been associated with actual motor processing (Ritter et al., 2009). That is, these "idle" or "resting" rhythms are of slower frequencies and are "blocked" or "suppressed" when sensory or motor events are called for, that is, when the accelerator is pressed. Interestingly, the occipital alpha rhythm and the precentral *mu* rhythm are both inversely related to the blood oxygen-level dependent (BOLD) contrast in functional magnetic resonance studies. This suggests that these rhythms are related to decreases in cerebral blood flow, which may be a general feature of "idle rhythms" (Ritter et al., 2009).

We should note that the *mu* rhythm is present at the vertex (EEG Cz electrode location) (also laterally at C3 or C4 depending on which hand or arm movement is being performed contralaterally). The vertex is where the P50 potential is maximal, which was discussed in Chapter 7 as the manifestation of RAS output after an auditory stimulus. That is, there may be a relationship between PPN projections to the intralaminar thalamus, specifically the Pf, which are expressed at the level of the cortex as a vertex-recorded P50 potential, and the *mu* rhythm. In Chapter 12, we will see that the vertex is also the region whose blood flow is most altered during deep brain stimulation (DBS) of the PPN in Parkinson's disease (PD) patients, that is, where PPN DBS has its most marked effects on cortical blood flow. Moreover, the vertex is precisely where the readiness potential is maximal, which is a wave in advance of a voluntary movement that will be discussed in detail in Chapter 10 in relation to voluntary movement. Voluntary movement itself is superimposed on a 10 Hz signal and it is called physiological tremor.

PHYSIOLOGICAL TREMOR

An excellent discussion of physiological tremor can be found in "I of the Vortex: From Neurons to Self" by Llinas (Llinas, 2001). Briefly, early work described a periodicity of 8–12 Hz to voluntary movement, an upper limit of 10–11 per second for voluntary movements from rest, and that movements were always initiated in phase with physiological tremor. Later, a 8–12 Hz rhythm was described during supported limbs at rest, during maintained posture, and during voluntary movements (Marsden et al., 1984). Recordings of muscle contractions using electromyography (EMG) found that, whether the movement of the same muscle was of slow, medium, or fast velocity, the EMG signal carried an underlying ~10 Hz signal regardless of velocity (Vallbo and Wallberg, 1993). These authors concluded that the signal underlying physiological tremor, on which voluntary movement was superimposed, was generated in the brain above the spinal cord. Llinas concluded that physiological tremor is a reflection of a descending command from the brain in a pulsatile fashion (Llinas, 2001). That is, that movement is discontinuous, not completely smooth.

What creates the rhythmic signal behind physiological tremor? How is the rhythm generated? Neurons in the inferior olivary nucleus manifest membrane oscillations at 8–12 Hz, and the cells fire action potentials at 1–2 Hz but always at the peaks of the oscillations (Llinas, 1981). Inferior olive cells are also electrically coupled with one of the highest coupling ratios (~10%) in the brain. This suggests that inferior olive neurons emanate a coherent 10 Hz signal to all Purkinje neurons, which are the main output of the cerebellar cortex, on which all resting muscle tone and voluntary movement are superimposed. At present, we do not know that, or even if, there is only one region that generates a 10 Hz that is manifested at various sites and subserves all of the ~10 Hz rhythms described in the occipital cortex, precentral cortex, and inferior olive. The inferior olive would seem a prime candidate for that function. Even if these regions generate their own 10 Hz signal, they appear to be coherent. In Chapter 9, we will discuss how gamma band activity is present not only in the cortex but also in the basal ganglia, cerebellum, hippocampus, and thalamus. We will also describe how these are not separate gamma band frequencies but they are coordinated. For example, when performing a precision grip task, gamma band activity in the monkey cortex and cerebellum is coherent (Soteropoulos and Baker, 2006), just as there is coherence between gamma band activity in the basal ganglia and cortex (Lalo et al., 2008; Litvak et al., 2012) and between the cerebellothalamic system and cortex (Timofeev and Steriade, 1997). The lower "resting state" 10 Hz rhythm is also represented in these structures and is probably coherent during rest across multiple regions, although evidence in support of this hypothesis is needed. Regardless of the origin, if there is only one, the 10 Hz frequency appears to be the one on which other activity is superimposed. The 10 Hz fulcrum lies at the transition between sleep and waking and carries little information if this "idle" rhythm decreases in frequency but is replaced by higher frequencies in keeping with the desired higher function.

THE 10 HZ FULCRUM

Figure 8.1 demonstrates the principle behind the "idle" or "resting" state 10 Hz frequency fulcrum. The top row contains examples of EEG recordings during various states. With eyes closed, occipital alpha is evident in the EEG, but if the eyes are opened and visual input is permitted, the resting frequency shifts to the right and increases to beta and even gamma depending on the task. The EEG recording becomes low amplitude and high frequency in keeping with synchronization of high-frequency rhythms. If the eyes are closed and the subject becomes drowsy, the resting frequency shifts to the left and becomes theta and of even lower frequency to delta upon reaching frank sleep. At the extreme end of the synchronization of low frequencies is epilepsy, marking maximal coherence across very large cell ensembles. At the fulcrum are also the precentral or motor cortex *mu* rhythm, the *tau* rhythm, and the inferior olive 10 Hz rhythm. Finally, the 10 Hz rhythm of physiological tremor is present in the power spectrum of EMGs of voluntary movements. If physiological tremor is decreased and loses its drive, a small decrease in frequency will result in a "tremor at rest" in the 6–8 Hz frequency, such as is manifested in PD. If there is further loss of drive, an even slower rhythm underlying the background activity

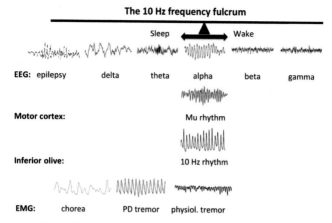

FIGURE 8.1 **The 10 Hz frequency fulcrum.** The top row shows examples of EEG recordings during various states. With eyes closed, occipital alpha is present in the EEG, but when eyes are opened, the resting frequency increases to beta and gamma depending on the task. If the subject becomes drowsy, the resting frequency decreases to theta and then to delta upon reaching frank sleep. The synchronization of low frequencies in epilepsy marks maximal coherence across very large cell ensembles. At the fulcrum are also the precentral or motor cortex *mu* rhythm and the inferior olive 10 Hz rhythm. The 10 Hz rhythm of physiological tremor is present in the power spectrum of EMGs of voluntary movements. If physiological tremor decreases, it results in a "tremor at rest" in the 6–8 Hz frequency, such as is manifested in PD. Even slower rhythm at rest will be manifested as choreoathetoid writhing in the 2–4 Hz range, which is present in HD.

at rest will be manifested as a choreoathetoid writhing in the 2–4 Hz range, which is present in Huntington's disease (HD). That is, we suggest that decreased drive in the 10 Hz "idle" rhythm will result in progressively slower involuntary movements "at rest," such as occur in PD and HD, with the elimination of pathways underlying specific rhythms. When the patient is asked to perform a voluntary movement, the rhythm will be "blocked" or "suppressed" in favor of higher-frequency activity, and the "tremor at rest" should disappear, as occurs in PD. Presumably, the PD patient can still voluntarily overcome the loss with greater exertion. It may become more difficult to overcome a greater loss of drive such as in HD, since these patients find it difficult to mask their involuntary movements as the disease progresses.

The implications of this proposal are far-reaching, suggesting that the 10 Hz frequency represents the "carrier" frequency of the brain at rest upon which other signals are superimposed. A carrier frequency is referred to as a wave that carries information, for example, a radio station has a carrier frequency that carries music and voice, the information conveyed by that carrier frequency. We assume that, when the 10 Hz frequency is "blocked," it is because higher frequencies are superimposed on it, but the 10 Hz rhythm is still at the center of the spectrum of higher frequencies. The 10 Hz carrier frequency is not so much "suppressed" as it is "modulated" by other frequencies that contain data. The 10 Hz fulcrum is needed as a basic central frequency to carry information contained in all sorts of higher frequencies. A similar role for theta oscillations may be present in the hippocampus (Kalauzi et al., 2012).

NATURAL FREQUENCY

Do brain circuits "at rest" have a natural frequency at which they resonate? Using sensory stimulation such as light flashes or auditory tone stimuli, various workers have found a thalamocortical resonant frequency ~10 Hz in humans and animals (Hermann, 2001; Narici and Romani, 1989; Rager and Singer, 1998). While these responses assess the tuning of thalamocortical circuits, they do so indirectly since they test the pathway that includes peripheral receptors, intervening synapses, induced sensory gating, and changes in attention levels. Others have used direct perturbations to detect the main rate of ensuing oscillations or the natural frequency of cortical regions (Rosanova et al., 2009). These authors used transcranial magnetic stimuli to perturb different cortical regions directly in order to measure their natural resonant frequency. We undertook parallel studies on the PPN. We used sagittal slices containing the rodent PPN and recorded population responses. We applied either neurochemical or electrical stimuli to determine the resonant frequencies induced by these stimuli applied directly to the PPN. We measured population responses using microelectrodes to detect multiunit activity, and we also used calcium imaging to detect changes in calcium levels in PPN neurons, especially since, as we will see in Chapter 9, calcium channels are responsible for PPN high-frequency activity.

Event-related spectral perturbation (ERSP)—For population studies, we used an interface chamber and recorded population responses with 1–2 MΩ pipettes filled with aCSF. Plots of the event-related spectral perturbation (ERSP) for each population response were generated with the EEGLAB Matlab Toolbox (Delorme and Makeig, 2004). An ERSP represents a measure of event-related brain dynamics induced in the EEG spectrum by a stimulus or event. It basically measures average dynamic changes in amplitude of the broadband EEG frequency spectrum as a function of time relative to an experimental event (Delorme and Makeig, 2004). These analyses generated power spectra for continuous points in time, for example, during and after the application of an agent or stimulus. These graphs plot frequency of activity over time, and the amplitude of the frequency shown is color-coded such that background (control) appears light green and higher amplitudes appear progressively more yellow and then red.

A convenient way of reading these graphs is to view an ERSP plot as a running power spectrum over time.

The application of transmitters—Previous studies have shown that injections of glutamate into the PPN induce increases in waking and REM sleep (Datta and Siwek, 1997; Datta et al., 2001), states that are marked by high-frequency activity. Glutamate activates NMDA, kainic acid, and AMPA receptors. Subsequent studies reported that injection of NMDA was found to increase waking but not REM sleep (Datta and Siwek, 1997), while injection of kainic acid (KA) was found to increase REM sleep but not waking (Datta, 2009). In addition, as we will see in Chapter 9, injection of the nonspecific cholinergic agonist carbachol (CAR) facilitates and maintains PPN activation (Kezunovic et al., 2013). Therefore, we first used superfusion of each agent during the same period of time (10 min) to determine the frequencies of activation induced by direct activation of NMDA, KA, or cholinergic receptors in the PPN on a steady-state level (Simon et al., 2010).

Figure 8.2 shows ERSPs following the application of NMDA, KA, or CAR. We first performed dose–response curves for each agent, so that the concentrations used in the studies described represent suprathreshold levels of activation. Since exposure to these agents was in the order of 10 min, the levels achieved were considered to be steady state. The application of NMDA induced oscillations; however, the oscillations were not at a specific frequency and the power was increased at almost every frequency (including gamma) (Figure 8.2 top left). The application of KA increased the overall activity but to a lesser extent than NMDA and specifically induced oscillations at frequencies in the theta and gamma ranges (Figure 8.2 top right). The effects of KA application over time suggest that specific peaks increased over time. Note the diagonally occurring increases in frequency of the darkest (highest-amplitude) peaks over time, suggestive of a free-running stepwise increase in frequencies from alpha to beta to gamma in succeeding steps. PPN neuronal population responses during the period of CAR exposure demonstrated that CAR induced oscillations at very specific frequency levels that were initiated after perfusion, beginning not only in the alpha range but also in the gamma range (Figure 8.2 bottom). Our population response studies showed that the application of specific transmitters to the PPN induced multiple frequency oscillations in population responses when pharmacologically stimulated with CAR (which induced activation at specific peaks in the alpha and gamma bands), NMDA (which induced overall increases in activity in alpha, beta, and gamma bands), and KA (which induced overall activation at theta and alpha and somewhat at beta and higher gamma frequencies). That is, regardless of the transmitter applied, there was always a resonant activation in the alpha frequency band, with various stepwise increases at other frequency bands.

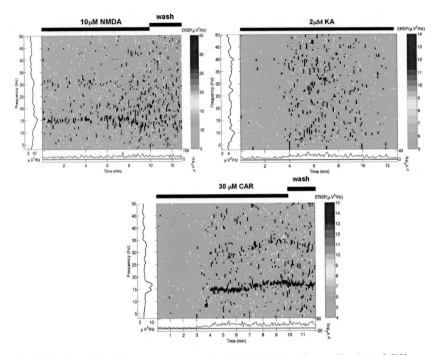

FIGURE 8.2   **Population responses in the PPN following the application of different transmitters.** Top left: The application of NMDA induced oscillations without specific frequencies and the power was increased at almost every frequency (including gamma). Top right: The application of KA increased the overall activity to a lesser extent than NMDA and induced oscillations in the theta and gamma ranges. Note the diagonally occurring increases in frequency of the darkest (highest-amplitude) peaks over time, suggestive of a free-running stepwise increase in frequencies from alpha to beta to gamma in succeeding steps. Bottom: CAR exposure induced oscillations at very specific frequency levels not only in the alpha range (10–20 Hz) but also in the gamma range.

Electrical stimulation and voltage-sensitive dye responses—We calculated ERSPs in the same manner as after transmitter application, except that instead of transmitter application, we delivered trains of 4 stimuli at either 1 Hz, 10 Hz, or 40 Hz, and recorded population responses for 500 ms following the last stimulus of each train. In the absence of activation with specific transmitters, these stimuli did not produce significant effects, therefore, we superfused CAR in order to raise the overall excitability of the slice. Voltage-sensitive dye recordings were performed by incubating slices with ANEPPS before superfusing CAR and applying electrical stimuli.

Figure 8.3 shows the effects of stimulation at various frequencies on the population responses of the PPN. Stimulation at 1 Hz led to synchronization of activity at alpha, beta, and gamma frequencies (left panel), suggesting that low-frequency stimuli induce a resonant frequency across all

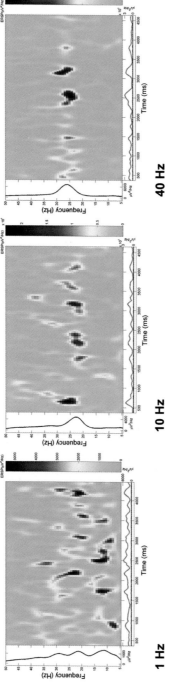

FIGURE 8.3 **Effects of stimulation at various frequencies on the population responses of the PPN.** Stimulation at 1 Hz led to synchronization of activity at alpha, beta, and gamma frequencies (left panel), suggesting that low-frequency stimuli induce a resonant frequency across all bands, but the most coherence occurred at alpha frequency. When the stimuli were delivered at 10 Hz, the population was synchronized to fire specifically at beta frequencies (middle panel), suggesting that external activation at alpha frequency induced coherence at beta frequency. When the stimuli were applied at 40 Hz, the population was entrained to fire specifically at gamma band frequency (right panel). That is, optimal coherence was achieved following stimulation at 40 Hz.

bands, but the most coherence occurred, albeit briefly, at alpha frequency. However, when the train of stimuli was delivered at 10 Hz, the population was synchronized to fire specifically at beta frequencies (middle panel), suggesting that external activation at alpha frequency induced coherence at beta frequency. When the stimuli consisted of a train at 40 Hz, the population was entrained to fire specifically at gamma band frequency (right panel). These results were interpreted to suggest that slightly above threshold activation has the potential to prime activity at alpha, beta, and gamma frequencies. When an external drive (in the form of a train of stimuli) at 10 Hz is applied, the PPN responded by coherence at a higher band, suggesting that its activation amplified the resultant response. When the external input was at gamma band activity, impressive coherence at gamma frequency was observed, suggesting that its optimal resonance when activated at gamma frequency is to synchronize at gamma band frequency.

These results were interpreted to suggest that, at threshold or minimal input, coherence, if any, occurs at alpha frequency. That is, the resonant frequency to threshold stimuli appears to be alpha. But, when external alpha frequency activation is applied, coherence results at beta, and if activation is applied at gamma, maximal coherence results in strong gamma frequency resonance. In Chapter 9, we will explore in more detail the maximal responsiveness of PPN cells at gamma band activity and the cellular mechanisms behind that maximal activation. However, the resonant frequency "at rest" appears to be closer to alpha or ~10 Hz.

CLINICAL IMPLICATIONS

In the chapters that follow, we will deal with a host of neurological and psychiatric disorders that have a common characteristic, the decrease or interruption of gamma band activity. That is, impairment in the generation and/or maintenance of high-frequency activity leads to a number of attentional, cognitive, perceptual, and mnemonic deficits. On the other hand, disorders that lead to a reduction in the "resting state" activity of the system will also produce deficits that impact higher-frequency induction and maintenance, as described above. More importantly, the proposal of a 10 Hz frequency fulcrum provides an understandable model of the manifestation of frequencies, especially in the EEG. The idea that the EEG represents a spectrum with a 10 Hz fulcrum at the transition between the lower frequencies in sleep and the higher frequencies in waking immediately reflects a dynamic scale across wake–sleep states. This allows for more rational assessment of disturbances in wake–sleep cycles as conditions that drive the manifestation of frequencies in one direction or the other. Any therapeutic manipulations that restore the spectrum towards its normal fulcrum are an indication of normalization of wake–sleep control, while manipulations that skew the spectrum towards one extreme or the other can be considered counterproductive to reestablishing normal wake–sleep control.

However, the transitions from the 10 Hz rhythm to higher frequencies do not seem to be a continuous spectrum but rather a stepwise increase from one stable state to the next. Studies measuring changes in activity during the recovery from anesthesia suggest that there is a series of transitions from discrete stable states (Hudson et al., 2014).

This suggests that higher functions step through individual levels during processing.

References

Cheyne, D., Gaetz, W., Garnero, L., Lachaux, J.P., Ducorps, A., Schwartz, D., Varela, F.J., 2003. Neuromagnetic imaging of cortical oscillations accompanying tactile stimulation. Brain Res. Cogn. Brain Res. 17, 599–611.

Datta, S., 2009. Regulation of neuronal activities within REM sleep-sign generators. Sleep 32, 1135–1147.

Datta, S., Siwek, D.F., 1997. Excitation of the brain stem pedunculopontine tegmentum cholinergic cells induces wakefulness and REM sleep. J. Neurophysiol. 77, 2975–2988.

Datta, S., Spoley, E.E., Patterson, E.H., 2001. Microinjection of glutamate into the pedunculopontine tegmentum induces REM sleep and wakefulness in the rat. Amer. J. Physiol. Reg. Integ. Comp. Physiol. 280, R752–R759.

Delorme, A., Makeig, S., 2004. EEGLAB: an open source toolbox for analysis of single-trial EEG dynamics including independent component analysis. J. Neurosci. Meth. 134, 9–21.

Gastaut, H., Dongier, M., Courtois, G., 1954. On the significance of "wicket rhythms" ("rhythmes en arceau") in psychosomatic medicine. Electroenceph. Clin. Neurophysiol. 6, 687–688.

Green, J.D., Arduini, A.A., 1954. Hippocampal electrical activity in arousal. J. Neurophysiol. 17, 533–537.

Hermann, C.S., 2001. Human EEG responses to 1–100 Hz flicker: response phenomena. Exp. Brain Res. 137, 346–353.

Hudson, A.E., Calderon, D.P., Pfaff, D.W., Proekt, A., 2014. Recovery of consciousness is mediated by a network of discrete metastable activity states. Proc. Natl. Acad. Sci. USA 111, 9283–9288.

Kalauzi, A., Spasic, S., Petrovic, J., Ciric, J., Saponjic, J., 2012. Cortico-pontine theta carrier frequency phase shift across sleep/wake states following monoaminergic lesion in rat. Gen. Physiol. Biophys. 31, 163–171.

Kezunovic, K., Simon, C., Hyde, J., Smith, K., Beck, P., Odle, A., Garcia-Rill, E., 2010. Arousal from slices to humans: Translational studies on sleep-wake control. Transl. Neurosci. 1, 2–8.

Kezunovic, N., Urbano, F.J., Simon, C., Hyde, J., Smith, K., Garcia-Rill, E., 2011. Mechanism behind gamma band activity in the pedunculopontine nucleus (PPN). Eur. J. Neurosci. 34, 404–415.

Kezunovic, N., Hyde, J., Simon, C., Urbano, F.J., Garcia-Rill, E., 2012. Gamma band activity in the developing parafascicular nucleus (Pf). J. Neurophysiol. 107, 772–784.

Kezunovic, N., Hyde, J., Goitia, B., Bisagno, V., Urbano, F.J., Garcia-Rill, E., 2013. Muscarinic modulation of high frequency activity in the pedunculopontine nucleus (PPN). Front. Neurol. 4 (176), 1–13.

Lalo, E., Thobois, S., Sharott, A., Polo, G., Mertens, P., Pogosyan, A., 2008. Patterns of bidirectional communication between cortex and basal ganglia during movement in patients with Parkinson disease. J. Neurosci. 28, 3008–3016.

Lehtela, L., Salmelin, R., Hari, R., 1997. Evidence for reactive magnetic 10-Hz rhythm in the human auditory cortex. Neurosci. Lett. 222, 111–114.

Litvak, V., Eusebio, A., Jha, A., Oosterveld, R., Barnes, G., Foltynie, T., 2012. Movement-related changes in local and long-range synchronization in Parkinson's disease revealed by simultaneous magnetoencephalography and intracranial recordings. J. Neurosci. 32, 10541–10553.

Llinas, R.R., 1981. Microphysiology of the cerebellum. In: Brooks, V.B. (Ed.), Handbook of Physiology, Vol. II, The Nervous System, Part IV. Amer Physiol Soc, Bethesda, MD, pp. 831–976.

Llinas, R.R., 2001. I of the Vortex, From Neurons to Self. MIT Press, Cambridge, MA.

Llinas, R.R., Ribary, U., Jeanmonod, D., Cancro, R., Kronberg, E., Schulman, J., Zonenshayn, M., Magnin, M., Morel, A., Siegmund, M., 2001. Thalamocortical dysrhythmia I. Functional and imaging aspects. Thal. Rel. Syst. 1, 237–244.

Marsden, C.D., Rothwell, J.C., Day, B.L., 1984. The use of peripheral feedback in the control of movement. Trends Neurosci. 7, 253–257.

Mena-Segovia, J., Sims, H.M., Magill, P.J., Bolam, J.P., 2008. Cholinergic brainstem neurons modulate cortical gamma activity during slow oscillations. J. Physiol. 586, 2947–2960.

Murakami, S., Okada, Y., 2006. Contributions of principal neocortical neurons to magnetoencephalography signals. J. Physiol. 575 (3), 925–936.

Narici, L., Romani, G.L., 1989. Neuromagnetic investigation of synchronized spontaneous activity. Brain Topogr. 2, 19–30.

Pfurtscheller, G., 1992. Event-related synchronization (ERS): An electrophysiological correlate of cortical areas at rest. Electroencephalogr. Clin. Neurophysiol. 83, 62–69.

Rager, G., Singer, W., 1998. The response of cat visual cortex to flicker stimuli of variable frequency. Eur. J. Neurosci. 10, 1856–1877.

Ritter, P., Moosmann, M., Villringer, A., 2009. Rolandic alpha and beta EEG rhythms' strengths are inversely related to fMRI-BOLD signal in primary somatosensory and motor cortex. Human Brain Map 30, 1168–1187.

Rosanova, M., Casali, A., Bellina, V., Resta, F., Mariotti, M., Massimini, M., 2009. Natural frequencies of human corticothalamic circuits. J. Neurosci. 29, 7679–7685.

Simon, C., Kezunovic, N., Ye, M., Hyde, J., Hayar, A., Williams, D.K., Garcia-Rill, E., 2010. Gamma band unit and population responses in the pedunculopontine nucleus. J. Neurophysiol. 104, 463–474.

Soteropoulos, D.S., Baker, S.N., 2006. Cortico-cerebellar coherence during a precision grip task in the monkey. J. Neurophysiol. 95, 1194–1206.

Steriade, M., Amzica, F., Contreras, D., 1996. Synchronization of fast (30–40 Hz) spontaneous cortical rhythms during brain activation. J. Neurosci. 16, 392–417.

Tattersall, T.L., Stratton, P.G., Coyne, T.J., Cook, R., Silbertsein, P., Silburn, P.A., Windels, F., Sah, P., 2014. Imagined gait modulates neuronal network dynamics in the human pedunculopontine nucleus. Nature Neurosci. 17, 449–454.

Thevathasan, W., Cole, M.H., Graepel, C.L., Hyam, J.A., Jenkinson, N., Brittain, J.S., Coyne, T.J., Silburn, P.A., Aziz, T.Z., Kerr, G., Brown, P., 2012. A spatiotemporal analysis of gait freezing and the impact of pedunculopontine nucleus stimulation. Brain 135, 1446–1454.

Timofeev, I., Steriade, M., 1997. Fast (mainly 30–100 Hz) oscillations in the cat cerebellothalamic pathway and their synchronization with cortical potentials. J. Physiol. 504, 153–168.

Tsang, E.W., Hamani, C., Moro, E., Mazzella, F., Poon, Y.Y., Lozano, A.M., Chen, R., 2010. Involvement of the human pedunculopontine nucleus region in voluntary movements. Neurology 75, 950–959.

Vallbo, A.B., Wallberg, J., 1993. Organization of motor output in slow finger movements in man. J. Physiol. 469, 673–691.

Gamma Band Activity

Nebojsa Kezunovic, PhD[*], *James Hyde, PhD*[†],
Francisco J. Urbano, PhD[‡], *and*
Edgar Garcia-Rill, PhD[§]

[*]Department of Neuroscience, Ichan School of Medicine at Mount Sinai,
New York, NY, USA
[†]Institute for Neuroscience, University of Pittsburgh, Pittsburgh, PA, USA
[‡]Laboratorio Fisiologia y Biologia Molecular, Facultad Ciencias Exactas,
Ciudad Universitaria, Buenos Aires, Argentina
[§]Department of Neurobiology and Developmental Sciences,
University of Arkansas for Medical Sciences, Little Rock, AR, USA

WHAT DOES THE EEG MEASURE?

How is gamma band activity detected? The most common method, of course, is the recording of amplified EEG at sites over the scalp. In the awake brain, especially when participating in perceptual or attentional tasks, the EEG manifests low-voltage, high-frequency activity. Power spectrum analysis of the EEG will show higher peaks in the gamma band range during such tasks. However, the EEG amplifier has inappropriate band-pass filters for detecting events as fast as action potentials, which occur in the 1–2 ms range. EEG amplifiers more faithfully reflect the activity of slow potentials such as dendritic postsynaptic potentials and intrinsic membrane oscillations, which occur in the 10-20 ms range. That is, the EEG is a measure of dendritic potentials and membrane oscillations, not of action potential frequencies (Murakami and Okada, 2006). Many investigators mistakenly assume that gamma band activity in the cortical EEG mainly reflects neuronal action potentials at 40–60 Hz or higher, but this is not the case.

Moreover, gamma band activity is mistakenly thought to occur only as a result of a circuit repeatedly cycling at 40–60 Hz, that is, as a result exclusively of synaptic activity. For example, a cell assembly of, say, six neurons, some of them in the cortex, is erroneously thought to be capable of maintaining gamma band activity cycling through the assembly over prolonged periods. That is, it is assumed that every synapse in the circuit will result in an action potential each time it cycles. However, cortical circuits are notoriously inadequate for maintaining such high-frequency responses over time. For example, repetitive stimulation of the auditory pathway is faithfully followed with responses following each stimulus at high frequencies but only at the level of the brain stem. However, at the level of the cortex, the primary auditory region hardly follows stimulation with simple clicks at 20 per second and certainly fails to respond to every stimulus at higher frequencies of stimulation. The same is true of all of the other primary sensory pathways that are presumed to be the highest synaptic security pathways. How, then, can the cortex manifest gamma frequency activity for prolonged periods? If circuits that involve cortical pathways cannot follow simple sensory stimuli at gamma frequencies, how does the cortical EEG exhibit high-frequency EEG? The answer again lies in membrane oscillations. Since synaptic circuits cannot follow gamma frequencies, it is the combination of dendritic potentials and subthreshold membrane oscillations, that is, intrinsic membrane properties, as well as the cell assembly circuit that serves to maintain gamma band activity. In such circuits, the volley of information flows faithfully around the circuit, but individual cells do not always maintain the rhythm, as long as some of the neurons during each cycle faithfully follow gamma frequencies. The likelihood that cells in the assembly will fire in tight relation to the rhythm is ensured by the synchronized membrane oscillations of the population at large. Every neuron is more likely to fire at the peaks of the oscillations than at the troughs. Thus, the EEG reflects the recurrent membrane oscillations of large assemblies of neurons.

Figure 9.1 shows a circuit of five regions linked sequentially in a circle. Each cell in each region projects to each cell in the next region and so on around the circuit. If the cells manifest membrane subthreshold oscillations as shown on the right, then action potentials are more likely to fire at the peaks of the oscillations since these are closest to action potential threshold. If any synapse fails to elicit an action potential in the next region, such as is denoted by the "X" on the left, the rhythm will still be maintained by the circuit as a whole. That is, any one synapse can fail as the rhythm travels through the circuit, but the rhythm will still be maintained since not every cell needs to fire on every cycle. Moreover, the frequency of the rhythm will be maintained because the subthreshold oscillation will entrain the circuit to fire at its preferred frequency. If there were no membrane oscillations, the circuit could not maintain the same frequency for any length of time, since any one failure would reset the firing rate of that region. With any failures in any of the circuits, the population as a whole would soon have little coherence and the circuit would not maintain the preferred frequency. Therefore, it is the synaptic circuitry along with membrane oscillations that help maintain specific rhythms for any duration.

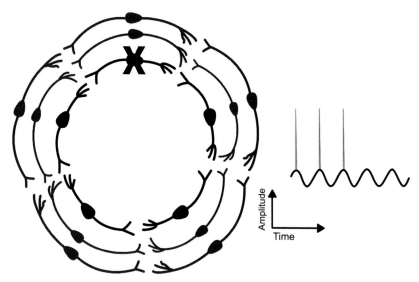

FIGURE 9.1 **Gamma band requires both synaptic links and subthreshold oscillations.** In this circuit with three regions, if the cells manifest membrane subthreshold oscillations as shown on the right, then action potentials are more likely to fire at the peak of the oscillation since it is closest to action potential threshold. If any synapse fails to elicit an action potential in the next region, such as is denoted by the "X" on the left, the rhythm will still be maintained by the circuit as a whole. That is, any one synapse can fail as the rhythm travels through the circuit, but the rhythm will still be maintained since not every cell needs to fire on every cycle.

BINDING AND PERCEPTION

Gamma frequency oscillations are thought to participate in sensory perception, problem solving, and memory (Eckhorn et al., 1988; Gray and Singer, 1989; Jones, 2007; Palva et al., 2009; Philips and Takeda, 2009; Voss et al., 2009). Coherence at gamma band frequencies can occur at cortical (Singer, 1993), or thalamocortical levels (Llinas et al., 1991). Coherent gamma band activation among thalamocortical networks (Llinas et al., 2002), and in other neuronal groups such as hippocampal and striatal afferents and efferents, is thought to contribute to the "binding" of information originating in separate regions (Llinás and Paré, 1991). The idea behind "binding" is that the perception of an object depends on a number of different properties. For example, for a visual stimulus, properties such as shape, color, size, texture, and movement are processed separately. The regions of the cortex subserving these different properties are dispersed, yet these properties are rapidly combined into a whole. Following the presentation of a complex stimulus such as a grating, oscillatory responses at separate cortical sites can synchronize to form neuronal assemblies in the visual cortex (Engel et al., 1990). Moreover, the presentation of a visual stimulus induces gamma oscillations with which single-cell activity is correlated (Gray and Singer, 1989). Similar findings have been reported for sensorimotor integration, learning, and memory (Miltner et al., 1999). That is, the presentation of a stimulus induces a sequence of activation of different regions that includes cortical and subcortical targets.

The perception of an object typically takes around 200 ms (Amano et al., 2006). Since the earliest evoked responses are elicited in the primary cortex at less than half that time, it is thought that, for perception to occur, the initial activation of the cortex must subsequently oscillate or rebound between the cortex and the thalamus. This thalamocortical resonance is closely linked to cognition and subjective experience (Ribary, 2005). Classic experiments by Libet demonstrated this principle for the somatosensory system (Libet and Kosslyn, 2004). In patients under local anesthesia prepared for epilepsy surgery, these experiments tested the response at the level of the somatosensory cortex of a cutaneous stimulus briefly applied to a contralateral finger. The stimulus elicited an evoked response on the corresponding finger region. However, when the finger region itself was stimulated briefly, the patient reported that no "sensation" was elicited by the brief stimulus. Only when the stimulus was extended to a duration ~200 ms did the patient detect that the "finger" had been stimulated. That is, although the initial sensory volley occurred at less than half that latency, the stimulus needed to induce a reverberating signal that lasted considerably longer, ~200 ms, for the stimulus to be perceived. More recently, elegant studies using magnetoencephalography have confirmed that perception of visual stimuli generally requires ~200 ms and occurs in an all-or-none fashion (Sekar et al., 2013). In summary, gamma band oscillations, reverberating between cortical and subcortical targets, are required to persist as long as 200 ms even for the perception of a simple sensory stimulus.

Gamma band activity is not only related to sensory perception. It has been proposed that gamma band activity supports the binding of distributed motor responses and the selection of specific motor parameters (Engel et al., 2006). EEG and MEG recordings in normal subjects (Ball et al., 2008; Cheyne et al., 2008; Hou et al., 2010; Waldert et al., 2008) and in patients with epilepsy (Crone et al., 2006; Mehring et al., 2004) showed that gamma oscillations occur in the contralateral motor cortex during movement onset. But, what are the cellular mechanisms that underlie gamma band oscillations in the cortex?

CORTICAL MECHANISMS OF GAMMA BAND ACTIVITY

Gamma oscillations emerge from the dynamic interaction between intrinsic neuronal and synaptic properties of thalamocortical networks (Steriade and Llinas, 1988). Cortical gamma band generation can be influenced by subcortical structures like the hippocampus and cerebellum

(Soteropoulos and Baker, 2006; Sirota et al., 2008). The neuronal networks behind such activity include (a) inhibitory cortical interneurons with intrinsic membrane potential oscillatory activity in the gamma range (Steriade and Llinas, 1988; Llinas et al., 1991; Steriade, 1999), many of which are electrically coupled (Gibson et al., 1999), and (b) rhythmically bursting pyramidal neurons (also electrically coupled) (Cunningham et al., 2004). At the thalamic level, thalamocortical excitatory neurons have intrinsic properties needed to generate subthreshold gamma band membrane potential oscillations (Pedroarena and Llinás, 1997). While cortical interneurons can generate membrane potential gamma oscillations through the activation of voltage-dependent, persistent sodium channel subthreshold oscillations (Llinas et al., 1991), and metabotropic glutamate receptors (Whittington et al., 1995), in thalamocortical neurons, the mechanism responsible for gamma band activity involves high-threshold P/Q-type voltage-gated calcium channels located in the dendrites (Pedroarena and Llinás, 1997; Whittington et al., 1995). Moreover, the same intrinsic properties mediating gamma band oscillations are present in the thalamus of several vertebrate species, indicating considerable evolutionary conservation of this mechanism (Llinas and Steriade, 2006).

Voltage-gated calcium channels are known to play a pivotal role in determining intrinsic properties and synaptic transmission throughout the central nervous system (Katz and Miledi, 1965; Llinas and Hess, 1976; Caterall, 1988; Llinas, 1988; Llinas et al., 2007). P/Q-type channels (also named $Ca(v)2.1$ channels) are present throughout the brain (Hillman et al., 1991; Uchitel et al., 1992; Jones, 2007; Llinas et al., 2007). N-type high-threshold, voltage-dependent calcium channels (also named $Ca(v)2.2$ channels) are found in the rat auditory brain stem, are restricted to the early postnatal period, and are replaced by P/Q-type channels later in development (Westenbroek et al., 1992). Immunocytochemical techniques have demonstrated the presence of N-type channels in brain stem structures (Shen et al., 2004). Importantly, P/Q-type mutant mice have deficient gamma band activity in the EEG, abnormal sleep-wake states, and ataxia; are prone to seizures (low-frequency synchrony); and die by 3 weeks of age (Llinas et al., 2007). N-type channel knockout animals show increased locomotor activity, deficits in long-term learning, and sensory gating (Nakagawasaki et al., 2010), as well as hyperaggression (Kim et al., 2009). N-type channels have been found to participate in nociception; in fact, a synthetic N-type calcium channel blocker derived from the marine snail ω-conotoxin, named ziconotide, is a powerful analgesic approved for the treatment of chronic pain (McGivern, 2007). In summary, the primary mechanisms for cortical gamma band activity production involve sodium-dependent persistent currents and P/Q-type (and perhaps also N-type) voltage-gated high-threshold calcium channels.

COORDINATION OF GAMMA BAND ACTIVITY

Before dealing with mechanisms behind gamma band activity in the RAS, we should note that a number of subcortical regions also manifest gamma band activity. For example, both the hippocampus and the cerebellum have the intrinsic and synaptic properties necessary to generate gamma band oscillatory activity. Hippocampal oscillatory activity in the gamma range (30–60 Hz) has been described and is functionally associated with entorhinal cortex input (Charpak et al., 1995). Neurons in the entorhinal cortex can oscillate at gamma band frequencies, suggesting a key role for such afferents in maintaining hippocampal gamma oscillations (Chrobak and Buzsáki, 1998). Moreover, gamma band activity in the CA1 area was divided into fast (~65–100 Hz) and slow (~25–60 Hz) frequency components that differentially couple CA1 and CA3 subfields, respectively (Colgin et al., 2009). Such differences have been proposed to "bind" CA1 fast gamma oscillations with very high-frequency activity from the entorhinal cortex, which provides information about object and place recognition in rodents (Bussey et al., 1999). On the other hand, slow gamma oscillations in CA1 appear locked to the slower frequencies present in the CA3 area, which is in charge of memory storage (Colgin et al., 2009; Colgin and Moser, 2010). That is, the two frequency ranges may have separate functions.

Gamma band activity also has been described in the Purkinje cell layer at the apex of the cerebellar cortex lobule and to a lesser extent in the distal white matter (Lang et al., 2006; Middleton et al., 2008). GABA$_A$ but not glutamate receptors are critical for gamma oscillation generation in Purkinje cells (Lang et al., 2006). Importantly, corticocerebellar coherence at gamma frequencies is evident in monkeys during performance of a manual precision grip task (Soteropoulos and Baker, 2006), and cerebellothalamic activity is synchronized with neocortical activity at gamma frequencies (Timofeev and Steriade, 1997). It has been proposed that cerebellar and thalamocortical networks oscillate at the same frequencies to enable information exchange among these brain areas (Middleton et al., 2008). Therefore, not only is gamma band activity present in various subcortical brain regions, but also there is coherence across these regions such that gamma oscillations are coordinated. For example, there is coherence across regions such that coherent gamma frequencies occur between the cortex and the cerebellum (Soteropoulos and Baker, 2006), the cerebellothalamic system and cortex (Timofeev and Steriade, 1997), and the hippocampus and cortex (Buzsaki, 2006). These findings suggest that gamma band activity in the cortex is not separate from that in other regions of the brain. That is, gamma band activity across cortical and subcortical regions is coherent.

In addition, gamma band activity has been described in the basal ganglia (Trottenberg et al., 2006). Such high-frequency activity has been recorded in the internal pallidum in patients with dystonia (Brucke et al., 2008; Liu et al., 2008) and from the subthalamic nucleus of patients with PD (Androulidakis et al., 2007; Kempf et al., 2007). Importantly, it was reported that gamma band activity in the motor cortex lags behind coherent activity in subcortical structures (Lalo et al., 2008; Litvak et al., 2012). This led to the suggestion that motor cortex gamma synchronization reflects a momentary arousal-related event for enabling the initiation of movement (Brucke et al., 2012; Cheyne and Ferrari, 2013; Jenkinson et al., 2013). That is, structures such as the RAS and thalamus may play an early permissive role in the control of movement. This issue will be addressed in more detail in Chapter 10; however, we must first address the cellular mechanisms behind gamma band activity in the RAS.

RAS MECHANISMS OF GAMMA BAND ACTIVITY

As far as the RAS is concerned, during waking and REM sleep, the EEG shows low-amplitude, high-frequency activity in the beta (20–30 Hz) and gamma (30–90 Hz) frequencies (Steriade and McCarley, 1990; Buzsáki and Draguhn, 2004). We discovered the presence of gamma band oscillations in three major centers of the RAS, (a) the pedunculopontine nucleus (PPN) (Simon et al., 2010; Kezunovic et al., 2010, 2011); (b) its major ascending target, the intralaminar parafascicular nucleus (Pf) (Kezunovic et al., 2010, 2012): and (c) its major descending target, the dorsal pontine subcoeruleus nucleus (SubCD) (Simon et al., 2011b). These were reviewed (Garcia-Rill et al., 2013; Urbano et al., 2012).

Mechanism for Gamma Activity in the Pedunculopontine Nucleus (PPN)— The PPN is most active during waking and REM sleep (Steriade and McCarley, 1990) and modulates ascending projections through the thalamus (modulating arousal and described in detail in Chapter 6) and descending projections through the pons and medulla (modulating REM sleep and posture and locomotion, and described in detail in Chapter 7). The PPN is made up of non-overlapping populations of cholinergic, glutamatergic, and GABAergic neurons (Wang and Morales, 2009). The PPN contains three cell types based on in vitro intrinsic membrane properties (Leonard and Llinas, 1990; Kamondi et al., 1992; Takakusaki and Kitai, 1997). These properties of PPN neurons are described in detail in Chapter 4. Recordings of PPN neurons in vivo identified multiple types of thalamic-projecting PPN cells distinguished by their firing properties relative to ponto–geniculo–occipital (PGO) wave generation (Steriade et al., 1990). Some neurons exhibited low spontaneous firing frequencies (<10 Hz), but most showed high rates of tonic firing in the beta/gamma range (20–80 Hz). In other in vivo studies, PPN neurons increased firing during REM sleep and were labeled "REM-on" cells or during both waking and

REM sleep and were called "wake/REM-on" cells, and the activity of both cell types decreased during SWS (Sakai et al., 1990; Steriade et al., 1990, 1991; Datta and Siwek, 2002). Stimulation of the PPN will potentiate the manifestation of fast (20–40 Hz) oscillations in the cortical EEG, outlasting stimulation by 10–20 s (Steriade et al., 1991). These results suggest that PPN cells do fire at gamma band frequencies in vivo and that its outputs can indirectly induce gamma band activity in its targets.

We were the first to report that almost all PPN cells fired maximally at gamma band frequency when depolarized using current steps (Figure 9.2) (Simon et al., 2010). We then tested the hypothesis that, in the presence

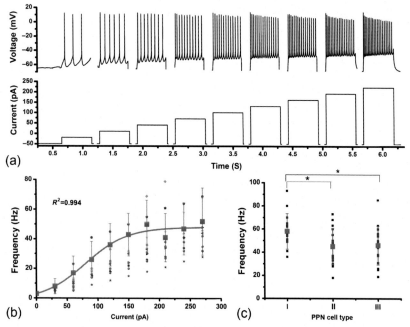

FIGURE 9.2   **Gamma band activity in whole-cell recorded PPN cells.** (a) Current steps increasing 30 pA per step, with each step 500 ms in duration, applied every 2.5 s (the record was truncated between current steps and spliced to show only the current steps) caused cells to fire action potentials at higher frequencies. This cell fired maximally at 54 Hz, which is within the gamma frequency range. (b) Graph showing the average firing frequency of the recorded cells (small characters) at the beginning, middle, and end of each current step. The average maximal firing frequency (large gray squares) was at the 180 pA current step, when cells fired at the average rate of 50+/−2 Hz at the beginning of the current step. Firing frequency decreased during the middle and end of the current step, and there was no significant difference between the firing rates during the middle and end of the step (ns, $p > 0.05$). (c) Graph showing the average firing frequency of each cell type at the beginning, middle, and end of the 180 pA current step. At the beginning of the current step, type I neurons fired significantly faster than type II or III cells (*$p < 0.05$), but type II and III neurons did not fire significantly faster than one another (ns, $p > 0.05$). Furthermore, there was no significant difference between the firing frequencies of the three cell types during the middle and end of the current step (ns, $p > 0.05$) (Simon et al., 2010).

of tetrodotoxin (TTX, to block action potential generation by blockade of sodium channels) and fast synaptic blockers (APV to block NMDA receptors, CNQX to block AMPA/KA receptors, gabazine to block GABA receptors, and strychnine to block glycine receptors), the remaining oscillatory activity observed in PPN neurons during current clamp square pulse depolarization was due to activation of intrinsic membrane properties in the form of voltage-dependent calcium channels (Kezunovic et al., 2011). Because of the potential activation of potassium channels during rapid depolarizing square steps, PPN neurons could not be depolarized beyond –25 mV. However, we used 1–2 s long depolarizing current ramps to gradually increase the membrane potential from resting values up to 0 mV. Standard square steps generated smaller amplitude gamma oscillations in PPN neurons compared to the amplitude of the oscillations generated by ramps (Figure 9.2). All PPN neurons tested exhibited membrane oscillations at both low (theta and alpha, 4–12 Hz) and beta/gamma band (>20 Hz) frequencies. Gamma band membrane oscillations were evident between –25 mV and –5 mV potentials and were absent at membrane potentials below –30 mV or above 0 mV, that is, within the voltage dependence window of high-threshold calcium channels. PPN neurons manifested statistically significant higher power spectrum amplitudes for gamma band oscillations when a depolarizing ramp was used (Figure 9.3), compared to square current steps (Kezunovic et al., 2011).

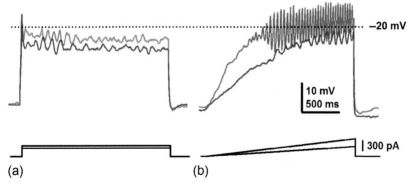

(a)   (b)

FIGURE 9.3 **Steps vs. ramps to reveal membrane oscillations.** Recordings in the PPN in the presence of tetrodotoxin (TTX, to block action potential generation by blockade of sodium channels) and fast synaptic blockers (APV to block NMDA receptors, CNQX to block AMPA/KA receptors, gabazine to block GABA receptors, and strychnine to block glycine receptors). (a) Current steps failed to depolarize the membrane potential sufficiently to activate high-threshold, voltage-dependent calcium channels. Note sag in membrane potential at the beginning of the current step, probably due to activation of potassium channels. (b) However, using ramps, membrane oscillatory activity was observed in PPN neurons due to activation of intrinsic membrane properties in the form of voltage-dependent calcium channels, usually at soma potentials between –30 mV and –10 mV. We used ω-CgTX and ω-Aga to determine that the depolarizing phase of these oscillations was manifested by N- and P/Q-type calcium channels, respectively (Kezunovic et al., 2011).

Using voltage clamp, no clear oscillatory currents were observed at holding potentials below −30 mV. However, clear beta/gamma band membrane oscillations were observed in the power spectra at both −20 mV and −10 mV holding potentials. The highest gamma band power amplitudes were observed at −10 mV holding potential (Kezunovic et al., 2011). The amplitudes of the peaks in the power spectrum were reduced after series resistance compensation (40–60% to reduce space clamp problems), suggesting that PPN neuronal oscillatory activity in voltage clamp was generated in neuronal compartments distant from the soma, as previously shown in a calcium imaging study of thalamic neurons (Pedroarena and Llinás, 1997). The N-type calcium channel blocker ω-conotoxin-GVIA (ω-CgTX) merely reduced gamma band oscillation amplitude and power spectrum, while the P/Q-type calcium channel blocker ω-agatoxin-IVA (ω-Aga) totally abolished them. Moreover, bath preapplication (>30 min) of ω-Aga prevented PPN neurons from oscillating at gamma band. In addition, application of ω-Aga and ω-CgTx together completely blocked all the oscillations in another group of cells tested. These results demonstrated that both voltage-dependent N- and P/Q-type calcium channels mediate the depolarizing phase of gamma band oscillations in the PPN. However, only P/Q-type channels appeared to be essential for gamma oscillation generation. Voltage clamp results suggested that calcium channels are located distally to the cell body, probably in PPN dendritic compartments (Kezunovic et al., 2011), as has been determined in thalamic neurons (Pedroarena and Llinás, 1997).

The average membrane oscillation frequency of PPN neurons was in the beta range at 23±1 Hz when perfused with synaptic blockers and TTX. When the nonspecific cholinergic agonist carbachol (CAR) was added, the average frequency of membrane oscillations was significantly higher and in the gamma range (47±2 Hz, $p < 0.001$). These oscillations were blocked by the application of ω-Aga. Different cell types did not show differential oscillatory frequencies after CAR exposure. The analysis of the power spectrum of oscillatory activity of all cells at different frequency ranges showed that CAR significantly reduced the power amplitude at 5–30 Hz frequencies, while there was no difference in average amplitude of oscillations at gamma (>30 Hz) frequency (Kezunovic et al., 2011).

We identified the channels responsible for the repolarizing phase of gamma frequency membrane oscillations in the PPN. High-voltage-gated potassium channels identified (Kv1.1, Kv1.2, and Kv1.6) belong to a class of delayed rectifier-like potassium channels. One of the functions of these channels is the regulation of state transitions and repetitive activity in striatal medium spiny neurons (Shen et al., 2004). We determined the effects of the specific potassium channel blockers on PPN neurons during ramp-induced oscillations: (a) α-dendrotoxin, a blocker of delayed rectifier-like voltage-gated potassium channels, (b) r-charybdotoxin, a blocker of big potassium (BK) calcium-dependent potassium channels, and (c) BDS-I, a blocker of the I_A current on PPN neurons. α-Dendrotoxin blocked gamma band oscillations (Figure 9.4) induced by 2s ramps, indicating that delayed rectifier-like voltage-gated potassium channels (Kv1.1, 1.2 and 1.6) play a role in the repolarizing phase of the PPN neuronal gamma oscillations. r-Charybdotoxin did not block PPN gamma oscillations in PPN (Figure 9.4), but did marginally increase their amplitude and frequency. We concluded that r-charybdotoxin increased the flow of K+ions by blocking large potassium outward currents through BK channels, which contributed to higher-amplitude and higher-frequency oscillations. BDS-I blocked the I_A current without affecting the oscillatory properties of PPN neurons.

These results (1) confirmed the presence of gamma band activity previously reported in PPN neurons (Simon et al., 2010) and (2) identified mechanisms underlying the generation of gamma band oscillations in the

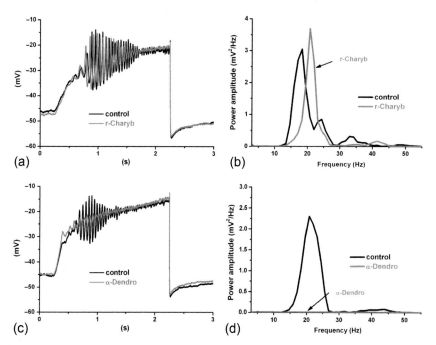

FIGURE 9.4 **Repolarization phase of membrane oscillations.** α-Dendrotoxin blocked ramp-induced oscillations, indicating that delayed rectifier-like voltage-gated potassium channels (Kv1.1, 1.2, and 1.6) play a role in the repolarizing phase of the PPN neuronal gamma oscillations. r-Charybdotoxin did not block PPN gamma oscillations in PPN, but marginally increased their amplitude and frequency. We concluded that r-charybdotoxin increased the flow of K+ions by blocking large potassium outward currents through BK channels, which contributed to higher-amplitude and higher-frequency oscillations. BDS-I blocked the I_A current without affecting the oscillatory properties of PPN neurons (Kezunovic et al., 2011).

PPN (Kezunovic et al., 2011). We also identified the roles of high-voltage-activated P/Q-type calcium and voltage-gated delayed rectifier-like potassium channels in mediating oscillations in PPN cells. Figure 9.5 is a diagram of the interplay between these channels in the PPN. PPN neurons manifest gamma band activity that is enhanced by CAR, such as that reported in cortical, thalamic, hippocampal, and cerebellar cells. In addition, PPN neurons appear to oscillate at gamma band through P/Q- and N-type calcium channels, as well as voltage-gated, delayed rectifier-like, potassium channels.

We undertook a series of studies to determine the location of high-threshold calcium channels in PPN neurons. We validated a novel ratiometric technique using Oregon Green BAPTA-1 (OGB1) with coinjections of a new long-Stokes-shift dye, Chromeo 494 (CHR). Fluorescent calcium transients were blocked with the nonspecific calcium channel blocker, cadmium, or by the combination of ω-agatoxin-IVA, a specific

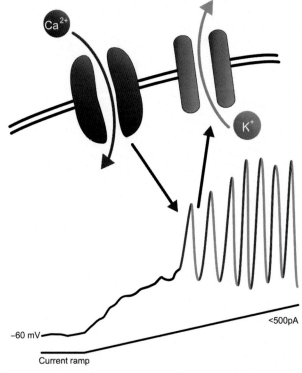

FIGURE 9.5 **Schematic of channels responsible for depolarizing and repolarizing phases of membrane oscillations.** PPN neurons appear to oscillate at gamma band frequencies through P/Q- and N-type calcium channels driving the depolarizing phase, as well as voltage-gated, delayed rectifier-like, potassium channels driving the repolarization phase (Kezunovic et al., 2011).

P/Q-type calcium channel blocker, and ω-conotoxin-GVIA, a specific N-type calcium channel blocker. The calcium transients were evident in different dendrites (suggesting channels are present throughout the dendritic tree), along the sampled length without interruption (suggesting channels are evenly distributed), and appeared to represent a summation of oscillations present in the soma (Figure 9.6). Thus, we confirmed that PPN calcium channel-mediated oscillations are due to P/Q- and N-type channels and revealed the fact that these channels are distributed along the dendrites of PPN cells (Hyde et al., 2013a).

Figure 9.7 provides a diagramed sequence of what occurs at the level of the soma and at the level of the dendrites when ramps are applied. In Figure 9.7(a), the left record is from the cell body before the ramp is applied, and the right record is from the dendrite. In Figure 9.7(b), the ramp has depolarized the cell body and induced oscillations at a higher membrane potential (left record) than at the dendrites (right record), which shows oscillations at or above action potential threshold (−50 mV to −40 mV). In Figure 9.7(c), the cell body has been depolarized beyond the window of activation (−30 mV to −10 mV) of high-threshold calcium channels, and oscillations are no longer evident at the level of the dendrites (less than −30 mV).

Mechanism for Gamma Activity in the Parafascicular Nucleus (Pf)—The Pf is part of the intralaminar thalamus (ILT), which is considered part of the "nonspecific" thalamocortical (TC) system. Pf cells differ from "specific" TC neurons in morphology, electrophysiological properties, and some

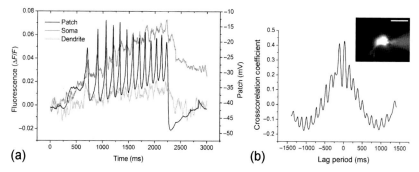

(a)   (b)

FIGURE 9.6   **Calcium imaging of PPN dendrites during ramp-induced oscillations.** (a) A novel ratiometric technique using Oregon Green BAPTA-1 with coinjections of a new long-Stokes-shift dye, Chromeo 494, revealed calcium oscillations in phase with electrically induced oscillations (black record) in the cell body (soma, dark gray record) and proximal dendrites (light gray record). The calcium transients were evident in different dendrites (suggesting channels are present throughout the dendritic tree), along the sampled length without interruption (suggesting channels are evenly distributed), and appeared to represent a summation of oscillations present in the soma. (b) Cross correlation analysis showed that the electrical and imaging signals were correlated with a 12 ms lag. Inset shows frame from inverted microscope during recording (Hyde et al., 2013a).

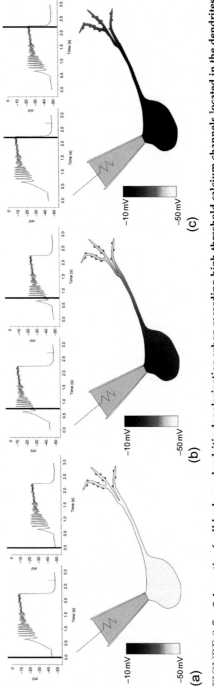

FIGURE 9.7 **Schematic of cell body vs. dendritic depolarization when recording high-threshold calcium channels located in the dendrites.** (a) The left record is from the cell body before the ramp is applied, and the right record is from the dendrite. (b) Application of the ramp depolarized the cell body and induced oscillations at a higher membrane potential (left record) than at the dendrites (right record), which shows oscillations at or above action potential threshold (−50 mV to −30 mV). (c) At the end of the ramp, the cell body has been depolarized beyond the window of activation (−30 mV to −10 mV) of high-threshold calcium channels, and oscillations are no longer evident at the level of the dendrites (less than −30 mV). Bottom diagram shows color-coded, gray to black, decreases in membrane potential.

synaptic connections. Pf neurons have long, sparsely branching processes in their proximal dendrites instead of compact bushy primary dendrites of TC relay cells (Deschenes et al., 1996a,b). TC relay neurons are present in the "specific" and some "nonspecific" thalamic nuclei and are bushy, multidendritic cells with stereotypical intrinsic properties, that is, bistable states of tonic vs. bursting patterns of activity due to the ubiquitous incidence of LTS mediated by T-currents (Llinas and Steriade, 2006). This mechanism is essential to inducing cortical synchronization of high-frequency rhythms during waking and REM sleep (tonic pattern) and synchronization of low-frequency rhythms during SWS (LTS + I_H oscillations) (McCormick and Pape, 1990). We previously demonstrated that "nonspecific" Pf cells showed reduced calcium-mediated LTS currents (Phelan et al., 2005), compared to "specific" TC neurons (Llinas and Jahnsen, 1982; Jahnsen and Llinas, 1984a,b). The Pf provides patterned inputs to separate striatal targets (Lacey et al., 2007). Thus, Pf neurons may play a different role in the modulation of TC activity compared to "specific" TC neurons.

The ILT is one of the targets of the cholinergic arm of the RAS and receives dense projections with symmetrical and asymmetrical terminals from the PPN and laterodorsal tegmental (LDT) nuclei (Erro et al., 1999; Kha et al., 2000; Capozzo et al., 2003; Kobayashi and Nakamura, 2003). The PPN, but not the LDT, is known to participate in the modulation of cortical arousal, sleep–wake cycles, and sensory awareness (Llinás and Paré, 1991; Steriade et al., 1991). Pf neurons were proposed to maintain the state of consciousness and selective attention in a study in primates (Minamimoto and Kimura, 2002; Raeva, 2006). Pf cells receive vagal input and participate in motor control, as well as pain modulation (Ito and Craig, 2005). High-frequency bursts recur rhythmically at ~40 Hz during wakefulness and sleep in ILT cells (Steriade et al., 1993). These authors suggested, "One of the most remarkable characteristics of these cells was the firing of high frequency bursts during natural states of wakefulness and REM sleep. At this time, this is the only known thalamic cell class exhibiting such behavior during brain-activated states" (Steriade et al., 1993).

We determined the maximal firing frequency of Pf neurons by using steps of increasing current amplitudes in current clamp mode (Kezunovic et al., 2012), as described above for the PPN. Figure 9.8(a) shows that the frequency of firing increased with increasing current steps but plateaued at gamma band frequencies. The initial firing frequency was higher than during the middle or end of the current step but within the gamma range (Figure 9.8(b)). Square steps generated lower-amplitude (and power of) oscillations compared to the amplitude and power of the oscillations generated by current ramps in the same cell. Statistical analysis showed that the amplitude of the oscillations generated by the ramps was significantly higher than that generated by the current steps. Oscillations were visible between –25 mV and –5 mV somatic membrane voltage range and

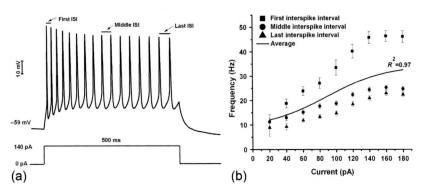

FIGURE 9.8 **Gamma band activity in whole-cell recorded Pf cells.** (a) Response of a Pf cell to depolarizing step showing that the initial (first interspike interval, ISI) firing frequency was higher than during the middle (middle ISI) or end (last ISI) of the current step. (b) Plot of the firing frequency during the first (squares), middle (circles), and end (triangles) ISIs of the current step, along with the average (line with R^2 value of 0.97), with increasing amplitude steps (Kezunovic et al., 2012).

were absent at membrane potentials below −30 mV or above 0 mV. Power spectrum analysis of the membrane oscillations induced during ramp recordings revealed that ~60% of the cells manifested the highest peak of oscillatory activity at the gamma range, while ~40% of cells had the highest peak at beta frequency. However, those cells showing frequencies below gamma band were typically at younger ages (Kezunovic et al., 2012).

We also showed that the maximal oscillatory frequency of Pf neurons began at lower ranges (alpha and beta) and gradually plateaued at gamma range with age (Figure 9.9). However, CAR could increase the firing frequency of younger cells to the gamma range, and their frequency plateaued within the gamma range, but no higher. In the oldest cells tested, CAR did not significantly increase frequency beyond the gamma range, suggesting these older cells were already "capped" at gamma band frequencies. In the Pf, the P/Q-type channel blocker ω-Aga abolished gamma band oscillations. The effect of the toxin was reversed during washout. Moreover, bath preapplication (>30 min) of ω-Aga prevented Pf neurons from oscillating at higher frequencies. The N-type blocker ω-CgTX reduced gamma band oscillation amplitude as evident in the power spectrum. The ω-CgTX effect was also reversed during washout and the amplitude of gamma band oscillations returned to the same level as in control. Bath preapplication of ω-CgTX did not prevent Pf neurons from oscillating. However, the amplitude of those oscillations was significantly lower than in control conditions. These results showed that both voltage-dependent P/Q- and N-type calcium channels mediate the depolarizing phase of gamma band oscillations in the Pf nucleus. However, only P/Q-type channels appeared to be essential, while N-type channels were permissive, for gamma band oscillation generation. The specific calcium channel blockers had the same

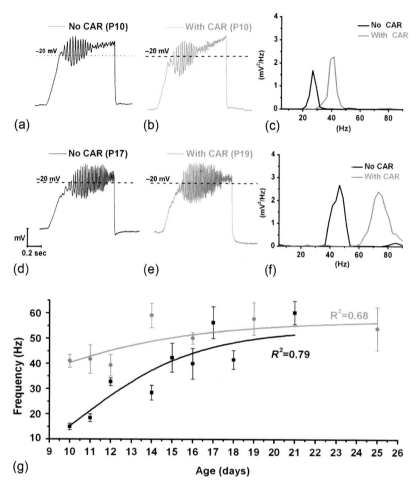

FIGURE 9.9   **Effect of carbachol (CAR) on ramp-induced oscillations in Pf neurons.**
(a) CAR increased the firing frequency of younger cells (P10) from the beta range (black record) to the gamma range (b), and their frequencies began at beta (black line) and plateaued within the gamma range (gray line) (c). In the oldest cells tested (P17), CAR did not significantly increase frequency beyond the gamma range, suggesting these older cells were already "capped" at gamma band frequencies (d, e). (f) Power spectrum of the oscillations without CAR (black line) and with CAR (gray line). (g) Maximum frequency manifested by Pf cells without CAR (black line, $R^2 = 0.79$) and with CAR (gray line, $R^2 = 0.68$) over age. CAR increased firing frequency in younger animals, but the intrinsic firing frequency was already established by 15 days in the Pf (Kezunovic et al., 2012).

type of blocking effect on Pf neuron oscillatory activity regardless of age (Kezunovic et al., 2012).

We analyzed calcium currents for both T-type and high-threshold components. We plotted current density values (pA/pF) against voltage measured at the beginning and at the end of square steps. T-type calcium channel-mediated currents peaked at −40 mV, while high-threshold

P/Q- and N-type mediated currents showed maximum values at −10 mV holding potential. The peak (holding potential −40 mV) T-type current density values were significantly increased after postnatal day 12, without affecting the peak density of P/Q- and N-type currents. Average *I-V* values from 12- to 25-day Pf cells showed an increase in T-type current density without changing N- and P/Q-type high-threshold current density (Kezunovic et al., 2012). Finally, we undertook a series of calcium imaging studies and determined that calcium transients generated during depolarizing current ramps could be visualized with a high-speed, wide-field fluorescence imaging system. Pf cells manifested calcium transients with oscillations in both somatic and proximal dendrite fluorescence recordings. Fluorescent calcium transients were blocked with the nonspecific calcium channel blocker, cadmium, or the combination of ω-Aga, the specific P/Q-type calcium channel blocker, and ω-CgTx, the specific N-type calcium channel blocker. We concluded that high-threshold calcium channels mediating gamma band oscillations in Pf cells were located in the dendrites (Hyde et al., 2013b).

In the striatum and subthalamic nucleus, depolarizing current steps linearly increased firing frequency to >500 Hz and >250 Hz, respectively (Azouz et al., 1997; Barraza et al., 2009), but did not plateau. In the monkey and rat prefrontal cortex, basket cells do not plateau, linearly increasing firing frequency (Povysheva et al., 2008). On the other hand, pyramidal cells in the mouse cortex plateaued at low frequencies, even after application of high-amplitude current steps (600–1000 pA) (Zhou et al., 2010). Our results showed that when Pf neurons were depolarized with increasing current steps, they fired initially at high frequency, but the firing frequency plateaued at gamma frequency (30–60 Hz). The only difference between our PPN results (Simon et al., 2010; Kezunovic et al., 2011) and our study on Pf cells (Kezunovic et al., 2012) was that PPN neurons did not show the initial high frequency rate. PPN neurons fired at ~50 Hz initially and later during the step at ~30–40 Hz, while Pf cells fired at ~80 Hz initially and then at ~50–60 Hz for the remainder of the step. Pf neurons may provide slightly differing initial signaling than PPN neurons, but both PPN and Pf ensure the maintenance of gamma band activity when maximally activated. These results suggest that the Pf generates gamma band activity when maximally activated and imparts gamma band activity on its targets. This newly discovered intrinsic membrane property of Pf neurons represents a novel mechanism for the induction of activated states such as waking by Pf efferents. As discussed above, the Pf projects to the cortex and to the striatum and may impart gamma band synchronization to both of these targets. The coherence between the RAS, basal ganglia, and motor cortex may well flow through the Pf and other intralaminar nuclei.

Mechanism for Gamma Activity in the Dorsal Subcoeruleus Nucleus (SubCD)—REM sleep is distinguished from other states by low-amplitude, high-frequency EEG activity, muscle atonia, and PGO waves in cats (P waves in the rat) (Aserinsky and Kleitman, 1953; Datta et al., 1998). Nuclei in the pons, including the SubCD, are critical for the generation of REM sleep (Baghdoyan et al., 1984; Boissard et al., 2002; Datta et al., 1998; Mouret et al., 1967; Marks et al., 1980; Yamamoto et al., 1990). The SubCD is most active during REM sleep (Boissard et al., 2002; Datta et al., 2009), and injection of the nonspecific cholinergic agonist CAR into this area induced a REM sleeplike state with muscle atonia and PGO waves (Baghdoyan et al., 1987; Boissard et al., 2002; Mitler and Dement, 1974; Vanni-Mercier et al., 1989; Yamamoto et al., 1990). Lesion of the SubCD produced REM sleep without muscle atonia or P waves (Karlsson et al., 2005; Mavanji et al., 2004; Mouret et al., 1967; Sanford et al., 1994) or diminished REM sleep (Lu et al., 2006). The SubCD receives input from several nuclei, including cholinergic, and perhaps glutamatergic, afferents from the PPN (Boissard et al., 2003; Datta et al., 1998; Mitani et al., 1988; Shiromani et al., 1988). The SubCD in turn projects to the thalamus, hippocampus, pons, and medulla (Datta and Hobson, 1994; Datta et al., 2004).

We showed that neurons in the SubCD fired APs at frequencies above gamma band (>100 Hz) at the beginning of a stimulus, but all neurons fired maximally (plateaued) at beta/gamma band during the remainder of the current step (Simon et al., 2011a) (Figure 9.10). Voltage and sodium channel-dependent subthreshold oscillations were identified as the mechanism involved in generating this activity. Subthreshold oscillations were isolated using APV, CNQX, GBZ, and STR to block fast inhibitory and excitatory spontaneous synaptic activity. At membrane potentials below AP threshold, subthreshold oscillations were observed and persisted at membrane potentials above AP threshold, where they were evident between APs. Subthreshold oscillations were observed following inactivation of sodium channels underlying APs with low levels of TTX, suggesting the existence of two populations of voltage-gated sodium channels, one related to AP generation and the other related to subthreshold oscillations (Simon et al., 2011a).

A sodium-dependent mechanism was revealed using TTX, an extracellular sodium channel blocker, and QX-314, an intracellular sodium channel blocker. Low concentrations of TTX (0.01 μM) completely blocked AP generation and reduced the power of gamma band oscillations but did not abolish subthreshold oscillations. On the other hand, high concentrations of TTX (10 μM) completely blocked the remaining subthreshold gamma oscillations. QX-314 in the intracellular recording solution blocked both APs and subthreshold gamma oscillations. These results showed that beta/gamma frequency, sodium-dependent subthreshold oscillations underlie the gamma frequency firing of SubCD neurons (Simon et al., 2011a).

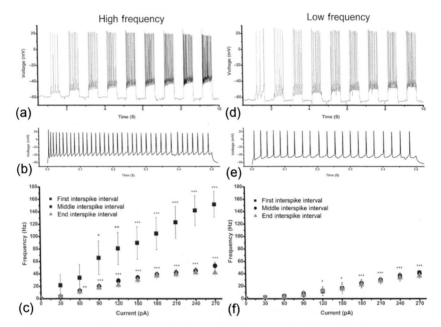

FIGURE 9.10   **Gamma band activity in whole-cell recorded SubCD cells.** SubCD cells showed two cell types based on the firing frequency achieved during the first ISI. (a) Current steps showed that some SubCD cells fired faster. (b) Response to a 180 pA depolarizing step) and reached higher firing frequencies. (c) Filled squares showed significantly higher frequencies after 120 pA) during the first ISI than during the middle and last ISI. (d) Other SubCD cells showed lower firing frequencies during increasing current steps and during a similar current (180 pA) step (e). (f) The firing frequency of low-frequency SubCD cells, like PPN and Pf cells, plateaued at 40–60 Hz (Simon et al., 2011a).

Previous studies in the SubCD hinted at the presence of gamma firing frequency in vivo and in vitro following depolarizing pulses (Brown et al., 2006; Datta and Hobson, 1994). One of these studies, however, sampled a region far wider than the SubCD since it included recordings of cholinergic and noradrenergic cells, which are not present in the SubCD (Brown et al., 2006). In any case, no previous study had demonstrated the ability of all SubCD neurons to fire maximally at or above gamma frequency, or the presence of two distinct populations of SubCD neurons that fire at different frequencies at the beginning of a current step, but plateau at gamma frequency following the initial burst of action potentials. Furthermore, some neurons showed membrane oscillations in the gamma range, which were not affected when APs were blocked by low concentrations of TTX, but were blocked by high concentrations of TTX, and thus appeared to be sodium channel-dependent subthreshold oscillations. In single-cell recordings, CAR increased activity in the beta/gamma range, and NMDA increased activity in the mid and high gamma range, while KA did not induce gamma frequency firing. In addition, population responses in the SubCD showed that NMDA increased activity at almost all frequencies, while CAR and KA induced specific peaks in the gamma range. These results suggest that the SubCD can generate beta (and to a lesser extent gamma) band activity via subthreshold oscillations and thereby may impart such activity on its targets when maximally activated (Simon et al., 2011a). Interestingly, a previous study reported the presence of subthreshold oscillations in type II PPN cells (Takakusaki and Kitai, 1997), and these oscillations were maximal in the beta range. Subthreshold oscillations in PPN and SubCD appear to peak at similar frequencies in the beta range, while P/Q-type calcium channel-mediated oscillations in the PPN and Pf appear to peak at similar frequencies in the gamma range. It is not clear if such a difference is present such that gamma band predominates during waking (when PPN and Pf are active) vs. beta band predominating during REM sleep (when PPN and SubCD are active).

Extracellular recordings from cells in the PGO wave-generating site in cats recorded the presence of "PGO-on" cells, which increased their firing rates before the first PGO wave until the end of REM sleep, but had low firing rates during waking and SWS without PGO waves (Datta and Hobson, 1994). These cells discharged high-frequency spike bursts (>500 Hz) during PGO-related states and fired tonically at 25–100 Hz. Perhaps, the cells recorded in vitro that showed high-frequency bursts of action potentials during the beginning of current steps could be putative "PGO-on" cells in the rat (Simon et al., 2011a,b).

GAMMA MAINTENANCE

We know that maintenance of gamma band activity must last at least 200 ms for the perception of a simple sensory stimulus. But how long must gamma band activity be maintained for an attentional task? How long do we maintain gamma band activity during learning or during a conversation lasting minutes to hours? It has been suggested that consciousness is associated with "continuous" gamma band activity rather than an interrupted pattern of activity (Vanderwolf, 2000a,b). The original description of the RAS specifically suggested that it participates in "tonic or continuous" arousal and that lesions of the RAS eliminated "tonic" arousal (Moruzzi and Magoun, 1949; Watson et al., 1974). RAS structures like the PPN, Pf, and SubCD, in which every cell in every nucleus manifests gamma band activity and in which a subgroup of cells are electrically coupled, then become a "gamma-making machine." We hypothesized that it is the activation of the RAS during waking and REM sleep that induces coherent activity (through electrically coupled cells) and high-frequency oscillations (through P/Q-type calcium channel and subthreshold oscillations). This leads to the generation of the background of gamma activity necessary to support a state capable of reliably assessing the world around us on a continuous basis. That is, these mechanisms may underlie preconscious awareness (Garcia-Rill et al., 2013; Urbano et al., 2012).

We, therefore, investigated the mechanism behind the maintenance of gamma band activity in the PPN. The PPN is known to receive cholinergic input from the contralateral PPN and LDT nuclei (Semba and Fibiger, 1992). PPN cholinergic neurons are hyperpolarized by the activation of postsynaptic muscarinic M_2 cholinergic receptors (Leonard and Llinas, 1994). We previously showed that ~73% of PPN output neurons were inhibited through M_2 cholinergic receptors, ~13% were excited through M_1 and nicotinic cholinergic receptors, and ~7% showed biphasic responses (Ye et al., 2010). The most abundant receptor type in the PPN is the M_2 muscarinic receptor, which is G_i protein-coupled. We found that short-duration application of CAR blocked PPN ramp-induced oscillations (Figure 9.11(a)) (Kezunovic et al., 2013). Acute or short-lasting CAR application blocked P/Q-type calcium channel-mediated oscillations in the PPN through M_2 muscarinic receptors, suggesting that brief cholinergic input to PPN cells induces inhibition of high-frequency oscillations (Kezunovic et al., 2013). This temporary membrane hyperpolarization caused by M_2 receptor activation explains the acute inhibitory effect of CAR on the oscillatory activity in PPN neurons.

On the other hand, when CAR was applied for prolonged periods of time, CAR increased the frequency (but not the amplitude) of the oscillations (Figure 9.11(b)). The specific P/Q-type calcium channel blocker, ω-Aga, abolished oscillatory activity in the PPN, suggesting that oscillations after tonic cholinergic activation were mediated by P/Q-type calcium channels. The increase in frequency of oscillations indicates that CAR altered the kinetic properties of these channels. We showed that M_2 (G_i protein-coupled) receptors not only effectively blocked calcium channel-mediated oscillations when administered acutely but also increased the frequency of oscillations after tonic activation of intracellular mechanisms. We proposed that persistent cholinergic activation of the PPN through the M_2 receptors plays a key role in maintaining high-frequency activity among PPN neurons.

We then tested the participation of G proteins in these effects. We showed that no oscillations were observed in the presence of the G protein blocker GDP-β-S, suggesting that availability of G proteins is key to generating high-frequency oscillations in the PPN (Figure 9.12(a) and (b)). However, the G protein stimulator GTP-γ-S did not change the ramp-induced oscillations (Figure 9.12(c) and (d)). These results suggested that G protein binding can saturate, beyond which no blockade of oscillations is possible (Kezunovic et al., 2013). Using a three-pulse protocol described by others

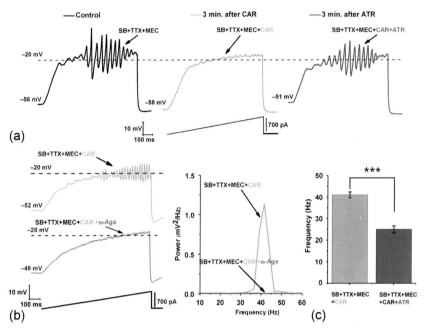

FIGURE 9.11 **Acute vs. persistent effects of CAR on PPN neurons.** (a) Ramp-induced oscillations (black left record) were blocked by short-duration application of CAR (light gray middle record), an effect eliminated by addition of atropine (ATR) (dark gray right record). Acute or short-lasting CAR application blocked P/Q-type calcium channel-mediated oscillations in the PPN through M_2 muscarinic receptors, suggesting that brief cholinergic input to PPN cells induces inhibition of high-frequency oscillations (Kezunovic et al., 2013). This temporary membrane hyperpolarization explains the acute inhibitory effect of CAR on the oscillatory activity in PPN neurons. (b) When CAR was applied for prolonged periods of time, it increased the frequency (but not the amplitude) of the oscillations, which were blocked by the P/Q-type calcium channel blocker ω-Aga (compare light gray vs. dark gray records and light gray vs. dark gray lines in power spectrum right side). (c) Graph of the peak frequency of oscillations induced by persistent application of CAR (light gray, left bar) in the gamma range was reduced to beta frequency after application of ATR (dark gray, right bar) (Kezunovic et al., 2013).

(Ikeda, 1996; Herlitze et al., 1996; Kammermeier et al., 2000), we tested the effects of specific calcium currents when G proteins are bound (prepulse) and compared them, after delivering a depolarizing step that removes G protein binding, to the response during a postpulse (Figure 9.13(a)). We found that CAR reduced calcium current amplitude of the prepulse more than the postpulse, suggesting a voltage-dependent G protein mechanism (Figure 9.13(b) and (d)). Moreover, acute CAR slowed the kinetics and reduced the amplitude of the calcium currents. The stimulatory effect of persistent CAR exposure on calcium currents was prevented by adding GDP-β-S and had similar effects on calcium currents only mediated by P/Q-type calcium channels (Figure 9.13(c) and (e)). This discovery

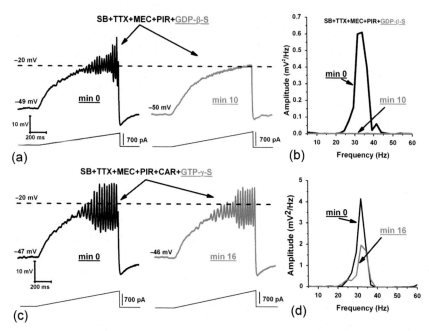

FIGURE 9.12   Role of G proteins in persistent effect of CAR. (a) In the presence of synaptic blockers (SB), TTX, MEC, and the M_1 cholinergic receptor blocker pirenzepine (PIR), ramps induced oscillations (black record, left side), but no oscillations were observed in the presence of the G protein blocker GDP-β-S (light gray record, right side), suggesting that availability of G proteins is key to generating high-frequency oscillations in the PPN. (b) The power spectrum shows that gamma band oscillations (black line) were blocked by GDP-β-S (light gray line). (c) The G protein stimulator GTP-γ-S did not change the ramp-induced oscillations (compare black left record and dark gray right record). (d) Power spectrum showing little change in the oscillations was effected by GTP-γ-S. These results suggested that G protein binding can saturate, beyond which no blockade of oscillations is possible (Kezunovic et al., 2013).

explains why the persistent CAR effect did not induce a total blockade of oscillations. Partial blockade of P/Q-type channels and slower activation/deactivation kinetics than in control conditions provided a mechanism for inducing faster frequencies of oscillations in the presence of CAR. G protein modulation of potassium channels provided the faster membrane potential repolarization necessary to sustain higher frequencies of oscillations (Kezunovic et al., 2013).

The differences between the manifestations of oscillations in PPN cells during acute vs. tonic exposure to CAR may be related to the presence of phasic vs. tonic cholinergic input. That is, short-duration cholinergic input to PPN neurons will block the capacity to oscillate at high frequencies; however, under persistent, long duration, or tonic cholinergic influence, PPN neurons oscillate at higher frequencies. Phasic cholinergic input is more characteristic of activity patterns during REM sleep, in which PPN

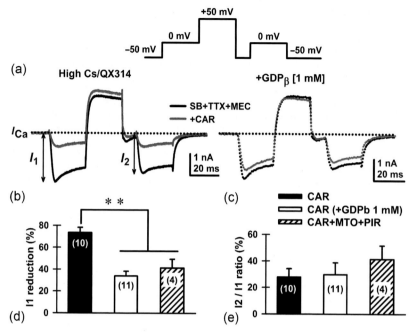

FIGURE 9.13   **Voltage-dependent G protein modulation by CAR of voltage-gated cal-cium currents in PPN.** (a) Three-step protocol used to study the voltage dependence of G protein modulation of calcium currents (I_{Ca}) in PPN neurons. (b) In the presence of SB, TTX, and MEC, calcium currents (I_{Ca}; black record) were reduced in amplitude when the three-pulse protocol was applied (left side I_1 vs. I_2). Adding CAR (dark gray record) reduced the total amount of current during both pulses without affecting calcium current amplitude in I_2/I_1 ratios. (c) The effects of CAR were prevented in recordings with GDP-β-S in the intra-cellular solution. Note that the black record (before CAR) was similar to the gray record (after CAR). (d) The effect of CAR (black column) was significantly reduced when either intracellular GDP-β-S (white column) or extracellular muscarinic receptor antagonists (M$_2$ antagonist methoctramine (MTO)+M$_1$ antagonist pirenzepine (PIR)) (hatched column) were used. (e) Calcium current I_2/I_1 ratio values (%) were unchanged by CAR under any of the experimental conditions. These results suggest that CAR reduced calcium currents through activation of M$_2$ muscarinic receptors that activate G proteins.

cells burst more, while persistent cholinergic tone is more characteristic of tonic patterns during waking. We proposed that, under phasic exposure to cholinergic input, G proteins will bind calcium channels, slowing their activation. This would reduce or prevent the induction of high-frequency oscillations usually mediated by P/Q-type calcium channels. However, under tonic cholinergic input, G proteins are bound to the persistent in-put, maximizing their utilization, thus freeing calcium channels to be acti-vated at their membrane potential threshold, leading to sustained gamma frequency oscillatory activity. This "permissive inhibition" of G proteins by tonic cholinergic input is a potential mechanism for the "maintenance" of gamma band activity (Kezunovic et al., 2013). It would be of interest to determine if this mechanism differs between gamma band activity during waking compared to that during REM sleep and if this mechanism is altered by total sleep or REM sleep deprivation.

GAMMA IN WAKING VS. REM SLEEP

The differences between gamma band activities during waking and during REM sleep are unknown. Why is this important? Because it is high-frequency, especially beta/gamma band, activity that drives our cognitive function not only during waking but also during REM sleep, two obviously different states of awareness. Our studies addressed the differential intracellular mechanisms assumed to subserve high-frequency activity during waking vs. REM sleep as a prelude to the selective pharmacological modulation of these states. We know that the two states are differentially regulated in the PPN. Injections of glutamate into the PPN were shown to increase both waking and REM sleep, but injections of NMDA increased only waking, while injections of kainic acid (KA) increased only REM sleep (Datta, 2002; Datta and Siwek, 1997; Datta et al., 2001a,b). Thus, the two states are independently activated by NMDA vs. KA receptors. Moreover, the intracellular pathways mediating the two states appear to differ. For example, the CaMKII activation inhibitor, KN-93, microinjected into the PPN of freely moving rats resulted in decreased waking but not REM sleep (Datta et al., 2010). Increased ERK1/2 signaling in the PPN is associated with maintenance of sleep via suppression of waking (Desarnaud et al., 2011) and that activation of intracellular protein kinase A (PKA) in the PPN instead contributed to REM sleep recovery following REM sleep deprivation (Datta and Desarnaud, 2010). These authors showed that during REM sleep, pCREB activation in PPN cholinergic neurons was induced by REM sleep and that PPN intracellular PKA activation and a transcriptional cascade involving pCREB occurred in cholinergic neurons (Datta et al., 2009). These results suggest that waking is modulated by the CaMKII pathway, while REM sleep is modulated by the cAMP/PKA pathway in the PPN. In vivo recording studies have shown that PPN neurons manifest two major types of cellular activity in relation to waking and REM sleep in the form of "wake/REM-on" and "REM-on" cells (Datta and Siwek, 2002). This suggests that some PPN cells fire in relation to waking and REM sleep and others only in relation to REM sleep, presumably through CaMKII and cAMP/PKA pathways vs. only the cAMP/PKA pathway, respectively. A caveat needs to be recalled, which is the lack of sampling of in vivo studies during active behavior and in enriched environments during single-cell recordings. This could reveal the presence of "wake-only" cells that would hardly exhibit activity under current experimental conditions.

Figure 9.14 shows results demonstrating that KN-93 blocked the ability of ramps to induce oscillations in PPN neurons (Figure 9.14(a)), an effect confirmed by the absence of oscillations in the power spectrum (Figure 9.14(b)). Recordings were carried out in the presence of synaptic blockers (SB), TTX, and mecamylamine (MEC), a nicotinic receptor antagonist. These findings suggest that blocking the activation of CaMKII using KN-93 eliminated the ability of ramps to induce gamma band oscillations in PPN neurons. These data suggest that CaMKII is necessary for the manifestation of ramp-induced oscillations in at least some PPN neurons (Garcia-Rill et al., 2014).

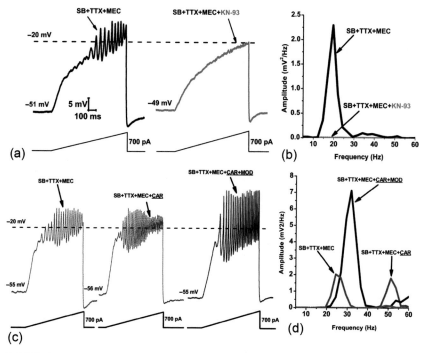

FIGURE 9.14 **Role of CaMKII and modafinil (MOD) in ramp-induced oscillations.** (a) Ramp-induced oscillations (black left record) were blocked by the CaMKII blocker KN-93 (green right record). (b) Power spectrum of records in A confirming the absence of oscillations after KN-93. These data suggest that CaMKII is necessary for the manifestation of ramp-induced oscillations in at least some PPN neurons (Garcia-Rill et al., 2014). (c, d) Figure shows ramp-induced oscillations in the presence of SB+TTX+MEC before CAR application (blue record, D blue line showing beta frequency oscillations at 25 Hz), but the addition of CAR for 20 min led to a significant increase in oscillation frequency (but not amplitude) to 50 Hz (C red record, D red line in power spectrum). Superfusion of MOD for 20 min in the presence of CAR significantly increased oscillation amplitude but decreased oscillation frequency to 30 Hz (C black record, D black line in power spectrum). These results showed that MOD potentiated CAR-induced oscillation amplitude but decreased their frequency from the gamma to the beta range in most cells, as was the case with MOD alone (Garcia-Rill et al., 2014).

The effects of the stimulant modafinil (MOD) are dependent on CaMKII, since its effects are blocked by the CaMKII activation blocker KN-93 (Urbano et al., 2007). Our studies using MOD (a) extensively characterized its effects on electrical coupling in the RAS and ILT, (b) established the concentration dependence of its effects in vitro and in vivo, (c) characterized its effects in animals and humans, and (d) determined that its effects were blocked specifically by gap junction blockers (Beck et al., 2008; Garcia-Rill et al., 2007, 2008; Heister et al., 2007; Kezunovic et al., 2010). MOD, which typically takes 10-20 min to induce changes in electrical coupling, by itself does not appear to change gamma band oscillation frequency or amplitude. However, MOD was found to enhance CAR-induced oscillation amplitude. We used SB plus the nicotinic receptor antagonist MEC to tonically activate only muscarinic receptors using CAR. Figure 9.14(c) and (d) shows ramp-induced oscillations in the presence of SB + TTX + MEC before CAR application (C blue record, D blue line showing beta frequency oscillations at 25 Hz), but the addition of CAR for 20 min led to a significant increase in oscillation frequency (but not amplitude) to 50 Hz (C red record, D red line in power spectrum). Superfusion of MOD for 20 min in the presence of CAR significantly increased oscillation amplitude but decreased oscillation frequency to 30 Hz (C black record, D black line in power spectrum). These results showed that MOD potentiated CAR-induced oscillation amplitude but decreased their frequency from the gamma to the beta range in most cells, as was the case with MOD alone (Garcia-Rill et al., 2014).

These data showed that MOD preferentially promotes high-frequency activity through the CaMKII ("waking") pathway, especially in the presence of tonic cholinergic input. Moreover, studies on cocaine abusers (Morgan et al., 2010) and on an animal model of sleep-disordered breathing (Panckeri et al., 1996), suggest that MOD may also decrease REM sleep. Further work will be needed to determine if the cAMP/PKA ("REM sleep") pathway is preferentially activated during REM sleep compared to the CaMKII pathway. These data may also have implications for the mechanisms behind the developmental decrease in REM sleep (Chapter 5). One possibility is that there is a gradual shift from the cAMP/PKA pathway to the CaMKII pathway with age. This would account for the more abundant generation of gamma band activity in the "waking gamma" compared to the "REM sleep gamma" pathway.

CLINICAL IMPLICATIONS

The implications of the discovery of gamma band activity as a ubiquitous mechanism in nuclei that modulate waking and REM sleep, and activate the cortex as well as postural and locomotor systems, are of critical importance. The fact that these centers can generate gamma band oscillations, however, should not be surprising given the descriptions of gamma activity in other subcortical structures such as the cerebellum, hippocampus, and especially basal ganglia. Moreover, gamma band activity between these regions is coherent, so that RAS gamma band activity is probably highly coordinated with cortical and other subcortical gamma band generators, depending on the task. This also should not be surprising, given the need for reverberating circuits essential for the persistent or maintenance of processes that mediate perception, movement, learning, and memory. The discovery of a permissive inhibition as one possible mechanism for the maintenance of gamma band activity has specific implications. The requirement for tonic cholinergic input to PPN neurons in order to occupy G proteins so that sensory inputs can make dendrites oscillate when activated probably needs considerable energy investment. That is, the process of maintaining gamma band activity is metabolically demanding. The cholinergic neurons must remain highly active during these states. Waking is hard work.

In addition, we know that glutamatergic input is important for the type of intracellular pathway that is activated to produce gamma band during waking vs. during REM sleep. While the cholinergic input must remain tonically active, the glutamatergic input must be superimposed on this background to shift the NMDA-responsive intracellular pathway towards waking or the KA-responsive intracellular pathway towards REM sleep. Interestingly, both cholinergic PPN and noncholinergic PPN, probably glutamatergic, neurons project to the Pf. This may mean that the cholinergic input must remain tonically active, while the glutamatergic input is superimposed, at the level of the Pf. However, it must first be established that a similar G protein-dependent mechanism is also at play in the Pf. Another implication is that if cholinergic cell firing wanes, then G proteins are not occupied and high-threshold calcium channels are bound, reducing the potential for high-frequency activity. One possibility is that the highest levels of activity by cholinergic cells are demanded during waking, but a partial reduction in the firing may shift the state from waking to REM sleep, while a significant reduction in cholinergic cell activity allows the induction of slow activity (no longer inhibits the reticular nucleus of the thalamus) to promote SWS.

The clinical implications of the presence of gamma band activity in the RAS are widespread. It may be possible to pharmacologically modulate the states of waking and REM sleep differentially. Findings cited above do suggest that modafinil promotes waking while decreasing REM sleep. These and the foregoing discussions suggest that gamma activity during waking is separate than that during REM sleep. Moreover, the two should not be manifested simultaneously, that is, promoting waking and REM sleep at the same time may be deleterious. The suggestion has been made that hallucinations represent REM sleep intrusion into waking (Dement, 1967; Mamelak and Hobson, 1989), that is, the manifestation of waking and REM sleep simultaneously. We hypothesize that the two types of gamma activity, through their respective intracellular pathways, CaMKII and cAMP/PKA, normally inhibit each other in order to prevent coincidence. This suggests that a potential therapeutic avenue for preventing hallucinations is downregulation of the cAMP/PKA pathway in psychosis (Chapter 11).

A number of neurological and psychiatric disorders are characterized by decreased or interrupted gamma band activity. These will be described individually in Chapters 11 and 12. Moreover, the maintenance of gamma band activity is critical for consciousness, learning and memory, and everyday survival that requires preconscious awareness, which will be addressed in Chapter 10. Interrupted or decreased gamma band activity will affect performance and cognitive behavior, from driving a car to flying an airplane, from performing surgery to performance in sports, from understanding directions to conducting traffic. Chapter 14 will address the issue of performance, intent, and novel ideas on the workings of the brain.

References

Amano, K., Nishida, S., Takeda, T., 2006. MEG responses correlated with the visual perception of velocity change. Vision Res. 46, 336–345.

Androulidakis, A.G., Kuhn, A.A., Chen, C.C., Blomstedt, P., Kempf, F., Kupsch, A., Schneider, G.H., Doyle, L., Dowsey-Limousin, P., Hariz, M.I., Brown, P., 2007. Dopaminergic therapy promotes lateralized motor activity in the subthalamic area in Parkinson's disease. Brain 130, 457–468.

Aserinsky, E., Kleitman, N., 1953. Regularly occurring periods of eye motility, and concomitant phenomena, during sleep. Science 118, 273–274.

Azouz, R., Gray, C.M., Nowak, L.G., McCormick, D.A., 1997. Physiological properties of inhibitory interneurons in cat striate cortex. Cereb. Cortex 7, 534–545.

Baghdoyan, H.A., Rodrigo-Angulo, M.L., McCarley, R.W., Hobson, J.A., 1984. Site-specific enhancement and suppression of desynchronized sleep signs following cholinergic stimulation of three brainstem regions. Brain Res. 306, 39–52.

Baghdoyan, H.A., Rodrigo-Angulo, M.L., McCarley, R.W., Hobson, J.A., 1987. A neuroanatomical gradient in the pontine tegmentum for the cholinoceptive induction of desynchronized sleep signs. Brain Res. 414, 245–261.

Ball, T., Demandt, E., Mutschler, I., Neitzel, E., Mehring, C., Vogt, K., Aertsen, A., Schulze-Bonhage, A., 2008. Movement related activity in the high gamma range in the human EEG. Neuroimage 41, 302–310.

Barraza, D., Kita, H., Wilson, C.J., 2009. Slow spike frequency adaptation in neurons of the rat subthalamic nucleus. J. Neurophysiol. 102, 3689–3697.

Beck, P., Odle, A., Wallace-Huitt, T., Skinner, R.D., Garcia-Rill, E., 2008. Modafinil increases arousal determined by P13 potential amplitude; an effect blocked by gap junction antagonists. Sleep 31, 1647–1654.

Boissard, R., Gervasoni, D., Schmidt, M.H., Barbagli, B., Fort, P., Luppi, P.H., 2002. The rat ponto–medullary network responsible for paradoxical sleep onset and maintenance: a combined microinjection and functional neuroanatomical study. Eur. J. Neurosci. 16, 1959–1973.

Boissard, R., Fort, P., Gervasoni, D., Barbagli, B., Luppi, P.H., 2003. Localization of the GABAergic and non-GABAergic neurons projecting to the sublaterodorsal nucleus and potentially gating paradoxical sleep onset. Eur. J. Neurosci. 18, 1627–1639.

Brown, R.E., Winston, S., Basheer, R., Thakkar, M.M., McCarley, R.W., 2006. Electrophysiological characterization of neurons in the dorsolateral pontine rapid-eye-movement sleep induction zone of the rat: Intrinsic membrane properties and responses to carbachol and orexins. Neuroscience 143, 739–755.

Brucke, C., Huebl, J., Kempf, F., Krauss, J.K., Yarrow, K., Kupsch, A., Schneider, G.H., Brown, P., Kuhn, A.A., 2008. Pallidal gamma activity is correlated to movement amplitude in patients with dystonia. Clin. Neurophysiol. 119 (S1), 49.

Brucke, C., Huebl, J., Schonecker, T., Wolf-Julian, N., Yarrow, K., Kupsch, A., Blahak, C., Lutjens, G., Brown, P., Krauss, J., Schneider, G.H., Kuhn, A., 2012. Scaling of movement is related to pallidal γ oscillations in patients with dystonia. J. Neurosci. 32, 1008–1019.

Bussey, T.J., Muir, J.L., Aggleton, J.P., 1999. Functionally dissociating aspects of event memory: the effects of combined perirhinal and postrhinal cortex lesions on object and place memory in the rat. J. Neurosci. 19, 495–502.

Buzsaki, G., 2006. Rhythms of the Brain. Oxford University Press, Oxford.

Buzsáki, G., Draguhn, A., 2004. Neuronal oscillations in cortical networks. Science 304, 1926–1929.

Capozzo, A., Florio, T., Cellini, R., Moriconi, U., Scarnati, E., 2003. The pedunculopontine nucleus projection to the parafascicular nucleus of the thalamus: an electrophysiological investigation in the rat. J. Neural Transm. 110, 733–747.

Caterall, W.A., 1988. Structure and function of neuronal Ca2+ channels and their role in neurotransmitter release. Cell Calcium 24, 307–323.

Charpak, S., Paré, D., Llinás, R.R., 1995. The entorhinal cortex entrains fast CA1 hippocampal oscillations in the anaesthetized guinea-pig: role of the monosynaptic component of the perforant path. Eur. J. Neurosci. 7, 1548–1557.

Cheyne, G., Ferrari, P., 2013. MEG studies of motor cortex gamma oscillations: evidence for a gamma "fingerprint" in the brain? Front. Hum. Neurosci. 7, 575.

Cheyne, D., Bells, S., Ferrari, P., Gaetz, W., Bostan, A.C., 2008. Self-paced movements induce high-frequency gamma oscillations in primary motor cortex. Neuroimage 42, 332–342.

Chrobak, J.J., Buzsáki, G., 1998. Gamma oscillations in the entorhinal cortex of the freely behaving rat. J. Neurosci. 18, 388–398.

Colgin, L.L., Moser, E.I., 2010. Gamma oscillations in the hippocampus. Physiology 25, 319–329.

Colgin, L.L., Denninger, T., Fyhn, M., Hafting, T., Bonnevie, T., Jensen, O., Moser, M.B., Moser, E.I., 2009. Frequency of gamma oscillations routes flow of information in the hippocampus. Nature 462, 353–357.

Crone, N.E., Sinai, A., Korzeniewska, A., 2006. High-frequency gamma oscillations and human brain mapping with electrocorticography. Prog. Brain Res. 159, 275–295.

Cunningham, M.O., Whittington, M.A., Bibbig, A., Roopun, A., LeBeau, F.E., Vogt, A., Monyer, H., Buhl, E.H., Traub, R.D., 2004. A role for fast rhythmic bursting neurons in cortical gamma oscillations in vitro. Proc. Natl. Acad. Sci. USA. 101, 7152–7157.

Data, S., O'Malley, M.W., Patterson, E.H., 2010. Calcium/calmodulin kinase II in the pedunculopontine tegmental nucleus modulates the initiation and maintenance of wakefulness. J. Neurosci. 31, 1700–17016.

Datta, S., 2002. Evidence that REM sleep is controlled by the activation of brain stem pedunculopontine tegmental kainite receptor. J. Neurophysiol. 87, 1790–1798.

Datta, S., Desarnaud, F., 2010. Protein kinase A in the pedunculopontine tegmental nucleus of rat contributes to regulation of rapid eye movement sleep. J. Neurosci. 30, 12263–12273.

Datta, S., Hobson, J.A., 1994. Neuronal activity in the caudolateral peribrachial pons: relationship to PGO waves and rapid eye movements. J. Neurophysiol. 71, 95–109.

Datta, S., Siwek, D.F., 1997. Excitation of the brain stem pedunculopontine tegmentum cholinergic cells induce wakefulness and REM sleep. J. Neurophysiol. 77, 2975–2988.

Datta, S., Siwek, D.F., 2002. Single cell activity patterns of pedunculopontine tegmentum neurons across the sleep–wake cycle in the freely moving rats. J. Neurosci. Res. 70, 79–82.

Datta, S., Siwek, D.F., Patterson, E.H., Cipolloni, P.B., 1998. Localization of pontine PGO wave generation sites and their anatomical projections in the rat. Synapse 30, 409–423.

Datta, S., Patterson, E.H., Spoley, E.E., 2001a. Excitation of pedunculopontine tegmental NMDA receptors induces wakefulness and cortical activation in the rat. J. Neurosci. Res. 66, 109–116.

Datta, S., Spoley, E.E., Patterson, E.H., 2001b. Microinjection of glutamate into the pedunculopontine tegmentum induces REM sleep and wakefulness in the rat. Am. J. Physiol. Regul. Integr. Comp. Physiol. 280, R752–R759.

Datta, S., Mavanji, V., Ulloor, J., Patterson, E.H., 2004. Activation of phasic pontine-wave generator prevents rapid eye movement sleep deprivation-induced learning impairment in the rat: a mechanism for sleep-dependent plasticity. J. Neurosci. 24, 1416–1427.

Datta, S., Siwek, D.F., Stack, E.C., 2009. Identification of cholinergic and non-cholinergic neurons in the pons expressing phosphorylated cyclic adenosine monophosphate response element-binding protein as a function of rapid eye movement sleep. Neuroscience 163, 397–414.

Dement, W.C., 1967. Studies on the effects of REM deprivation in humans and animals. Res. Publ. Assoc. Res. Nerv. Ment. Dis. 43, 456–467.

Desarnaud, F., Macone, B.W., Datta, S., 2011. Activation of extracellular signal-regulated kinase signaling in the pedunculopontine tegmental cells is involved in the maintenance of sleep in rats. J. Neurochem. 116, 577–587.

Deschenes, M., Bourassa, J., Doan, V.D., Parent, A., 1996a. A single-cell study of the axonal projections arising from the posterior intralaminar thalamic nuclei in the rat. Eur. J. Neurosci. 8, 329–343.

Deschenes, M., Bourassa, J., Parent, A., 1996b. Striatal and cortical projections of single neurons from the central lateral thalamic nucleus in the rat. Neuroscience 72, 679–687.

Eckhorn, R., Bauer, R., Jordan, W., Brosch, M., Kruse, W., Munk, M., Reitboeck, H.J., 1988. Coherent oscillations: a mechanism of feature linking in the visual system? Biol. Cybern. 60, 121–130.

Engel, A.K., Konig, P., Gray, C.M., Singer, W., 1990. Stimulus-dependent neuronal oscillations in cat visual cortex: inter-columnar interaction as determined by cross-correlation analysis. Eur. J. Neurosci. 2, 588–606.

Engel, A.K., Moll, C.K., Fried, I., Ojemann, G.A., 2006. Invasive recordings from the human brain: clinical insights and beyond. Nat. Rev. Neurosci. 6, 35–47.

Erro, E., Lanciego, J.L., Gimenez-Amaya, J.M., 1999. Relationships between thalamostriatal neurons and pedunculopontine projections to the thalamus: a neuroanatomical tract-tracing study in the rat. Exp. Brain Res. 127, 162–170.

Garcia-Rill, E., Heister, D.S., Ye, M., Charlesworth, A., Hayar, A., 2007. Electrical coupling: novel mechanism for sleep–wake control. Sleep 30, 1405–1414.

Garcia-Rill, E., Charlesworth, A., Heister, D.A., Ye, M., Hayar, A., 2008. The developmental decrease in REM sleep: the role of transmitters and electrical coupling. Sleep 31, 1–18.

Garcia-Rill, E., Kezunovic, N., Hyde, J., Beck, P., Urbano, F.J., 2013. Coherence and frequency in the reticular activating system (RAS). Sleep Med. Rev. 17, 227–238.

Garcia-Rill, E., Kezunovic, N., D'Onofrio, S., Luster, B., Hyde, J., Bisagno, V., Urbano, F.J., 2014. Gamma band activity in the RAS- intracellular mechanisms. Exp. Brain Res. 232, 1509–1522.

Gibson, J.R., Beierlein, M., Connors, B.W., 1999. Two networks of electrically coupled inhibitory neurons in neocortex. Nature 402, 75–79.

Gray, C.M., Singer, W., 1989. Stimulus-specific neuronal oscillations in orientation columns of cat visual cortex. Proc. Natl. Acad. Sci. USA 86, 1698–1702.

Heister, D.S., Hayar, A., Charlesworth, A., Yates, C., Zhou, Y., Garcia-Rill, E., 2007. Evidence for electrical coupling in the SubCoeruleus (SubC) nucleus. J. Neurophysiol. 97, 3142–3147.

Herlitze, S., Garcia, D.E., Mackie, K., Hille, B., Scheuer, T., Catterall, W.A., 1996. Modulation of Ca2+ channels by G-protein beta gamma subunits. Nature 380, 258–262.

Hillman, D., Chen, S., Aung, T.T., Cherksey, B., Sugimori, M., Llinas, R.R., 1991. Localization of P-type calcium channels in the central nervous system. Proc. Natl. Acad. Sci. USA 88, 7076–7080.

Hou, X., Xiang, J., Wang, Y., Kirtman, E.G., Kotecha, R., Fujiwara, H., Hamsilpin, N., Rose, D.F., Degrauw, T., 2010. Gamma oscillations in the primary motor cortex studied with MEG. Brain Dev. 32, 619–624.

Hyde, J., Kezunovic, N., Urbano, F.J., Garcia-Rill, E., 2013a. Spatiotemporal properties of high speed calcium oscillations in the pedunculopontine nucleus. J. Appl. Physiol. 115, 1402–1414.

Hyde, J., Kezunovic, N., Urbano, F.J., Garcia-Rill, E., 2013b. Visualization of fast calcium oscillations in the parafascicular nucleus. Pflugers Arch. 465, 1327–1340.

Ikeda, S.R., 1996. Voltage-dependent modulation of N-type calcium channels by G-protein beta gamma subunits. Nature 380, 255–258.

Ito, S., Craig, A.D., 2005. Vagal-evoked activity in the parafascicular nucleus of the primate thalamus. J. Neurophysiol. 94, 2976–2982.

Jahnsen, H., Llinas, R.R., 1984a. Electrophysiological properties of guinea-pig thalamic neurones: an in vitro study. J. Physiol. 349, 205–226.

Jahnsen, H., Llinas, R.R., 1984b. Ionic basis for the electro-responsiveness and oscillatory properties of guinea-pig thalamic neurones in vitro. J. Physiol. 349, 227–247.

Jenkinson, N., Kuhn, A.A., Brown, P., 2013. Gamma oscillations in the human basal ganglia. Exp. Neurol. 245, 72–76.

Jones, E.G., 2007. Calcium channels in higher-level brain function. Proc. Natl. Acad. Sci. USA 14, 17903–17904.

Kammermeier, P.J., Ruiz-Velasco, V., Ikeda, S.R., 2000. A voltage-independent calcium current inhibitory pathway activated by muscarinic agonists in rat sympathetic neurons requires both Gα q/11 and Gβγ. J. Neurosci. 20, 5623–5629.

Kamondi, A., Williams, J., Hutcheon, B., Reiner, P., 1992. Membrane properties of mesopontine cholinergic neurons studied with the whole-cell patch-clamp technique: implications for behavioral state control. J. Neurophysiol. 68, 1359–1372.

Karlsson, K.A., Gall, A.J., Mohns, E.J., Seelke, A.M., Blumberg, M.S., 2005. The neural substrates of infant sleep in rats. PLoS Biol. 3 (e143), 891–901.

Katz, B., Miledi, R., 1965. The effect of calcium on acetylcholine release from motor nerve terminals. Proc. R. Soc. London, Ser. B 161, 483–495.

Kempf, F., Kuhn, A.A., Kupsch, A., Brucke, C., Weise, L., Schneider, G.H., Brown, P., 2007. Premovement activities in the subthalamic area of patients with Parkinson's disease and their dependence on task. Eur. J. Neurosci. 25, 3137–3145.

Kezunovic, N., Simon, C., Hyde, J., Smith, K., Beck, P., Odle, A., Garcia-Rill, E., 2010. Arousal from slices to humans: translational studies on sleep–wake control. Transl. Neurosci. 1, 2–8.

Kezunovic, N., Urbano, F.J., Simon, C., Hyde, J., Smith, K., Garcia-Rill, E., 2011. Mechanism behind gamma band activity in the pedunculopontine nucleus (PPN). Eur. J. Neurosci. 34, 404–415.

Kezunovic, N., Hyde, J., Simon, C., Urbano, F.J., Garcia-Rill, E., 2012. Gamma band activity in the developing parafascicular nucleus (Pf). J. Neurophysiol. 107, 772–784.

Kezunovic, N., Hyde, J., Goitia, B., Bisagno, V., Urbano, F.J., Garcia-Rill, E., 2013. Muscarinic modulation of high frequency activity in the pedunculopontine nucleus (PPN). Front. Neurol. Sleep Chronobiol. 4, 176 (1-13).

Kha, H.T., Finkelstein, D.I., Pow, D.V., Lawrence, A.J., Horne, M.K., 2000. Study of projections from the entopeduncular nucleus to the thalamus of the rat. J. Comp. Neurol. 426, 366–377.

Kim, C., Jeon, D., Young-Hoon, K., Lee, J.C., Kim, H., Hee-Sup, S., 2009. Deletion of N-type Ca2+ channel Cav2.2 results in hyperaggressive behaviors in mice. J. Biol. Chem. 284, 2738–2745.

Kobayashi, S., Nakamura, Y., 2003. Synaptic organization of the rat parafascicular nucleus, with special reference to its afferents from the superior colliculus and the pedunculopontine tegmental nucleus. Brain Res. 980, 80–91.

Lacey, C.J., Bolam, J.P., Magill, P.J., 2007. Novel and distinct operational principles of intralaminar thalamic neurons and their striatal projections. J. Neurosci. 27, 4374–4384.

Lalo, E., Thobois, S., Sharott, A., Polo, G., Mertens, P., Pogosyan, A., 2008. Patterns of bidirectional communication between cortex and basal ganglia during movement in patients with Parkinson disease. J. Neurosci. 28, 3008–3016.

Lang, E.J., Sugihara, I., Llinás, R.R., 2006. Olivocerebellar modulation of motor cortex ability to generate vibrissal movements in rat. J. Physiol. 571, 101–120.

Leonard, C.S., Llinas, R.R., 1990. Electrophysiology of mammalian pedunculopontine and laterodorsal tegmental neurons in vitro: implications for the control of REM sleep. In: Steriade, M., Biesold, D. (Eds.), Brain Cholinergic Systems. Oxford Science, Oxford, pp. 205–223.

Leonard, C.S., Llinas, R., 1994. Serotonergic and cholinergic inhibition of mesopontine cholinergic neurons controlling REM sleep: an in vitro electrophysiological study. Neurosci. 59, 309–330.

Libet, B., Kosslyn, S.M., 2004. Mind Time: The Temporal Factor in Consciousness. Harvard Press, Cambridge, MA.

Litvak, V., Eusebio, A., Jha, A., Oosterveld, R., Barnes, G., Foltynie, T., 2012. Movement-related changes in local and long-range synchronization in Parkinson's disease revealed by simultaneous magnetoencephalography and intracranial recordings. J. Neurosci. 32, 10541–10553.

Liu, X., Wang, S., Yianni, J., Nandi, D., Bain, P.G., Gregory, R., Stein, J.F., Aziz, T.Z., 2008. The sensory and motor representation of synchronized oscillations in the globus pallidus in patients with primary dystonia. Brain 131, 1562–1573.

Llinas, R.R., 1988. The intrinsic electrophysiological properties of mammalian neurons: insights into central nervous system function. Science 242, 1654–1664.

Llinas, R.R., Hess, R., 1976. Tetrodotoxin-resistant dendritic spikes in avian Purkinje cells. Proc. Natl. Acad. Sci. USA 73, 2520–2523.

Llinas, R. R., Jahnsen, H., 1982. Electrophysiology of mammalian thalamic neurones in vitro. Nature 297, 406–408.

Llinas, R.R., Steriade, M., 2006. Bursting of thalamic neurons and states of vigilance. J. Neurophysiol. 95, 3297–3308.

Llinás, R.R., Paré, D., 1991. Of dreaming and wakefulness. Neuroscience 44, 521–535.

Llinas, R.R., Grace, A.A., Yarom, Y., 1991. In vitro neurons in mammalian cortical layer 4 exhibit intrinsic oscillatory activity in the 10- to 50-Hz frequency range. Proc. Natl. Acad. Sci. USA 88, 897–901.

Llinas, R.R., Ribary, U., Jeanmonod, D., Kronberg, E., Mitra, P.P., 1999. Thalamocortical dysrhythmia: a neurological and neuropsychiatric syndrome characterized by magnetoencephalography. Proc. Natl. Acad. Sci. USA 96, 15222–15227.

Llinas, R.R., Leznik, E., Urbano, F.J., 2002. Temporal binding via cortical coincidence detection of specific and nonspecific thalamocortical inputs: a voltage-dependent dye-imaging study in mouse brain slices. Proc. Natl. Acad. Sci. USA 99, 449–454.

Llinas, R.R., Soonwook, C., Urbano, F.J., Hee-Sup, S., 2007. γ-Band deficiency and abnormal thalamocortical activity in P/Q-type channel mutant mice. Proc. Natl. Acad. Sci. U. S. A. 104, 17819–17824.

Lu, J., Sherman, D., Devor, M., Saper, C.B., 2006. A putative flip-flop switch for control of REM sleep. Nature 441, 589–594.

Mamelak, A.N., Hobson, J.A., 1989. Dream bizarreness as the cognitive correlate of altered neuronal brain in REM sleep. J. Cogn. Neurosci. 1, 201–222.

Marks, G.A., Farber, J., Roffwarg, H.P., 1980. Metencephalic localization of ponto–geniculo–occipital waves in the albino rat. Exp. Neurol. 69, 667–677.

Mavanji, V., Ulloor, J., Saha, S., Datta, S., 2004. Neurotoxic lesions of phasic pontine-wave generator cells impair retention of 2-way active avoidance memory. Sleep 27, 1282–1292.

McCormick, D.A., Pape, H.C., 1990. Properties of a hyperpolarization-activated cation current and its role in rhythmic oscillation in thalamic relay neurons. J. Physiol. 431, 291–318.

McGivern, J.G., 2007. Ziconotide: a review of its pharmacology and use in the treatment of pain. Neuropsychiatr. Dis. Treat. 3, 69–85.

Mehring, C., Nawrot, M.P., de Oliveira, S.C., Vaadia, E., Schulze-Binhage, A., Aertsen, A., Ball, T., 2004. Comparing information about arm movement direction in single channels of local and epicortical field potentials from monkey and human motor cortex. J. Physiol. 98, 498–506.

Middleton, S.J., Racca, C., Cunningham, M.O., Traub, R.D., Monyer, H., Knöpfel, T., Schofield, I.S., Jenkins, A., Whittington, M.A., 2008. High-frequency network oscillations in cerebellar cortex. Neuron 58, 763–774.

Miltner, W.H., Braun, C., Arnold, M., Witte, H., Taub, E., 1999. Coherence of gamma-band EEG activity as a basis for associative learning. Nature 397, 434–436.

Minamimoto, T., Kimura, M., 2002. Participation of the thalamic CM-Pf complex in attentional orienting. J. Neurophysiol. 87, 3090–3101.

Mitani, A., Ito, K., Hallanger, A.E., Wainer, B.H., Kataoka, K., McCarley, R.W., 1988. Cholinergic projections from the laterodorsal and pedunculopontine tegmental nuclei to the pontine gigantocellular tegmental field in the cat. Brain Res. 451, 397–402.

Mitler, M.M., Dement, W.C., 1974. Cataplectic-like behavior in cats after micro-injections of carbachol in pontine reticular formation. Brain Res. 68, 335–343.

Morgan, P.T., Pace-Schott, E., Pittman, B., Stickgold, R., Malison, R.T., 2010. Normalizing effect of modafinil on sleep in chronic cocaine users. Am. J. Psychiatry 167, 331–340.

Moruzzi, G., Magoun, H.W., 1949. Brainstem reticular formation and activation. Electroencephalogr. Clin. Neurophysiol. 1, 455–473.

Mouret, J., Delorme, F., Jouvet, M., 1967. Lesions of the pontine tegmentum and sleep in rats. C. R. Seances Soc. Biol. Fil. 161, 1603–1606.

Murakami, S., Okada, Y., 2006. Contributions of principal neocortical neurons to magnetoencephalography signals. J. Physiol. 575 (3), 925–936.

Nakagawasaki, O., Onogi, H., Mitazaki, S., Sato, A., Watanabe, K., Saito, H., Murai, S., Nakaya, K., Murakami, M., Takahashi, E., Tan-No, K., Tadano, T., 2010. Behavioral and neurochemical characterization of mice deficient in the N-type Ca2+ channel alpha1B subunit. Behav. Brain Res. 208, 224–230.

Palva, S., Monto, S., Palva, J.M., 2009. Graph properties of synchronized cortical networks during visual working memory maintenance. Neuroimage 49, 3257–3268.

Panckeri, K.A., Schotland, H.M., Pack, A.I., Hendricks, J.C., 1996. Modafinil decreases hypersomnolence in the English bulldog, a natural animal model of sleep-disordered breathing. Sleep 19, 626–631.

Pedroarena, C., Llinás, R.R., 1997. Dendritic calcium conductances generate high-frequency oscillation in thalamocortical neurons. Proc. Natl. Acad. Sci. USA 94, 724–728.

Phelan, K.D., Mahler, H.R., Deere, T., Cross, C.B., Good, C., Garcia-Rill, E., 2005. Postnatal maturational properties of rat parafascicular thalamic neurons recorded in vitro. Thalamus Relat. Syst. 1, 1–25.

Philips, S., Takeda, Y., 2009. Greater frontal-parietal synchrony at low gamma-band frequencies for inefficient then efficient visual search in human EEG. Int. J. Psychophysiol. 73, 350–354.

Povysheva, N.V., Zaitsev, A.V., Rotaru, D.C., Gonzalez-Burgos, G., Lewis, D.A., Krimer, L.S., 2008. Parvalbumin-positive basket interneurons in monkey and rat prefrontal cortex. J. Neurophysiol. 100, 2348–2360.

Raeva, S.N., 2006. The role of the parafascicular complex (CM-Pf) of the human thalamus in the neuronal mechanisms of selective attention. Neurosci. Behav. Physiol. 36, 287–295.

Ribary, U., 2005. Dynamics of thalamo-cortical network oscillations and human perception. Prog. Brain Res. 150, 127–142.

Sakai, K., El Mansari, M., Jouvet, M., 1990. Inhibition by carbachol microinjections of presumptive cholinergic PGO-on neurons in freely moving cats. Brain Res. 527, 213–223.

Sanford, L.D., Morrison, A.R., Mann, G.L., Harris, J.S., Yoo, L., Ross, R.J., 1994. Sleep patterning and behaviour in cats with pontine lesions creating REM without atonia. J. Sleep Res. 3, 233–240.

Sekar, K., Findley, W.M., Poeppel, D., Llinas, R.R., 2013. Cortical response tracking the conscious experience of threshold duration visual stimuli indicates visual perception is all or none. Proc. Natl. Acad. Sci. USA 110, 5642–5647.

Semba, K., Fibiger, H.C., 1992. Afferent connections of the laterodorsal and the pedunculopontine tegmental nuclei in the rat: a retro- and antero-grade transport and immunohistochemical study. J. Comp. Neurol. 323, 387–410.

Shen, W., Hernandez-Lopes, S., Tkatch, T., Held, E., Surmeier, J., 2004. Kv1.2-containing K+channels regulate subthreshold excitability of striatal medium spiny neurons. J. Neurophysiol. 91, 1337–1349.

Shiromani, P.J., Armstrong, D.M., Gillin, J.C., 1988. Cholinergic neurons from the dorsolateral pons project to the medial pons: a WGA-HRP and choline acetyltransferase immunohistochemical study. Neurosci. Lett. 95, 19–23.

Simon, C., Kezunovic, N., Ye, M., Hyde, J., Hayar, A., Williams, D.K., Garcia-Rill, E., 2010. Gamma band unit and population responses in the pedunculopontine nucleus. J. Neurophysiol. 104, 463–474.

Simon, C., Hayar, A., Garcia-Rill, E., 2011a. Responses of developing pedunculopontine neurons to glutamate receptor agonists. J. Neurophysiol. 105, 1918–1931.

Simon, C., Kezunovic, N., Williams, D.K., Urbano, F.J., Garcia-Rill, E., 2011b. Cholinergic and glutamatergic agonists induce gamma frequency activity in dorsal subcoeruleus nucleus neurons. Am. J. Physiol. Cell Physiol. 301, C327–C335.

Singer, W., 1993. Synchronization of cortical activity and its putative role in information processing and learning. Annu. Rev. Physiol. 55, 349–374.

Sirota, A., Montgomery, S., Fujisawa, S., Isomura, Y., Zugaro, M., Buzsáki, G., 2008. Entrainment of neocortical neurons and gamma oscillations by the hippocampal theta rhythm. Neuron 60, 683–697.

Soteropoulos, D.S., Baker, S.N., 2006. Cortico-cerebellar coherence during a precision grip task in the monkey. J. Neurophysiol. 95, 1194–1206.

Steriade, M., 1999. Cellular substrates of oscillations in corticothalamic systems during states of vigilance. In: Lydic, R., Baghdoyan, H.A. (Eds.), Handbook of Behavioral State Control. Cellular and Molecular Mechanisms. CRC Press, New York, pp. 327–347.

Steriade, M., Llinás, R.R., 1988. The functional states of the thalamus and the associated neuronal interplay. Physiol. Rev. 68, 649–742.

Steriade, M., McCarley, R.W., 1990. Brainstem Control of Wakefulness and Sleep. Plenum Press, New York, NY.

Steriade, M., Paré, D., Datta, S., Oakson, G., Curro Dossi, R., 1990. Different cellular types in mesopontine cholinergic nuclei related to ponto–geniculo–occipital waves. J. Neurosci. 10, 2560–2579.

Steriade, M., Curro Dossi, R., Pare, D., Oakson, G., 1991. Fast oscillations (20–40 Hz) in thalamocortical systems and their potentiation by mesopontine cholinergic nuclei in the cat. Proc. Natl. Acad. Sci. USA 88, 4396–4400.

Steriade, M., Curro-Dossi, R., Contreras, D., 1993. Electrophysiological properties of intralaminar thalamocortical cells discharging rhythmic (~40Hz) spike-bursts at ~1000Hz during waking and rapid eye movement sleep. Neuroscience 56, 1–9.

Takakusaki, K., Kitai, S.T., 1997. Ionic mechanisms involved in the spontaneous firing of tegmental pedunculopontine nucleus neurons of the rat. Neuroscience 78, 771–794.

Timofeev, I., Steriade, M., 1997. Fast (mainly 30–100Hz) oscillations in the cat cerebellothalamic pathway and their synchronization with cortical potentials. J. Physiol. 504, 153–168.

Trottenberg, T., Fogelson, N., Kuhn, A.A., Kivi, A., Kupsch, A., Schneider, G.H., Brown, P., 2006. Subthalamic gamma activity in patients with Parkinson's disease. Exp. Neurol. 200, 56–65.

Uchitel, O.D., Protti, D.A., Sanchez, V., Cherkesey, B.D., Sugimori, M., Llinas, R.R., 1992. P-type voltage-dependent calcium channel mediates presynaptic calcium influx and transmitter release in mammalian synapses. Proc. Natl. Acad. Sci. USA 89, 3330–3333.

Urbano, F.J., Leznik, E., Llinas, R.R., 2007. Modafinil enhances thalamocortical activity by increasing neuronal electrotonic coupling. Proc. Natl. Acad. Sci. USA 104, 12554–12559.

Urbano, F.J., Kezunovic, N., Hyde, J., Simon, C., Beck, P., Garcia-Rill, E., 2012. Gamma band activity in the reticular activating system (RAS). Front. Neurol. Sleep Chronobiol. 3 (6), 1–16.

Vanderwolf, C.H., 2000a. What is the significance of gamma wave activity in the pyriform cortex? Brain Res. 877, 125–133.

Vanderwolf, C.H., 2000b. Are neocortical gamma waves related to consciousness? Brain Res. 855, 217–224.

Vanni-Mercier, G., Sakai, K., Lin, J.S., Jouvet, M., 1989. Mapping of cholinoceptive brainstem structures responsible for the generation of paradoxical sleep in the cat. Arch. Ital. Biol. 127, 133–164.

Voss, U., Holzmann, R., Tuin, I., Hobson, J.A., 2009. Lucid dreaming: a state of consciousness with features of both waking and non-lucid dreaming. Sleep 32, 1191–1200.

Waldert, S., Preissl, H., Demandt, E., Braun, C., Birbaumer, N., Aertsen, A., Mehring, C., 2008. Hand movement direction decoded from MEG and EEG. J. Neurosci. 28, 1000–1008.

Wang, H.L., Morales, M., 2009. Pedunculopontine and laterodorsal tegmental nuclei contain distinct populations of cholinergic, glutamatergic and GABAergic neurons in the rat. Eur. J. Neurosci. 29, 340–358.

Watson, R.T., Heilman, K.M., Miller, B.D., 1974. Neglect after mesencephalic reticular formation lesions. Neurology 24, 294–298.

Westenbroek, R.E., Hell, J.W., Warner, C., Dubel, S.J., Snutch, T.P., Catterall, W.A., 1992. Biochemical properties and subcellular distribution of an N-type calcium channel alpha 1 subunit. Neuron 9, 1099–1115.

Whittington, M.A., Traub, R.D., Jefferys, J.G., 1995. Synchronized oscillations in interneuron networks driven by metabotropic glutamate receptor activation. Nature 373, 612–615.

Yamamoto, K., Mamelak, A.N., Quattrochi, J.J., Hobson, J.A., 1990. A cholinoceptive desynchronized sleep induction zone in the anterodorsal pontine tegmentum: spontaneous and drug-induced neuronal activity. Neuroscience 39, 295–304.

Ye, M., Hayar, A., Strotman, B., Garcia-Rill, E., 2010. Cholinergic modulation of fast inhibitory and excitatory transmission to pedunculopontine thalamic projecting neurons. J. Neurophysiol. 103, 2417–2432.

Zhou, L., Gall, D., Qu, Y., Prigogine, C., Cheron, G., Tissir, F., Schiffmann, S.N., Goffinet, A.M., 2010. Maturation of "neocortex isole" in vivo in mice. J. Neurosci. 30, 7928–7939.

10

Preconscious Awareness

Edgar Garcia-Rill, PhD

Center for Translational Neuroscience, Department of Neurobiology
and Developmental Sciences, University of Arkansas for Medical Sciences,
Little Rock, AR, USA

THE ROLE OF GAMMA BAND ACTIVITY IN THE RAS

In Chapter 9, based on the fact that every neuron in each of the three major nuclei modulating waking and rapid eye movement (REM) sleep manifests gamma band activity, the reticular activating system (RAS) was referred to as a "gamma-making machine." We also learned that in the various regions of the CNS that exhibit gamma band activity, such as the cortex, thalamus, cerebellum, basal ganglia, hippocampus, and RAS, they all demonstrate coherence between regions. That is, these gamma band generators are not isolated but correlated, and in some cases, subcortical oscillations lead cortical oscillations. Based on the presence of electrical coupling, intrinsic membrane properties, and circuitry capable of generating and maintaining gamma band activity, we proposed a novel role for gamma band activity in the RAS. The usual role for gamma band activity in the cortex is that of sensory or motor binding. However, we hypothesized that it is the continued activation of the RAS during waking that allows the maintenance of the background of gamma activity necessary to support a state capable of reliably assessing the world around us on a continuous basis (Urbano et al., 2012; Garcia-Rill et al., 2013). Based on our results, we suggested that a similar mechanism to that in the cortex for achieving temporal coherence at high frequencies is present in the pedunculopontine nucleus (PPN) and its subcortical targets such as the Pf and SubCD nuclei (Garcia-Rill et al., 2013). We suggested that gamma band activity and electrical coupling generated in the PPN may help stabilize coherence related to arousal and provide a stable activation state. We identified not only the intrinsic membrane properties but also the intracellular mechanisms that allow this generation and maintenance for prolonged periods (see Chapter 9). Our overall hypothesis is that sensory input will induce gamma band activity in the RAS that participates in preconscious awareness. William James proposed that the "stream of consciousness" is "a river flowing forever through a man's conscious waking hours" (James, 1890/2007). The stream of consciousness flowing into our brains supports our conscious waking hours, but we do not pay attention to that stream; rather, we are preconsciously aware of the information; and only when we pay attention to a particular event are we fully "conscious" of it. Most of the information is available so that we are aware of it "preconsciously." The RAS seems the ideal site for preconscious awareness since it provides a continuous flow of internal and external information, is phylogenetically conserved, and modulates wake–sleep cycles, the startle response, and fight-or-flight responses that include changes in muscle tone and locomotion.

Sigmund Freud identified three different parts of the mind based on our level of awareness, the "conscious" mind, the "preconscious" mind, and the "subconscious" mind. Freud suggested that the "subconscious" mind has a will and purpose of its own that cannot be known to the conscious mind and also that it is a storage depot for socially unacceptable ideas, traumatic memories, desires, and painful emotions put out of mind by the mechanism of psychological repression. As we will see below, there is little evidence that the "subconscious" state actually exists, yet the term is used widely. The "preconscious" mind includes those things of which we are aware, but we are not paying attention to them. If we choose to pay attention, we bring them to the "conscious" mind (Civin and Lombardi, 1990). We speculate that it is the activation of the RAS during waking that induces coherent activity (through electrically coupled cells) and high-frequency oscillations (through P/Q-type calcium channel and subthreshold oscillation activity) and leads to the maintenance (through activation of G proteins) of the background of gamma activity necessary to support a state capable of reliably assessing the world around us on a continuous basis. That is, these mechanisms underlie the process of preconscious awareness in the RAS.

LEVELS OF AWARENESS

Slow-wave sleep (SWS)—Thus, we propose that there are two states of awareness, "conscious" and "preconscious," the latter being mediated by the RAS. One question that arises is, is there a "subconscious" process? Given the discussion in Chapter 8, in which the 10 Hz frequency fulcrum is at the transition between waking and sleep, and that both "conscious" and "preconscious" processes are mediated by higher-frequency oscillations, is there a "subconscious" process present during the low frequencies observed during SWS? As discussed in earlier chapters, as the electroencephalogram (EEG) becomes more high amplitude and low frequency, coactivation of adjacent cortical columns produces little lateral inhibition and therefore little perception of specific sensory events. As more and more global low-frequency synchronization occurs, less and less discrimination is likely. Therefore, if there is a "subconscious" state that occurs during SWS, it has little chance of producing perception, learning, or memory. Instead, it is likely that SWS represents an "unconscious" state, rather than a "subconscious" state. The ultimate slow-wave state in which numerous columns are synchronized is epilepsy, which is characterized by a lack of consciousness and of memory of the event.

REM sleep—The other state during sleep is REM sleep. A number of authors have suggested that REM sleep plays a role in memory formation or consolidation. One hypothesis suggests that dreams delete unimportant memories through modulation of synapses, inhibition of sensory information, and/or suppression of movement (Jouvet, 1999; Crick and Mitchison, 1983). Others have suggested a role for REM sleep in the consolidation of memories (Skaggs and McNaughton, 1996) and that cognitive activity experienced during daily learning is "replayed" to promote long-term storage (Stickgold et al., 2001; Wilson and McNaughton, 1994). Alternative theories suggest a role for REM sleep in experience-dependent changes (Frank et al., 2001; Gais et al., 2002; Maquet et al., 2000). This suggestion has been questioned because of (a) the absence of "replays" in REM sleep *after* the behavioral episode involving the novel task, (b) the absence of dream reports linked to recent experiences, (c) the lack of effect of REM sleep deprivation on consolidation, and (d) the lack of effects on consolidation by monoamine oxidase inhibitors that are known to block REM sleep (Maquet, 2001; Siegel, 2001).

Another critical concern is that blood flow is increased during REM sleep in the brain stem and limbic areas but decreased in the frontal cortex (Maquet et al., 1966). These findings suggest that metabolism decreases during SWS and increases during REM sleep in the brain stem, but the frontal cortex remains depressed in both SWS and REM sleep. This "hypofrontality" is possibly why there is a lack of critical judgment during dreaming, which occurs during REM sleep. Critical judgment is thought to be a function of the frontal lobes. The decreased blood flow to the frontal lobes essentially deafferents the frontal cortex and may explain why dream content is accepted at face value. That is, when we dream, we never question the surreal conditions such as flying overhead or finding ourselves in outrageous circumstances. We never seem to say, "Time out, let's have a realistic dream!" It is difficult to imagine how this state would promote the consolidation of accurate memories.

Another discrepancy is the role of REM sleep during development that was described in Chapter 5. Even if the memory consolidation hypothesis has some merit, it would mean that the role of REM sleep is different after, compared to before, birth. That is, there is a dissociation between the prenatal function and the postnatal function of REM sleep. The prenatal period marks the most abundant age exhibiting REM sleep. What memories does the fetus consolidate? We have few to no memories of our infancy. As described in Chapters 8 and 9, one theory of conscious perception suggests that thalamocortical projections from specific thalamic nuclei provide the "content" of sensory experience, while thalamocortical projections from nonspecific (ILT) thalamic nuclei provide the "context," and their coincidence leads to binding or conscious perception (Llinas et al., 1994). Since there is no eye or ear opening before birth, it would seem that the "content" of memories to be consolidated by REM sleep would be highly reduced. Therefore, why is there a need for so much REM sleep? This would suggest that the majority of the activity to be consolidated in the fetus would be related to "context." In summary, a role for REM sleep in memory consolidation may not apply prenatally, even if it is a function in the adult.

A final concern is that, during REM sleep, descending signals induce the atonia of REM sleep. This is a decrease in muscle tone, particularly in extensor muscles, caused by descending inhibition of motoneurons. The only motoneurons that are not actively inhibited are eye movement-related motoneurons, thus the name "REM sleep." That is, only our eyes act out our dreams. The brain paralyzes the remaining muscles because it does not want us to act out our dreams, especially given their weak link to reality. Simply from a heuristic viewpoint, REM sleep would seem a state that the brain obviously requires, but given its hypofrontality and atonia, do we really want to consolidate important memories during such a state?

While the interpretation of dreams during REM sleep has been deemed important, or at least informative, we believe that it is likely that dreams are simply an epiphenomenon of REM sleep. That is, dreams occur during a necessary state, but the body is prevented from acting out their content, and for good reason. In the disorder REM sleep behavior disorder, the patient acts out dream content, with serious and dangerous repercussions (see Chapter 7).

Transition states—During the state of lucid dreaming, one is aware that one is dreaming and in some cases may even influence dream content. While there are opinions that it is a state of sleep, others believe it represents the beginnings of wakefulness. EEG studies tell us that the transitions between states such as waking and SWS, and between SWS and REM sleep, are stepwise. For example, as long as 1–2min before the beginning of REM sleep with atonia, ponto–geniculo–occipital (PGO) waves are present in the EEG during SWS, and they increase in frequency the closer the time is to REM sleep (Steriade, 1999). This suggests that the transition from one state to another takes time, but the transitions appear to shift from one stable state to the next until full consciousness is reached (Hudson et al., 2014). Just as the 10Hz fulcrum is a "resting state" at the transition between sleep and frank waking, and increased synchronization of slower frequencies leads to SWS, while increased synchronization of faster frequencies leads to higher states of arousal, these transitions take time before the full state is recruited. It is likely that lucid dreaming is a transition state toward waking, in which the stepwise shift to full consciousness has not been reached. Aside from gradual EEG changes, what happens to blood flow when we awaken?

WAKING AND PRECONSCIOUS AWARENESS

Functionally, the process of preconscious awareness would need to be fairly continuous during waking in order to provide the sensory foundation for planned behavior. Importantly, this "stream of preconsciousness" would need to begin upon waking. As mentioned previously, it has been shown that increases in blood flow in the thalamus and brain stem begin within 5 min of waking but as much as 15 min elapse before significant changes in frontal cortex blood flow are observed (Balkin et al., 2002). This is more surprising when considering that waking follows the last REM sleep episode of the night, during which frontal cortex blood flow is low (Garcia-Rill, 2010). The sudden onset of waking after the last REM sleep episode of the night does not instantly increase frontal lobe blood flow. Thus, upon waking, significant increases in the brain stem and thalamus precede restoration of blood flow to the frontal lobes. The importance of subcortical structures in the determination of states of awareness is being growingly emphasized. Damasio proposed that the brain stem is critical to the formulation of the self, which is critical to the formulation of feelings (Damasio, 2010). The fact that we awaken immediately as ourselves, despite low levels of frontal cortical blood flow, supports the view that subcortical structures are essential to the process of the formulation of the self. Wilder Penfield is a famous American neurosurgeon who became Canadian and did his most important work at the Montreal Neurological Institute in the 1940s and 1950s. He created maps of the cortical sensory and motor homunculus during surgery on patients with epilepsy. He extensively described the responses to cerebral cortex stimulation in a large number of patients. Penfield arrived at the conclusion that there was a mechanism for consciousness that was subcortical. He stated, "There is no place in the cerebral cortex where electrical stimulation will cause a patient to believe or to decide" (Penfield, 1975). In addition, he emphasized that, while cortical seizures localized to specific cortical regions elicit sensory or motor effects but maintain consciousness, *petit mal* seizures in "mesothalamic" (the midbrain and thalamus) regions always eliminate consciousness. Based on these results, Penfield proposed the presence of a "centrencephalic integrating system" that fulfills the role of sensorimotor integration necessary for consciousness.

Cortical function has been proposed to take place through reentrant signaling or reverberating oscillations (Olafspons et al., 1989). We assume that a similar process of cycling is present within brain stem–thalamus interactions, given the reciprocal nature of projections between the PPN and the intralaminar thalamus and between the PPN and the SubCD. These oscillations can be relayed trough the intralaminar thalamus to the cortex and basal ganglia and through the SubCD to the hippocampus and posterior brain stem. That is, RAS oscillations are probably coherent with those in other cortical and subcortical systems. Much additional research is needed to substantiate this speculation, which is proposed only as a starting point for discussion of the nature of the potential process of preconscious awareness by the RAS. From a pathological point of view to be discussed in Chapters 11 and 12, a similar process to thalamocortical dysrhythmia (TCD) (to be described in those chapters) could occur at subcortical levels, in which the timing of brain stem–thalamus oscillations would be disturbed. The P50 potential is dysregulated in a number of neurological and psychiatric disorders (Garcia-Rill and Skinner, 2001; Skinner et al., 2004). The deficits produced by such dysregulation could be severe, perhaps resulting in the release of automatic behaviors such as fixed action patterns and tics, in addition to arousal and sensory gating deficits. That is, the deficits could arise from a lack of cortical regulation of these automatisms, resulting in involuntary movements, unregulated arousal, and exaggerated fight-or-flight responses. There would also be an impact on voluntary movements.

VOLITION AND FREE WILL

A common definition of consciousness is awareness, such that if there is no awareness, there is no consciousness (Searle, 1998). This is fundamentally intertwined with the concept of free will. When choosing to make a movement, the implication is that there is a will that decides to engage the motor system to then induce the movement. The problem is that the brain initiates a movement before there is conscious awareness of volition (Hallett, 2007). That is, there is a process that takes place before we have the subjective "conscious" awareness that we intend to act. That process likely involves preconscious awareness. Preconscious awareness amounts to all that information that is available and influences our actions, but we do not consider it as determining our actions. In other words, we often act in order to meet desired goals and feel that conscious will is the cause of our behavior. The results of considerable scientific research suggest that this may not be the case. Under some conditions, actions are initiated even though we are "unconscious" of the goal. But, are we really "unconscious" of the goal, or are we just "preconscious"?

Pioneering studies by Libet first showed that when people consciously set a goal to engage in a behavior, their conscious will to act begins "unconsciously" (Libet et al., 1983). These authors studied the readiness potential (RP), a negative DC shift present long before the execution of a voluntary movement. The RP was originally described by Kornhuber as a negative DC shift that began 600–800 ms before a voluntary, uncued movement (Kornhuber and Deecke, 1965). The RP was present at maximal amplitude at the vertex in the region of the supplementary motor cortex and precentral cortex (Deecke et al., 1976; Kornhuber and Deecke, 1965). That is, the RP is manifested not only over the same region as the midlatency auditory-evoked P50 potential but also in the same regions as the *mu* rhythm. As described in Chapter 6, the P50 potential is an arousal-related waveform that is generated by PPN outputs to the intralaminar thalamus and the cortex in response to an auditory stimulus. We will see in Chapter 12 that the RP is not only decreased in Parkinson's disease (PD), but stimulation of the PPN in PD patients implanted for deep brain stimulation (DBS) manifests the most significant changes in blood flow in the same region, that is, at the vertex. It would be interesting to determine if the RP represents preparatory activity in the RAS and intralaminar thalamus and if it is subserved by the *mu* rhythm. That is, as mentioned above, the RP is localized in the same region as the "resting-state" *mu* rhythm described in Chapter 8. Parallel studies showed that intense acoustic stimuli sufficient to elicit a startle response can release patterned ballistic movements (Valls-Sole et al., 1995). Further work showed that an intense sensory stimulus, such as would be expected to activate the RAS and PPN, can induce early latency, unintentional planned movements (McKinnon et al., 2013). In Chapter 7, we saw how the PPN modulates the startle response. That is, the localization of the RP, the auditory P50 potential and the magnetic M50 response, the blood flow changes resulting from PPN DBS, the modulation of the startle responses, and the resting-state *mu* rhythm are all colocalized. It is highly unlikely that these disparate functions all exist in the same region but are mediated by different cells. The more parsimonious explanation is that the same process is present during the resting state, during the preparation for movement, and during the induction of the movement by sensory, especially supramaximal, stimuli. This strongly suggests that it would be worth determining if indeed the RP represents preparatory, arousal-related activity in the RAS and intralaminar thalamus.

Regardless of the confluence of these measures, the implication we draw is that the same processes that underlie preconscious awareness may be released when voluntary movements are executed, and this process occurs before the conscious control of movement. Let us consider again that preconscious awareness is the information that is available and influences our actions, but we do not consider it as determining our actions. The interpretation of the Libet studies becomes markedly different by changing the term "subconscious" to "preconscious." The studies of Libet employed the RP, which is known to have an early component that precedes the movement by as much as 1–2 s and a late component that precedes the movement by 400 ms (Shibasaki and Hallett, 2006). Libet's subjects were asked to move voluntarily and were also asked to subjectively time the moment at which they felt the "will" to move as well as the onset of the actual movement. The early and late phases of the RP preceded the "consciously" determined will to move by hundreds of milliseconds. These authors concluded that cerebral initiation of spontaneous, freely voluntary acts can begin "unconsciously," before there is any subjective "conscious" awareness that a decision to act was initiated cerebrally. Even simple movements appear to be generated "subconsciously," and the "conscious" sense of volition comes later (Hallett, 2007). Hallett described the details of studies showing that voluntary movements can be triggered with stimuli that are not "consciously" perceived, that movement may well occur prior to the apparent planning of the movement, and that not only the sense of the movement having occurred but also the sense of willing the movement happens before the actual movement (Hallett, 2007). Libet suggested that voluntary acts begin "unconsciously," before there is subjective "conscious" awareness that a decision to act was initiated by the brain. This conclusion has been extrapolated to suggest that there is no free will; however, Libet suggested that, although the movement was indeed initiated "subconsciously," it was subject to veto once it reached consciousness (Libet, 1999). This has been regarded as unsatisfactory and not answering the question of whether there is free will. The question is complex because so many factors influence the sense of volition such as the perception of time, the conditions under which the movement is executed, and the perception of volition (Hallett, 2007). We propose an alternative view simply by concluding that it is the interpretation of the results that assumes that the process preceding the movement is "unconscious." There is no evidence that this is the case. The preparation for movement generates brain processes that are clearly related to the intent, but especially occurs during waking, and therefore should be labeled "preconscious," not "subconscious." The replacement of the word "preconscious" for the word "subconscious" significantly alters the conclusions of these studies. That is, the conclusion should have been: "voluntary acts begin preconsciously, before there is subjective conscious awareness that a decision to act was initiated by the brain."

Figure 10.1 is a representation of an RP and the timing reported by subjects performing an uncued voluntary movement. The estimation of the sense of will (signified by the "W") or intent to move occurred well after the beginning of the RP, and the sensation of movement (signified by the "M") occurred even later and well after the beginning of the RP as in the Libet

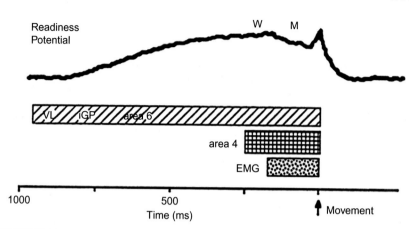

FIGURE 10.1 **The RP, will, and volition.** Tracing of the vertex-recorded RP showing the locations of the subjective timing of the will (W) and perception of movement (M) in subjects asked to perform a voluntary movement and denote the timing of these events. Note that the RP usually begins hundreds of milliseconds before the subjective feeling of intention or movement. The rectangles below the RP are approximate latencies of changes in activity recorded in animals (cats and monkeys) performing an uncued handle movement. Activity in the VL, internal pallidum (iGP), and anterior motor cortex (area 6) begins firing very early during the RP, while activity in the hand area (area 4) begins just before the EMG of the arm muscles. That is, several subcortical circuits are involved in drafting a voluntary movement long before the primary motor cortex.

study. Included in this figure are data from animal studies trained to perform uncued, voluntary movements showing that single-cell activity in various cortical and subcortical areas began well in advance of the movement. Recordings in the ventrolateral thalamic nucleus, which receives input from the cerebellum and basal ganglia; the internal pallidum, which is a major basal ganglia output; and the premotor cortex (cortical area 6) all showed activity correlated with the beginning of the RP. Neurons in the pallidum and thalamus showed onsets of firing over 500 ms before the movement, as did premotor/axial cortex cells, but cells in the forelimb (precentral) cortex were active immediately before the onset of electromyographic activity (Neafsey et al., 1978). Therefore, while there is no "prime mover," that is, a single region that triggers the RP, certain specific circuits involving the basal ganglia, thalamus, and cortex may all conspire to build up the RP.

Our conclusion of the Libet studies is that the subjects performing the task had plenty of information available "preconsciously" that helped determine the subsequent voluntary action. We question the use of the word "subconscious"; in fact, as discussed above, we have little convincing evidence that there is a "subconscious" state. What most people refer to as "subconscious" may actually be "preconscious" information available that helps determine our actions. Under this umbrella, the results of Libet are not surprising. In other words, the idea that the background information is available if we wish to pay attention to it, yet participates in the formulation of voluntary movements and free will, eliminates the argument that the Libet results suggest the absence of free will. The question is not "do we have free will?"; it should be "how much of our free will is preconscious?"

The RAS is also involved in the preparation for movements that are triggered by external stimuli. For example, the delivery of a startling stimulus immediately before the performance of a reaction time task will decrease reaction time (Castellote et al., 2013). These studies show that the RAS influences the execution of self-generated voluntary movements as well as preprogrammed reaction time tasks.

MECHANISMS FOR PRECONSCIOUS AWARENESS

So how does preconscious awareness arise? An attractive model of conscious perception is based on the presence of gamma band activity in thalamocortical regions. During activated states (waking and paradoxical sleep), EEG responses are characterized by low-amplitude, high-frequency oscillatory activity in the gamma band range (~30–90 Hz). Gamma frequency oscillations have been proposed to participate in sensory perception, problem solving, and memory, and it has been suggested that such coherent events occur at cortical and/or thalamocortical levels. The mechanisms behind such activity, as described in Chapter 9, include the presence of inhibitory GABAergic cortical interneurons that exhibit intrinsic oscillatory activity in the gamma band frequency, many of which are electrically coupled, of neurons with sodium-dependent subthreshold oscillations, as well as of fast rhythmic bursting pyramidal neurons that are also electrically coupled. That is, the rhythm is generated initially and maintained, not by the circuit alone, but by the combination of a specific circuit, electrical coupling-induced coherence, and the high-frequency intrinsic membrane properties of cortical neurons.

Cognition has been proposed to arise from "specific" thalamocortical projections that carry the *content* of conscious experience, interacting with "nonspecific" thalamocortical projections that carry the *context* of conscious events (Llinas et al., 1999). The cortical sites peaking at gamma band frequency via "specific" thalamocortical projections are thought to reverberate and summate with coincident "nonspecific" gamma band activity (Linas and Pare, 1991; Llinas et al., 2002). This summation, along with the coherence provided by electrical coupling, is proposed to provide global binding. Binding is the mechanism whereby the different aspects of sensation, say, color, motion, and shape, are combined into a unified perceptual experience. Disturbance in this mechanism results in "TCD" and is thought to be involved in a number of neurological and psychiatric disorders (Llinas et al., 1999).

Is there a mechanism, perhaps based on electrical coupling and intrinsic oscillatory activity, that generates a similar process but at *subcortical* levels? As we saw above, gamma band activity has been reported in the thalamus, basal ganglia, hippocampus, and cerebellum. Now, three nuclei of the RAS exhibit electrical coupling, providing a novel mechanism for wake–sleep control based on coherence driven by electrical coupling (Garcia-Rill et al., 2007). Moreover, virtually every neuron in these nuclei, regardless of cell or transmitter type, exhibited gamma band activity generated by intrinsic membrane properties. Regardless of depolarizing level or input, these cells plateau at gamma band frequency (30–60 Hz) (Kezunovic et al., 2011, 2012; Simon et al., 2010). This is a very unique property. Taken together, these results suggest that a similar mechanism to that in the cortex for achieving temporal coherence at high frequencies is present in the PPN and in its subcortical targets such as the Pf and SubCD nuclei. We suggested that gamma band activity and electrical coupling generated in the PPN may help stabilize coherence related to arousal, providing a stable activation state during waking and paradoxical sleep that participates in preconscious awareness (Garcia-Rill et al., 2013; Urbano et al., 2012).

EVERYDAY PRECONSCIOUS AWARENESS

On a daily basis, we awaken to receive a host of afferent sensory information, which is superimposed on a level of excitability that is generated by simply waking up. The background or "idle" 10 Hz activity upon waking is overlaid with afferent volleys generated by visual, auditory, vestibular, proprioceptive, cutaneous, and other sensory and motor information. Stimuli strong enough to reach arousal threshold are typically of higher amplitude, while sensory thresholds for merely perceiving the stimuli are typically of lower amplitude. For example, an auditory stimulus in the order of only 15 dB elicits a response in the primary auditory cortex, but the P50 potential arousal-related response requires an auditory stimulus in the order of 70 dB for the RAS to be activated (Garcia-Rill and Skinner, 2001). That is, while low-amplitude sensory events may not demand "attention" or break into consciousness, they nevertheless produce brain responses. This multitude of afferent information forms a continuous background of activity, basically the stream of information we require for preconscious awareness, but we are not entirely conscious of it as long as we do not direct attention to specific events.

Under these circumstances, firing patterns in an awake individual who is walking and talking should be quite high and very variable.

EEG recordings, however, are never performed under such conditions, mainly from fear of movement artifacts. Human subjects are typically recorded in a sitting or lying position, with little sensory activation aside from that delivered by specific stimuli if the protocol calls for such. It is a wonder that human subjects manifest gamma band activity given such a sedentary, understimulated state. Animal studies are similar. Why do so many studies, especially in rats, report fairly low rates of neuronal activity in RAS neurons? In contrast, investigators using the same strains of rats report gamma frequency activity in corticostriatal networks (Berke, 2009); in cortical (Yu et al., 2012) and striatal (Howe et al., 2011) networks during learning; in the hippocampus during waking (Montgomery et al., 2008), exploration (Jacobson et al., 2013), place-seeking (Senior et al., 2008), learning (Bauer et al., 2007), and memory tasks (Montgomery and Buzsaki, 2007); and in the cerebellum during simple behaviors in the rat (Cao et al., 2012) and precision grip studies in monkeys (Soteropoulos and Baker, 2007). The main difference is that most studies on RAS cells (a) record from single neurons and not field or population activity and (b) record only during "quiet waking." This means that the animal is understimulated and almost asleep. There is little stimulation; certainly, only a minimal "stream of consciousness" is passing through that brain, that is, there is little to activate the RAS. Under these conditions, it would indeed be surprising if RAS neurons fired at rates much higher than at "idle" frequencies ~10 Hz. However, RAS neurons, in particular PPN cells, have been studied for their roles in behavior, which can inform on participation in complex actions.

THE PPN AND BEHAVIOR

Lesions of the PPN in the cat were found to reduce REM sleep events as well as PGO waves and REMs (Shouse and Siegel, 1992). Significantly, lesions of the laterodorsal tegmental nucleus failed to change sleep parameters, again stressing that the function of this medial cholinergic partner of the PPN may not involve direct wake-sleep control. Other studies using PPN lesions were performed in rats and were confirmed to affect sleep architecture, mainly increasing SWS and fractionating REM sleep episodes, especially disturbing transitions between SWS, REM sleep, and waking, but without major effects on waking duration (Hernandez-Chan et al., 2011; Petrovic et al., 2013). Behavioral studies showed that excitotoxic lesions of the PPN impaired the acquisition of several learning tasks including spatial navigation in the Morris water maze with a submerged platform (Dellu et al., 1991), one-trial passive avoidance and two-way shuttle box active avoidance (Fujimoto et al., 1992), externally cued reinforced bar pressing (Florio et al., 1999), delayed spatial win-shift task in an eight-arm radial maze (Keating and Winn, 2002), and spatial delayed matching to position and nonmatching to position in a T-maze (Leri and Franklin, 1998; Satorra-Marin et al., 2005). In general, then, waking and REM sleep are disturbed but not eliminated, and complex interactions ensue when unilateral vs. bilateral lesions are executed. This is not surprising when lesioning a homeostatic system with such global function as coordinating exteroceptive and internal information and modulating responses to environmental conditions. Most studies have employed behavioral tasks that require narrow responsiveness or ask limited questions about sleep architecture or behavioral reactions, but few assess the types of activity that would entail preattentional functions and sensory gating. The reader is directed to studies on the P50 potential in humans and P13 potential in the rat that measure sensory gating or habituation to repetitive stimulation. This process is impaired in a number of neurological and psychiatric disorders to be discussed in Chapters 11 and 12. It is worth remembering that injections of GABA or GABA agonists into the PPN reduced and eliminated the vertex-recorded P13 potential (Miyazato et al., 1999). This sleep state-dependent waveform is a manifestation of PPN outputs through the intralaminar thalamus to the cortex.

Studies using excitotoxic lesions of the PPN in the rat found that there were no motivational deficits or lack of spontaneous activity. However, lesioned animals retained incentive information for reward learned before the lesion, but after the lesion manifested disruption in the formation of stimulus-reward associations (Inglis et al., 1994, 2000). The authors concluded that there was a deficit in appetitive learning that could include attentional problems and sensory gating deficits. Another excitotoxic lesion study showed that PPN lesions disrupted the acquisition of copulation and the rewarding consequences of copulation in experienced animals (Kippin and van der Kooy, 2003). These studies also concluded that the association between stimulus and reward was impaired. This is the sort of function that would be impaired by damage to substrates of preconscious awareness; however, tasks more directly dedicated to such a mechanism need to be designed and implemented in PPN-lesioned animals. Such tasks should investigate the effects on PPN function in the processing of interfering or distracting stimuli such as are used in prepulse inhibition paradigms or masked stimuli in priming tasks.

In a visually guided saccade task, PPN neurons in the monkey brain were found to respond to cues that predicted reward or the actual delivery of the reward, and the amplitude of the activity was related to reward magnitude (Okada et al., 2009, 2011). The idea that PPN activity is related to actions (responses to cues) and outcomes (rewards) has been explored in a number of studies using lesions. As mentioned above, lesions of the PPN in rats impaired learning a radial maze task (Keating and Winn, 2002), or to lever press for amphetamine reward, but not if the animals learned the reward association to lever press prior to the lesion (Alderson et al., 2004). The same lab reported that such lesions impaired learning by slowing the rate of acquisition and the rate of altering behavior to changes in reinforcement (Wilson et al., 2009). Another study suggested that pharmacological inactivation of the PPN impaired sustained attention, but not motor performance, in the omissions committed in a sustained attention task (Rostron et al., 2008). These findings were extended in recent experiments showing that PPN lesions were essential for updating associations between action and outcome (MacLaren et al., 2013). That is, the main deficit induced by PPN inactivation was the inability to update the causal relationship between the requirements of performing actions in order to obtain an outcome. In order to perform such action-outcome learning, the PPN must be able to receive a multitude of sensory inputs and select one out of many competing actions. Therefore, it must be able to analyze sensory events (in keeping with its role as part of the ascending RAS, Chapter 6) and produce a rapid response (as part of the descending RAS, Chapter 7).

These results imply that the PPN circuit has the ability to be aware of events, objects, or sensory patterns without necessarily attending to particular events and that it can sense and act or react to a condition or event. This circuit appears to integrate sensations from the environment with desired goals and to guide behavior in a goal-oriented manner, such as in volition. In other words, these findings imply that the PPN is highly likely, at least in part, responsible for preconscious awareness.

CLINICAL IMPLICATIONS

Given the foregoing, the number of potential functions affected by PPN dysregulation is numerous and involves multiple functions from perception to response to goal-oriented behavior to voluntary movements to just about anything that we are aware of but to which we are not attending. Under these circumstances, a large number of neurological and psychiatric disorders will certainly include PPN dysregulation, and, since the PPN is part of the RAS, these disorders will also suffer from wake–sleep problems. In fact, as we will see in Chapters 11 and 12, wake–sleep dysregulation may presage the onset of a number of disorders by many years. By the same token, normalization of wake–sleep regulation will likely presage normalization of function following successful treatment in some of these disorders. Since this is the main highway to our sensory information, and the first level of processing of our reactions to environmental conditions, it should not be surprising that the PPN is involved in so many functions and disorders.

References

Alderson, H.L., Latimer, M.P., Blaha, C.D., Phillips, A.G., Winn, P., 2004. An examination of d-amphetamine self-administration in pedunculopontine tegmental nucleus-lesioned rats. Neuroscience 125, 349–358.

Balkin, T.J., Braun, A.R., Wesensten, N.J., Jeffries, K., Varga, M., Baldwin, P., Belenky, G., Herscovitch, P., 2002. The process of awakening: a PET study of regional brain activity patterns mediating the re-establishment of alertness and consciousness. Brain 125, 2308–2319.

Bauer, E.P., Paz, R., Pare, D., 2007. Gamma oscillations coordinate amygdalo-rhinal interactions during learning. J. Neurosci. 27, 9369–9379.

Berke, J.D., 2009. Fast oscillations in cortico-striatal networks switch frequency following rewarding events and stimulant drugs. Eur. J. Neurosci. 30, 848–859.

Cao, Y., Maran, S.K., Dhamala, M., Jaeger, D., Heck, D.H., 2012. Behavior related pauses in simple spike activity of mouse Purkinje cells are linked to spike rate modulation. J. Neurosci. 32, 8678–8685.

Castellote, J.M., Van den Berg, M.E.L., Valls-Sole, J., 2013. The StartReact effect on self-initiated movements. BioMed. Res. Int. 2013, 471792.

Civin, M., Lombardi, K.L., 1990. The preconscious and potential space. Psychoanal. Rev. 77, 573–585.

Crick, F., Mitchison, G., 1983. The function of dream sleep. Nature 304, 111–114.

Damasio, A., 2010. Self Comes to Mind: Constructing the Conscious Brain. Pantheon Books, New York, NY.

Deecke, L., Grozinger, B., Kornhuber, H.H., 1976. Voluntary finger movement in man: cerebral potentials and theory. Biol. Cybern. 23, 99–119.

Dellu, F., Mayo, W., Cherkaoui, J., Le Moal, M., Simon, H., 1991. Learning disturbances following excitotoxic lesion of cholinergic pedunculo-pontine nucleus in the rat. Brain Res. 544, 126–132.

Florio, T., Capozzo, A., Puglielli, E., Pupillo, R., Pizzuti, G., Scarnati, E., 1999. The function of the pedunculopontine nucleus in the preparation and execution of an externally-cued bar pressing task in the rat. Behav. Brain Res. 104, 95–104.

Frank, M.G., Issa, N.P., Stryker, M.P., 2001. Sleep enhances plasticity in the developing visual cortex. Neuron 30, 275–287.

Fujimoto, K., Ikeguchi, K., Yoshida, M., 1992. Impaired acquisition, preserved retention and retrieval of avoidance behavior after destruction of pedunculopontine nucleus areas in the rat. Neurosci. Res. 13, 43–51.

Gais, S., Molle, M., Helms, K., Born, J., 2002. Learning-dependent increases in sleep spindle density. J. Neurosci. 22, 6830–6834.

Garcia-Rill, E., 2010. Reticular activating system. In: Stickgold, R., Walker, M. (Eds.), The Neuroscience of Sleep. Academic Press, London, UK, pp. 133–138.

Garcia-Rill, E., Skinner, R.D., 2001. The sleep state-dependent P50 midlatency auditory evoked potential. In: Lee-Chiong, T.L., Carskadon, M.A., Sateia, R. (Eds.), Sleep Medicine. M.J. Hanley & Belfus, Philadelphia, PA, pp. 697–704.

Garcia-Rill, E., Heister, D.S., Ye, M., Charlesworth, A., Hayar, A., 2007. Electrical coupling: novel mechanism for sleep–wake control. Sleep 30, 1405–1414.

Garcia-Rill, E., Kezunovic, N., Hyde, J., Beck, P., Urbano, F.J., 2013. Coherence and frequency in the reticular activating system (RAS). Sleep Med. Rev. 17, 227–238.

Hallett, M., 2007. Volitional control of movement: the physiology of free will. Clin. Neurophysiol. 118, 1179–1192.

Hernandez-Chan, N.G., Gongora-Alfaro, J.L., Alvarez-Cervera, F.J., Solis-Rodriguez, F.A., Heredia-Lopez, F.J., Arankowsky-Sandoval, G., 2011. Quinolinic acid lesions of the pedunculopontine nucleus impair architecture, but not locomotion, exploration, emotionality or working memory in the rat. Behav. Brain Res. 225, 482–490.

Howe, M.W., Atallah, H.E., McCool, A., Gibson, D.J., Graybiel, A.M., 2011. Habit learning is associated with major shifts in frequencies of oscillatory activity and synchronized spike firing in striatum. Proc. Natl. Acad. Sci. USA 108, 16801–16806.

Hudson, A.E., Calderon, D.P., Pfaff, D.W., Proekt, A., 2014. Recovery of consciousness is mediated by a network of discrete metastable activity states. Proc. Natl. Acad. Sci. USA 111 (25), 9283–9288.

Inglis, W.L., Dunbar, J.S., Winn, P., 1994. Outflow from the nucleus accumbens to the pedunculopontine tegmental nucleus: a dissociation between locomotor activity and the acquisition of responding for conditioned reinforcement stimulated by d-amphetamine. Neurosci. 62, 51–64.

Inglis, W.L., Olmstead, M.C., Robbins, T.W., 2000. Pedunculopontine tegmental nucleus lesions impair stimulus–reward learning in autoshaping and conditioned reinforcement paradigms. Behav. Neurosci. 114, 285–294.

Jacobson, T.K., Howe, M.D., Schmidt, B., Hinman, J.R., Escabi, M.A., Markus, E.J., 2013. Hippocampal theta, gamma, and theta-gamma coupling: effects of aging, environmental change, and cholinergic innervation. J. Neurophysiol. 109, 1852–1865.

James, W., 1890/2007. The Principles of Psychology. Cosimo Classics, New York (Original work published 1890).

Jouvet, M., 1999. The Paradox of Sleep: The Story of Dreaming. MIT Press, Cambridge, MA.

Keating, G.L., Winn, P., 2002. Examination of the role of the pedunculopontine tegmental nucleus in radial maze tasks with or without a delay. Neuroscience 112, 687–696.

Kezunovic, N., Urbano, F.J., Simon, C., Hyde, J., Smith, K., Garcia-Rill, E., 2011. Mechanism behind gamma band activity in the pedunculopontine nucleus (PPN). Eur. J. Neurosci. 34, 404–415.

Kezunovic, N., Hyde, J.R., Simon, C., Urbano, F.J., Willimas, D.K., Garcia-Rill, E., 2012. Gamma band activity in the developing parafascicular nucleus (Pf). J. Neurophysiol. 107, 772–784.

Kippin, T.E., van der Kooy, D., 2003. Excitotoxic lesions of the tegmental pedunculopontine nucleus impair copulation in naive male rats and block the rewarding effects of copulation in experienced male rats. Eur. J. Neurosci. 18, 2581–2591.

Kornhuber, H.H., Deecke, L., 1965. Hirnpotentiala underungen bei Willkurbewegungen und passiven bewegungen des menschen: bereitschaftspotential und reafferente potential. Pflugers Arch. 284, 1–17.

Leri, F., Franklin, K.B., 1998. Learning impairments caused by lesions to the pedunculopontine tegmental nucleus: an artifact of anxiety? Brain Res. 807, 187–192.

Libet, B., 1999. Do we have free will? J. Conscious. Stud. 9, 47–57.

Libet, B., Gleason, C.A., Wright, E.W., Pearl, D.K., 1983. Time of conscious intention to act in relation to onset of cerebral activity (readiness-potential). The unconscious initiation of a freely voluntary act. Brain 106, 623–642.

Linas, R.R., Pare, D., 1991. Of dreaming and wakefulness. Neuroscience 44, 521–535.

Llinas, R.R., Ribary, U., Joliot, M., Wang, X.J., 1994. Content and context in temporal thalamocortical binding. In: Buzsaki, G., Christen, Y. (Eds.), Temporal Coding in the Brain. Springer-Verlag, Berlin, pp. 251–272.

Llinas, R.R., Ribary, U., Jeanmonod, D., Kronberg, E., Mitra, P.P., 1999. Thalamocortical dysrhythmia: a neurological and neuropsychiatric syndrome characterized by magnetoencephalography. Proc. Natl. Acad. Sci. USA 96, 15222–15227.

Llinas, R.R., Leznik, E., Urbano, F.J., 2002. Temporal binding via cortical coincidence detection of specific and nonspecific thalamocortical inputs: a voltage-dependent dye-imaging study in mouse brain slices. Proc. Natl. Acad. Sci. USA 99, 449–454.

MacLaren, D.A.A., Wilson, D.I.G., Winn, P., 2013. Updating action-outcome associations is prevented by inactivation of the posterior pedunculopontine tegmental nucleus. Neurobiol. Learn. Mem. 102, 28–33.

Maquet, P., 2001. The role of sleep in learning and memory. Science 294, 1048–1052.

Maquet, P., Peters, J.M., Aerts, J., Delfiore, G., Degueldre, C., Luxen, A., Franck, G., 1966. Functional neuroanatomy of human rapid-eye-movement sleep and dreaming. Nature 383, 163–166.

Maquet, P., Laureys, S., Peigneux, P., Fuchs, S., Petiau, C., Phillips, C., Aerts, J., Del Fiore, G., Degueldre, C., Meulemans, T., Luxen, A., Franck, G., Van Der Linden, M., Smith, C., Cleeremans, A., 2000. Experience-dependent changes in cerebral activation during human REM sleep. Nat. Neurosci. 3, 831–836.

McKinnon, C.D., Allen, D.P., Shiratori, T., Rogers, M.W., 2013. Early and unintentional release of planned motor actions during motor cortical preparation. PLoS One 8, e63417.

Miyazato, H., Skinner, R.D., Garcia-Rill, E., 1999. Neurochemical modulation of the P13 mid-latency auditory evoked potential in the rat. Neuroscience 92, 911–920.

Montgomery, S.M., Buzsaki, G., 2007. Gamma oscillations dynamically couple hippocampal CA3 and CA1 regions during memory task performance. Proc. Natl. Acad. Sci. USA 104, 14495–14500.

Montgomery, S.M., Sirota, A., Buzsaki, G., 2008. Theta and gamma coordination of hippocampal networks during waking and REM sleep. J. Neurosci. 28, 6731–6741.

Neafsey, E.J., Hull, C.D., Buchwald, N.A., 1978. Preparation for movement in the cat. II. Unit activity in the basal ganglia and thalamus. Electroencephalogr. Clin. Neurophysiol. 44, 714–723.

Okada, K., Toyama, K., Inoue, Y., Isa, T., Kobayashi, Y., 2009. Different pedunculopontine tegmental neurons signal predicted and actual task rewards. J. Neurosci. 29, 4858–4870.

Okada, K., Nakamura, K., Kobayashi, Y., 2011. A neural correlate of predicted and actual reward-value information in monkey pedunculopontine tegmental and dorsal raphe nucleus during saccade tasks. Neural Plast. 2011, ID 570840.

Olafspons, J., Gally, G., Reeke, J., Edelman, G., 1989. Reentrant signaling among simulated neuronal groups leads to coherency in their oscillatory activity. Proc. Natl. Acad. Sci. USA 86, 7265–7269.

Penfield, W., 1975. The Mystery of the Mind. Princeton University Press, Princeton, NJ.

Petrovic, J., Lazic, K., Ciric, J., Kalauzi, A., Saponjic, J., 2013. Topography of the sleep/wake states related EEG microstructure and transitions structure differentiates the functionally distinct cholinergic innervation disorders in rat. Behav. Brain Res. 256, 108–118.

Rostron, C.L., Farquhar, M.J., Latimer, M.P., Winn, P., 2008. The pedunculopontine tegmental nucleus and the nucleus basalis magnocellularis: do both have a role in sustained attention? BMC Neurosci. 9, 16.

Satorra-Marin, N., Homs-Ormo, S., Arévalo-García, R., Morgado-Bernal, I., Coll-Andreu, M., 2005. Effects of pre-training pedunculopontine tegmental nucleus lesions on delayed matching- and non-matching-to-position in a T-maze in rats. Behav. Brain Res. 160, 115–124.

Searle, J.R., 1998. How to study consciousness scientifically. Philos. Trans. R. Soc. Lond. B Biol. Sci. 353, 1935–1942.

Senior, T.J., Huxter, J.R., Allen, K., O'Neill, J., Csicsvari, J., 2008. Gamma oscillatory firing reveals distinct populations of pyramidal cells in the CA1 region of the hippocampus. J. Neurosci. 28, 2274–2286.

Shibasaki, H., Hallett, M., 2006. What is Bereitschaftspotential? Clin. Neurophysiol. 117, 2341–2356.

Shouse, M.N., Siegel, J., 1992. Pontine regulation of REM sleep components in cats: integrity of the pedunculopontine tegmentum (PPT) is important for phasic events but unnecessary for atonia during REM sleep. Brain Res. 571, 50–63.

Siegel, J.M., 2001. The REM sleep-memory consolidation hypothesis. Science 294, 1058–1063.

Simon, C., Kezunovic, N., Ye, M., Hyde, J., Hayar, A., Williams, D.K., Garcia-Rill, E., 2010. Gamma band unit and population responses in the pedunculopontine nucleus. J. Neurophysiol. 104, 463–474.

Skaggs, W.E., McNaughton, B.L., 1996. Replay of neuronal firing sequences in rat hippocampus during sleep following spatial experience. Science 271, 1870–1873.

Skinner, R.D., Miyazato, H., Garcia-Rill, E., 2004. Arousal mechanisms related to posture and movement. I. Ascending modulation. In: Mori, S., Stuart, D.G., Wiesendanger, M. (Eds.), Brain Mechanisms for the Integration of Posture and Movement. Prog. Brain Res. 143, 291–298.

Soteropoulos, D.S., Baker, S.N., 2007. Different contributions of the corpus callosum and cerebellum to motor coordination in monkey. J. Neurophysiol. 98, 2962–2973.

Steriade, M., 1999. Cellular substrates of oscillations in corticothalamic systems during states of vigilance. In: Lydic, R., Baghdoyan, H.A. (Eds.), Handbook of Behavioral State Control. Cellular and Molecular Mechanisms. CRC Press, New York, NY, pp. 327–347.

Stickgold, R., Hobson, J.A., Fosse, R., Fosse, M., 2001. Sleep, learning, and dreams: off-line memory reprocessing. Science 294, 1052–1057.

Urbano, F.J., Kezunovic, N., Hyde, J., Simon, C., Beck, P., Garcia-Rill, E., 2012. Gamma band activity in the reticular activating system (RAS). Front. Neurol. Sleep Chronobiol. 3 (6), 1–16.

Valls-Sole, J., Sole, A., Valldeoriola, F., Munoz, E., Gonzalez, L.E., Tolosa, E.S., 1995. Reaction time and acoustic startle in normal human subjects. Neurosci. Lett. 195, 97–100.

Wilson, M.A., McNaughton, B.L., 1994. Reactivation of hippocampal ensemble memories during sleep. Science 265, 676–679.

Wilson, D.I., MacLaren, D.A.A., Winn, P., 2009. Bar pressing for food: differential consequences of lesions to the anterior versus posterior pedunculopontine. Eur. J. Neurosci. 30, 504–513.

Yu, C., Fan, D., Lopez, A., Yin, H.H., 2012. Dynamic changes in single unit activity and γ oscillations in a thalamocortical circuit during rapid instrumental learning. PLoS One 7, e50578.

Psychiatric Disorders and the RAS

Stasia D'Onofrio, BS*, *Erick Messias, MD, PhD†,
and Edgar Garcia-Rill, PhD**

*Center for Translational Neuroscience, Department of Neurobiology and
Developmental Sciences, University of Arkansas for Medical Sciences,
Little Rock, AR, USA
†Department of Psychiatry, University of Arkansas for Medical Sciences,
Little Rock, AR, USA

AROUSAL, WAKING, AND PSYCHIATRIC DISORDERS

Given the information outlined in the previous chapters, what aspects of reticular activating system (RAS) function will be involved in psychiatric disorders? (1) We know that the RAS participates in fight-or-flight responses; therefore, we would expect that responses to sudden alerting stimuli will be abnormal. For disorders in which the RAS is overactive, this would mean that such stimuli will produce exaggerated responses that would be manifested as *exaggerated startle responses or hyperactive reflexes* such as the blink reflex (see Chapter 7). (2) Another property of the RAS is its rapid habituation to repetitive stimuli. This is reflected in its lack of responsiveness to rapidly repeating stimuli, that is, its habituation. This endows the RAS with its capacity for sensory gating, the property of decreasing responsiveness of repetitive events in favor of novel or different stimuli. For disorders in which this property is affected, we expect a *decrease in habituation or a sensory gating deficit* (see Chapter 6). (3) The RAS controls waking and sleep, so that sleep patterns would be dysregulated. If the RAS is downregulated by a disorder, we expect an inability to remain awake, the presence of *excessive daytime sleepiness*, and an excess of total sleep time, especially an increase in slow-wave sleep (SWS). If, on the other hand, the RAS is upregulated, we expect difficulty in getting to sleep and maintaining sleep. This would be reflected in decreased SWS, *insomnia*, and fragmented sleep during the night, as well as *increased rapid eye movement (REM) sleep drive*, which is characterized by nightmares and frequent awakenings (see Chapters 2 and 5). In Chapter 5, we considered what would happen if the developmental decrease in REM sleep drive did not occur. Such a condition would lead to increased REM sleep drive during sleep (resulting in intense dreaming) but perhaps also during waking, resulting in dreaming while awake or hallucinations, along with *hypervigilance* during waking. (4) The RAS also modulates the maintenance of waking, a property ignored by many but one that affects a host of cognitive functions. The inability to maintain a steady waking state, in the form of *maintained gamma band activity*, will interfere with attention, learning, and memory, to name a few processes (see Chapter 9).

Another factor in all of these disorders is the level of frontal lobe blood flow. Decreased frontal lobe blood flow, or *hypofrontality*, is present in all of these disorders to some extent. This state is normally present during REM sleep and may account for the lack of critical judgment present in our dreams; however, such a state during waking would lead to impulsive "knee-jerk" or reflexive reactions with lack of consideration of consequences. This condition is probably involved in the lack of habituation to repetitive stimuli, or sensory gating deficit. Under the condition of decreased cortical modulation, fight-or-flight responses and reflexes would be exaggerated. Whether hypofrontality is a cause of RAS dysregulation or RAS dysregulation leads to hypofrontality remains to be determined. However, successful treatment of any of these disorders will be marked by normalization of wake–sleep rhythms and reflexes, as well as frontal lobe blood flow.

Therefore, in the assessment of the role of the RAS in psychiatric disorders, we need to consider (a) responses to reflexes such as the startle response and the blink reflex using electromyographic (EMG), (b) sensory gating using paired stimuli or prepulse inhibition paradigms of sensory responses such as the P50 potential, (c) sleep patterns or somnography using nighttime electroencephalogram (EEG), (d) the maintenance or interruption of gamma band activity during waking using EEG or magnetoencephalography (MEG), and, finally, (e) the level of frontal lobe blood flow. These measures will allow us to determine in detail the role of the RAS in specific psychiatric disorders.

SCHIZOPHRENIA

Symptoms—Schizophrenia is a heterogeneous disorder marked by psychotic symptoms such as delusions and hallucinations, as well as attentional impairment, emotional withdrawal, apathy, and cognitive impairment (Andreasen and Flaum, 1991). More specifically, the positive symptoms (not present normally but experienced by most patients) include hallucinations, delusions, thought disorder, and agitation, while negative symptoms (present normally but less experienced by patients) include lack of affect, anhedonia, and withdrawal. Cognitive symptoms include poor executive function, lack of attention, and disturbed working memory. In addition, abnormal movements have been described. Equally heterogeneous mechanisms have been advanced to explain the disease, including cortical atrophy, catecholaminergic abnormalities, and early brain injury.

Etiology—The factors involved in schizophrenia include genetic, environmental, and developmental. Some stimulants, specifically amphetamine and cocaine, can induce some of the symptoms of schizophrenia. Since these agents increase dopaminergic drive and the use of dopamine (DA) receptor blockers have been found to alleviate some of the symptoms of schizophrenia, the "dopamine theory" of schizophrenia has some validity. In about 80% of patients, the disorder develops after puberty, between the ages of 15 and 25. It is not clear that some aspects of the disorder were not present well before puberty in severe cases, but due to the reluctance of placing the diagnostic label of schizophrenia on children, the diagnosis is not offered even with fairly clear positive and negative symptoms. On the other hand, some cases of schizophrenia are postpubertal without prior history of related symptomatology. The incidence of schizophrenia is 0.5–1.0% worldwide, with a male to female ratio of 1.4:1. In addition, schizophrenia is accompanied by comorbid depression (in 50% of patients), anxiety and posttraumatic stress disorder (PTSD) (in 30%), and obsessive compulsive disorder (OCD) (in 25%) (Buckley et al., 2009).

As far as genetic factors are concerned, there is an increased risk (~40%) of developing the disorder in monozygotic twins and an even greater risk if both parents are schizophrenic (Craddock and Owen, 2010; Picchioni and Murray, 2007). A number of different genes have been implicated in schizophrenia, many of which are also implicated in bipolar disorder and autism, yet there are enough differences to warrant separate diagnostic categories (Craddock and Owen, 2010). Because of our discovery of gamma band activity in the RAS being mediated by high-threshold, voltage-dependent calcium channels, we will concentrate on what is known regarding the genetics of calcium channels in schizophrenia (see Section "Bipolar Disorder").

As far as environmental factors are concerned, schizophrenia is more common in cities, stressful environments, and exposure to drug abuse (Picchioni and Murray, 2007). As far as developmental factors are concerned, abnormal fetal development or perinatal injury can increase risk, as well as developmental changes at puberty (Feinberg, 1969; Weinberger, 1982). Epidemiological studies in six countries have established that in about 20% of schizophrenics, the mother had an influenza attack in the second trimester (Mednick et al., 1988). Such a viral insult would induce increased levels of gamma interferon, a potent mitogen that can influence such agents as tumor necrosis factor and fibroblast growth factor (FGF). We identified FGF as the survival molecule for cholinergic pedunculopontine nucleus (PPN) neurons (Garcia-Rill et al., 1991). In cell culture studies, nerve growth factor did not influence the survival of PPN cells but maintained basal forebrain cholinergic cells, whereas PPN neurons were maintained by FGF but not by nerve growth factor. This is particularly important because schizophrenics have simplified whorls in their fingerprints (Bracha et al., 1991) and they show disturbed fibroblast metabolism (Hashimoto et al., 2003). This indicates a developmental dysregulation of FGF metabolism in at least some patients with schizophrenia. Interestingly, FGF is the survival molecule for subsets of cells in the same regions that have been implicated in schizophrenia, including the cortex (Akbarian et al., 1993), cerebellum (Nasrallah et al., 1991), hippocampus (Kovelma and Scheibel, 1984), and RAS (Garcia-Rill et al., 1991). That is, dysregulation of FGF metabolism during development may be responsible for the subsequent symptoms in at least some schizophrenic patients. Moreover, FGF receptors can act as portals to viral infection (Kaner et al., 1990), and exposure to viruses has been proposed in the etiology of schizophrenia (Yolken, 2004). Whether or not the involvement of FGF in schizophrenia is through developmental (FGF and calcium control axon growth in the brain) or genetic dysregulation and/or via viral infection remains to be determined.

EEG, reflexes, and P50 potential—Schizophrenia is characterized by abnormalities in wake–sleep control, including hypervigilance, decreased SWS especially deep sleep stages, increased REM sleep drive, and fragmented sleep (Caldwell and Domino, 1967; Feinberg et al., 1969; Itil et al., 1972; Jus et al., 1973; Zarcone et al., 1975). These wake–sleep abnormalities reflect increased vigilance and REM sleep drive, that is, overactive RAS output. The increased REM sleep drive has been proposed to account for REM sleep intrusion during waking, that is, in eliciting hallucinations (Dement, 1967; Mamelak and Hobson, 1989). This suggests that, normally, gamma band activity during waking is separated from gamma band activity during REM sleep. In fact, gamma band activity during waking may be modulated by the CaMKII pathway, and gamma band activity during REM sleep may be mediated by the cAMP/PKA pathway (Garcia-Rill et al., 2014). We hypothesize that these two pathways are activated individually and may, in fact, inhibit each other. However, in schizophrenia, the two pathways are activated simultaneously, thus leading to REM sleep intrusion during waking, that is, hallucinations.

The sleep disturbances are correlated better with negative rather than positive symptoms (Ganguli et al., 1987). Other studies emphasized the relationship between negative symptoms in schizophrenia and increased cholinergic output (Janowsky et al., 1979; Tandon et al., 1993). There are also postural and motor abnormalities (King, 1974; Manschrek, 1986), as well as eye movement dysregulation (Holzman et al., 1973; Karson et al., 1990). These findings are in keeping with the modulation of motor control by the RAS described in Chapters 7 and 10. In addition, these patients suffer from sensory gating deficits determined using habituation of the blink reflex (Geyer and Braff, 1982). One of the first studies using the P50 potential in clinical conditions showed that habituation of the P50 potential using a paired stimulus paradigm was decreased in schizophrenia (Freedman et al., 1983). These sensory gating deficits demonstrate a lack of inhibition of responses to repetitive stimuli, which can be manifested in exaggerated fight-or-flight responses upon sudden arousal. The P50 potential, as established in human and animal studies, is generated by the PPN and is manifested at the vertex (see Chapters 6 and 10).

In addition, aberrant gamma band activity and coherence during cognitive tasks or attentional load have been reported in schizophrenic patients (Uhlhaas and Singer, 2010). Several human studies demonstrated frequency-specific deficits in the coherence and maintenance of gamma oscillations in patients with schizophrenia (Spencer et al., 2003). These results suggest that the generation and maintenance of gamma band activity may be abnormal in schizophrenia. As described in Chapter 9, the RAS generates gamma band activity related to awareness. We discovered a potential mechanism for the decreases in gamma band activity and maintenance in schizophrenia that is described below (see Section "Bipolar Disorder"). In addition, schizophrenic patients suffer from hypofrontality, which contributes not only to the sensory gating deficits but also to the lack of critical judgment. This sign may also exacerbate the intensity of hallucinations, such that the patient is deluded into obeying the voices, with dangerous and unfortunate repercussions.

The results of EEG, reflex, and P50 potential testing all point to increased arousal and increased REM sleep drive. That is, the PPN is overactive in schizophrenia, but it is overactive in a specific manner. Responses to repetitive stimuli are increased and reflexes are exaggerated suggesting that phasic responses to brief stimuli are dysregulated. However, the decreased and interrupted gamma band activity also suggests that gamma oscillations are not properly maintained on a tonic basis. This combination of short-term hyperexcitability and long-term diminution of RAS activity is functionally devastating. This is in agreement with findings described above showing that anticholinergic agents appear to alleviate some of the negative symptoms of schizophrenia (Janowsky et al., 1979; Tandon et al., 1993).

Treatment—We should point out that the treatments for schizophrenia are only marginally effective, with as many as one-half of patients not responsive to even the latest-generation agents. Few patients, even those responding, ever regain a normal living and social life. Therefore, while we have done well to do away with padded rooms and lobotomy, modern therapies have a long way to come. Early treatment was directed at the positive symptoms, specifically psychosis. The first-generation antipsychotics included chlorpromazine and haloperidol, which showed DA D_2 receptor blockade (Snyder, 2006). A second-generation antipsychotic named clozapine was found to act on the same DA receptors in addition to muscarinic cholinergic and serotonergic receptors. In fact, clozapine was initially developed as an antimuscarinic cholinergic agent intended to balance the decrease in DA present in Parkinson's disease, that is, by decreasing cholinergic tone, the idea was to rebalance the striatum. This proved untenable due to its side effects of agranulocytosis (decreasing white blood cells) but became an effective antipsychotic that not only reduced negative symptoms but also affected positive ones. Figure 11.1 is a diagram of the sites of action at which clozapine can act in the RAS. While increased DA drive to the striatum is a major component of the disease, the substantia nigra (SN) is activated by the PPN, especially by muscarinic cholinergic input. Clozapine appears to partially (~40% penetrance) block muscarinic input to the SN as well as DA input to the striatum. In addition, clozapine acts as a partial serotonin reuptake receptor blocker, thereby increasing inhibition of the PPN, further downregulating the RN-PPN-SN-striatum pathway. Drug companies have attempted to eliminate the side effects of clozapine, but only olanzapine has retained antimuscarinic properties and is the most widely used third-generation antipsychotic.

A large number of other transmitter modulators are being tested without clear success. However, it is likely that more than one drug, along with intensive therapy, will probably become the treatment of choice in the future. Hopefully, the intensive research being performed on this disease will render better options sooner than later. One approach that seems to be lacking is the more extended use of physiological measures such as EEG to determine changes in wake–sleep patterns, startle response and

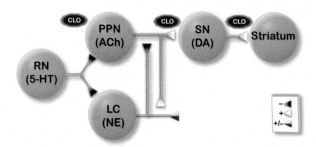

FIGURE 11.1   **Sites of action of clozapine.** As shown in Figure 4.1, the dorsal raphe nucleus (RN) generally inhibits the PPN and LC, while the LC inhibits the PPN and the PPN excites the LC. The muscarinic input to the SN is excitatory, and the DA input to the striatum is mixed. The symptoms of psychosis suggest that hypervigilance and increased REM sleep drive are mediated by overactive PPN output that overdrives SN output. The suggestion is that this pathway drives many of the negative symptoms of the disorder. Clozapine was designed as a muscarinic antagonist that also has partial penetrance at serotonin reuptake receptors (acting to increase inhibitory drive to the PPN, thus downregulating its activity). This drug also acts to decrease activation of the SN by PPN output and by decreasing DA input to the striatum from the SN.

P50 potential habituation, and frontal lobe blood flow. These quantitative assessments are rarely used in clinical trials, which themselves are hampered by differences in dosage and testing instruments.

Some investigators believe that nicotinic agonists may be a potential treatment for the symptoms of schizophrenia since these patients tend to chain smoke. The idea was advanced that this is tantamount to self-medication and nicotine was found to normalize a rodent model of sensory gating. However, the fact is that the P50 potential and the sleep dysregulation are not sensitive to nicotinic agents, but respond to muscarinic agents. In Chapter 13, we discuss the effects of tobacco on the RAS. Basically, smoking induces a powerful, short latency and short-duration anxiolytic response. It is this brief respite from the overwhelming intrusion wrought by decreased sensory gating that the patients are seeking. Unfortunately, tobacco then induces a longer latency anxiogenic response, so that, by the end of the cigarette, the smoker is prompted to light up again. This is a very addictive mechanism in smokers, which is even less resisted by patients. In fact, most of the disorders described that include hypofrontality will be less likely to kick the habit than normal individuals.

BIPOLAR DISORDER

Symptoms—Bipolar disorder is characterized by periods of elevated mood or mania and periods of depression that significantly impair function at work and socially. Mania can be milder and is termed hypomania, although it can progress to full mania. In many cases with hypomania, patients live for many years before a diagnosis is made, with recurrent adverse consequences. The depressive episodes are longer-lasting and the cycling between elevated and depressed moods can repeat regularly, with more rapid cycling with age being indicative of increased disease severity. Stress and anxiety can trigger symptoms, which are similar to those in attention deficit hyperactivity disorder (ADHD), but the label of bipolar disorder is generally avoided in children, unless they manifest intense episodes of mania or depression. Bipolar disorder patients also show comorbid anxiety disorder and may even manifest psychotic symptoms. Bipolar disorder patients are more likely to commit suicide than major depression patients, emphasizing the need for effective treatment (Nordentoft et al., 2011). The symptoms of bipolar disorder are very similar to those in schizophrenia, with the major difference being a history of mania or hypomania (Anderson et al., 2012). The presence of psychotic symptoms in bipolar disorder can be mistaken for schizophrenia. The incidence of bipolar disorder is ~1% and also has a postpubertal age of onset, with most patients being diagnosed in their early 20s, with symptoms present as teenagers (Merikangas et al., 2011).

Etiology—Physical or sexual abuse and environmental factors all contribute to increase incidence (Etain et al., 2008). However, genetic factors contribute the most to the incidence of bipolar disorder, which also has genetic components overlapping with schizophrenia (Sullivan et al., 2012). Interestingly, FGF metabolism is also disturbed in bipolar disorder and major depression (Liu et al., 2014; Persson et al., 2009).

EEG, reflexes, and P50 potential—As expected, the wake–sleep patterns manifested in the EEG of bipolar disorder patients are similar to those in schizophrenia, including fragmented sleep, decreased SWS, increased vigilance, and increased REM sleep drive (Kadrmas and Winokur, 1979; Kupfer et al., 1978). As in schizophrenia, bipolar disorder patients show decreased habituation of the P50 potential in a paired stimulus paradigm (Olincy and Martin, 2005; Schulze et al., 2007), along with exaggerated startle response (Perry et al., 2001), and dysregulation of blink reflexes (Depue et al., 1990).

Reduced gamma band activity has been reported in bipolar disorder patients (Ozerdem et al., 2011), similarly to that reported in schizophrenia (Uhlhaas and Singer, 2010). Decreased gamma band activity can account for many of the symptoms in these disorders, including wake–sleep, arousal, and cognitive symptoms. The mechanism behind the decrease in gamma band activity and maintenance was unknown until recently. Human post-mortem studies reported increased expression of neuronal calcium sensor protein-1 (NCS-1) in the brains of bipolar disorder and schizophrenic patients compared to normal controls and major depression patients (Bergson et al., 2003; Koh et al., 2003). That is, gamma band activity is reduced or disrupted in precisely the same disorders that show brain NCS-1 overexpression. We tested the hypothesis that NCS-1 modulates calcium channels in PPN neurons that generate gamma band oscillations (see Chapter 9) and that excessive levels of NCS-1, as would be expected with overexpression, reduce or block gamma band oscillations in these cells.

Recordings in PPN neurons using $1\,\mu M$ NCS-1 were found to increase the amplitude and frequency of ramp-induced oscillations within ~25 min of diffusion into the cell. Figure 11.2(a) is a representative example of ramp-induced membrane potential oscillations in a PPN neuron in the presence of synaptic blockers and tetrodotoxin. Shortly after patching, the ramp typically induced low-amplitude oscillations in the beta/gamma range (light gray record in Figure 11.2(a) and light gray line in Figure 11.2(b)). Figure 11.2(a) dark gray record shows that, after 10 min of recording, some increase in the oscillation amplitude and frequency was present (also evident in Figure 11.2(b) as a dark gray line in the power spectrum). After 25 min of recording, NCS-1 at $1\,\mu M$ significantly increased the amplitude and frequency of oscillations (back record in Figure 11.2(a) and as the black line in the power spectrum). Control cells recorded without NCS-1 in the pipette manifested no significant changes in amplitude or

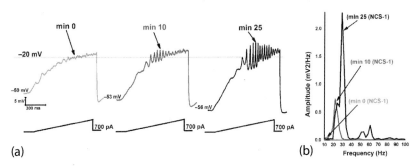

FIGURE 11.2   **Effects of NCS-1 on ramp-induced oscillations in PPN neurons.** (a) Using 1 s long current ramps, we induced oscillations in a PPN neuron in synaptic blockers and TTX in the extracellular solution and 1 µM NCS-1 in the recording pipette (left record, light gray). After 10 min of NCS-1 diffusing into the cell, the oscillatory activity increased slightly (middle record, dark gray). And after 25 min of NCS-1 diffusion, both oscillation amplitude and frequency were increased (right record, black). (b) Power spectrum of the records shown in (a) before band-pass filtering, showing the increased amplitude and frequency of oscillations after 25 min exposure to 1 µM NCS-1.

frequency throughout the 30 min recording period. These values were not significantly different from each of the 0 min recordings using pipettes with NCS-1, that is, before NCS-1 induced significant effects, so that the 0 min recordings are an accurate representation of control levels.

We then carried out a study to determine the effects of NCS-1 concentration on PPN cell ramp-induced oscillations. Figure 11.3 shows the changes in the amplitude of the oscillations over time when the pipette contained different concentrations of NCS-1. When the pipette contained 0.5 µM NCS-1, no changes in amplitude were observed throughout the recording, suggesting that this concentration does not significantly affect oscillation amplitude. When using 1 µM NCS-1, however, the oscillation amplitude increased significantly by 20 min and thereafter, suggesting a gradual effect in tripling amplitude as the NCS-1 diffused into the cell. When using 5 µM NCS-1, there was a significant increase in amplitude at 5 min but not afterward. There were no further changes observed, so that we conclude that the effect at 5 min was not consistent, or briefly increased amplitude and the effect saturated. When using 10 µM NCS-1, the oscillation amplitude immediately increased to four times the levels and gradually decreased until it was significantly reduced by 30 min. These effects suggest an immediate effect on amplitude by very high levels of NCS-1 that ultimately led to partial blockade. Based on these results, 1 µM NCS-1 seems to be the most critical concentration for promoting gamma oscillations (Garcia-Rill et al., 2014; D'Onofrio et al., in press).

In order to evaluate the effects of NCS-1 on high-threshold voltage-dependent calcium currents (I_{Ca}) present in PPN neurons, square voltage steps were used in combination with high cesium/QX314 intracellular

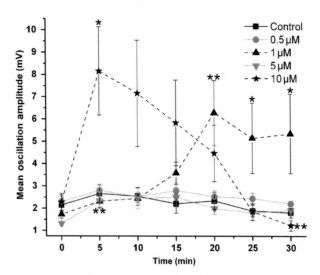

FIGURE 11.3   **Effects of NCS-1 on peak oscillation amplitude.** The graph shows the mean peak amplitude in mV of control cells recorded (black squares) that demonstrated no significant changes over time. Cells recorded using 0.5 µM NCS-1 also showed no significant changes over time (gray circles). Cells recorded using 1 µM NCS-1 (black upward triangles) showed significant increases in mean peak oscillation amplitude at 20 min and beyond. Cells recoded using 5 µM NCS-1 (gray downward triangles) showed no significant changes over time, but cells recorded using 10 µM NCS-1 (black stars) showed a significant increase in mean peak oscillation amplitude at 10 min, but not thereafter. *$p < 0.05$; **$p < 0.01$.

pipette solution and synaptic receptor blockers. Figure 11.3 shows the results of the calcium current study carried out using NCS-1 concentrations of 0.5 and 1 µM NCS-1. Figure 11.4(a) shows the voltage step protocol and representative calcium currents at min 0 through min 15. Subsequent time points were similar to 15 min and are not shown. Figure 11.4(b) shows a graph of the time course of mean reduction of I_{Ca} by either 0.5 µM (black circles) or 1 µM NCS-1 (gray circles). Both curves were well fitted to a single exponential ($R^2 > 0.99$, dotted lines), yielding *tau* (τ) values of and 9.4 or 7.8 min for 0.5 or 1 µM NCS-1 curves, respectively. These results suggest that NCS-1 decreased calcium currents and increased series resistance (Figure 11.4(c)), suggesting long-term effects on intracellular calcium metabolism that require further study.

The postmortem results previously described (Koh et al., 2003) suggest that only some patients with schizophrenia may suffer from significant overexpression of NCS-1, which may be manifested as decreased gamma band activity only in a subpopulation of patients. No human study has measured gamma band activity and correlated it with NCS-1 levels. Unfortunately, serum sampling does not reflect brain levels and, in fact, NCS-1 levels in leukocytes are actually decreased in schizophrenic patients (Torres et al., 2009). However, future clinical trials in patients with

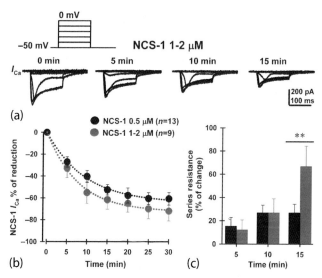

(a)

(b)

(c)

FIGURE 11.4 **Effects of NCS-1 on calcium currents in PPN cells.** (a) Example of calcium currents (I_{Ca}) recorded using depolarizing square steps from −40 to 0 mV from −50 mV holding potential over time (0, 5, 10, and 15 min of exposure). (b) Mean time course of I_{Ca} block by intracellular 0.5 μM (black circles) and 1 μM (gray circles). Data were fitted to a single exponential decay yielding tau (τ) values of 9.4 or 7.8 min for 0.5 or 1 μM NCS-1 curves, respectively. (c) Series resistance change (in % of 0 min values) observed at 5, 10, and 15 min after exposure to either 0.5 μM (black bars) or 1 μM NCS-1 (gray bars). Series resistance values reached a plateau after 15 min, when using 1 μM NCS-1. **$p < 0.001$.

schizophrenia or bipolar disorder may benefit from prior determination of a significant decrease in gamma band activity, which may also help address the heterogeneity of schizophrenia and facilitate the process of identifying more homogeneous groups within the syndrome (Picardi et al., 2012). It is to those patients that pharmacological targeting to increase gamma band activity may be of benefit. We have preliminary evidence suggesting that the stimulant modafinil may indeed compensate to some extent for excessive amounts of NCS-1. We found a partial return of gamma oscillations that were suppressed by high levels of NCS-1 after exposure to modafinil (Garcia-Rill et al., 2014).

Treatment—Serendipitously, the mood disturbances in this disorder were treated effectively using lithium, an ion that remains one of the best treatment options, although it is limited by side effects (Brown and Tracy, 2012). Lithium has also been used with some success for regulating mood in schizophrenia. Lithium has also been proposed as a neuroprotective agent. Lithium may act by inhibiting the interaction between NCS-1 and inositol 1,4,5-triphosphate receptor protein (InsP) (Schleker et al., 2006), which is overexpressed in bipolar disorder and schizophrenia (Bergson et al., 2003; Koh et al., 2003). A number of other agents are also used successfully, including anticonvulsants and antipsychotics. Fortunately, for many patients, such therapy does allow them to function more normally and lead productive lives. We should note that NCS-1 enhances the activity of InsP (Kasri et al., 2004), which is present in the PPN (Rodrigo et al., 1993). Lithium, as mentioned above, inhibits the effects of NCS-1 on InsP (Schleker et al., 2006). A diagram of these intracellular pathways appears in Figure 11.5. That is, lithium may reduce the effects of overexpressed NCS-1 in bipolar disorder (and schizophrenia), thereby normalizing gamma band oscillations mediated by N- and P-/Q-type calcium channels modulated by NCS-1. That is, the effects of overexpression of NCS-1 in schizophrenia and bipolar disorder may be decreased by lithium. We found that excessive NCS-1 decreased gamma oscillations; therefore, lithium may prevent the downregulation of gamma band activity and restore normal levels of gamma band oscillations. These findings taken together resolve the 60-year mystery of how lithium works in bipolar disorder and schizoaffective disease. An interesting observation is that NCS-1 downregulates N-type calcium channels, at least in some cell lines (Gambino et al., 2007). This may mean that under some circumstances, NCS-1 may inhibit N-type channel function while promoting P-/Q-type channel function.

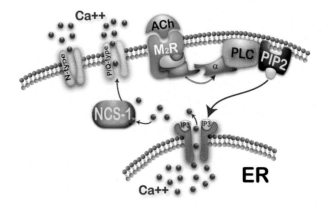

FIGURE 11.5 **Intracellular pathways for NCS-1 and inositol trisphosphate (IP3).** In the case of muscarinic cholinergic receptors (M_2R blue structure in the membrane) activated by acetylcholine (ACh), as would occur in a PPN neurons, G proteins are activated to trigger the phospholipase C (PLC)-phosphatidylinositol bisphosphate (PIP2) pathway to release IP3 into the cytoplasm. The IP3 binds to the IP3 receptor in the endoplasmic reticulum (ER) to release calcium (red spheres) intracellularly. NCS-1 (green cytoplasmic protein) binds calcium that flows from the ER and through P-/Q-type channels (and may inhibit N-type channels) in the membrane. The NCS-1 facilitates or potentiates gamma band oscillations mediated by P-/Q-type calcium channels. If NCS-1 is overexpressed such as in schizophrenia and bipolar disorder, lithium may act to inhibit the role of NCS-1 in accumulating intracellular calcium, thus protecting the neuron.

MAJOR DEPRESSION

Symptoms—The most striking symptom in major depression is the persistent low mood, which is accompanied by anhedonia and low self-esteem. In severe cases, patients experience psychosis, which can include delusions and hallucinations. Changes in wake–sleep patterns and appetite are common, suggesting that other homeostatic systems besides wake–sleep control are impaired. About half of patients show comorbid anxiety (Kessler et al., 1996), which can increase the risk of suicide (Hirschfeld, 2001), and patients with major depression can have a number of systemic disorders like cardiovascular disease, chronic pain, and diabetes. The incidence of depression ranges from 8% to 12% worldwide. Suicidal ideation is perhaps the most astounding delusion. The person believes that the world is better off without them. That means that the survival instinct has been completely abrogated. If someone tries to throw you off the top of a building, you fight for your life. Yet a deeply depressed person is willing to jump off the building voluntarily. That this disorder can eliminate one of our strongest instincts, the survival instinct, makes major depression among the most serious medical conditions. Such delusions have to be taken very seriously, because the sufferer is probably deeply incapacitated. Under such circumstances, decision-making must be suspect. Because terminally ill patients can develop depression and seek assisted suicide, regardless of the level of acceptance for the procedure, it is advisable to ensure that the person is not clinically depressed and hypofrontal when deciding to end their existence.

Etiology—Genetic factors are perhaps the most common (Kendler et al., 2006), with indications of dysregulation in serotonin transporter metabolism (Caspi et al., 2003). The monoamine hypothesis for major depression has been proposed (Nutt, 2008). In addition, poverty, social isolation, and early abuse all are linked with developing major depression later in life (Heim et al., 2008). A number of metabolic disturbances produce depressive symptoms, including thyroid problems, calcium imbalance, and chronic disease.

EEG, reflexes, and P50 potential—Insomnia is a hallmark of major depression and, in addition to wake–sleep cycle dysregulation, includes decreased SWS, increased REM sleep drive, and frequent awakenings (Arfken et al., 2014; Kudlow et al., 2013; Seifritz, 2001). The startle response (Carroll et al., 2007; Kohl et al., 2013) and blink reflexes (Kohl et al., 2013) are exaggerated in major depression. We found that the P50 potential exhibited decreased habituation in depression (Garcia-Rill et al., 2002). Briefly, the habituation at the 250 ms ISI as well as the 500 ms ISI was decreased compared to normal, age, and gender matched controls (Figure 11.6). These results suggest that the changes in wake–sleep, startle/reflex responses, and P50 potential in major depression are all in the same

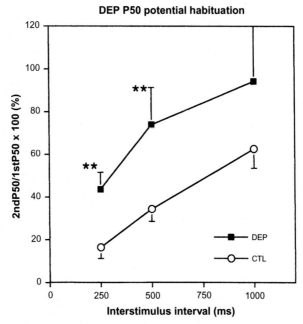

FIGURE 11.6   **Effects of depression on P50 potential habituation.** Graph of the mean percent (±S.E.) P50 potential habituation (sensory gating) using a paired stimulus paradigm in depressed (DEP) (filled squares) and control (CTL) (open circles) groups. Three ISIs were tested, 250, 500, and 1000 ms. For each ISI, the amplitude of the P50 potential following the second stimulus was calculated as a percent of the amplitude of the P50 potential following the first stimulus. The percent habituation was significantly (**$p < 0.01$) higher in the DEP compared to the CTL group at the 250 and 500 ms ISI. There was no statistically significant difference in the percent habituation of the groups at the 1000 ms ISI. Data from Garcia-Rill et al., 2002.

direction as in schizophrenia and bipolar disorder. Interestingly, major depression patients do not show a decrement in gamma band activity like bipolar disorder and schizophrenic patients (Liu et al., 2012). Postmortem studies, as described above, showed that schizophrenic and bipolar disorder patients manifested overexpression of NCS-1, while major depression patients were not different from control (Koh et al., 2003). This suggests that NCS-1 may be critical for the expression of effective levels of gamma band activity in schizophrenia and bipolar disorder.

Treatment—The most common form of treatment aside from psychotherapy is antidepressant intervention, usually in the form of specific serotonin reuptake inhibitors (SSRIs). The site of action of SSRIs, as far as the RAS is concerned, is an increase in serotonergic inhibition of the PPN and locus coeruleus (LC). That is, as evident on the left side of Figure 11.7 (DEP), the inhibitory inputs to both the PPN and LC would increase, thereby decreasing the outputs of these nuclei. This would have the effect

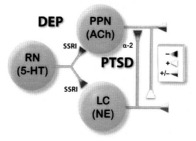

FIGURE 11.7 **RAS diagram showing sites of dysregulation in major depression and PTSD.** In depression (DEP, left side), decreased inhibition of the PPN and LC by the RN appears to be a determinant of wake–sleep symptomatology. By administering SSRIs, the serotonergic inhibition of PPN will help reduce the increased arousal, hypervigilance, and increased REM sleep drive in major depression. On the other, in PTSD (right side), decreased LC inhibition of the PPN appears to be a determinant of the hypervigilance and increased REM sleep drive, especially following LC cell loss. Administration of α-2 adrenergic receptor agonists such as clonidine will inhibit PPN, downregulating these symptoms, while α-2 adrenergic receptor antagonists such as yohimbine will exacerbate arousal and REM sleep drive dysregulation.

of decreasing some of the hypervigilance and increased REM sleep drive present in major depression. Electroconvulsive therapy and, more recently, transcranial magnetic stimulation have been approved for the treatment of depression. Unfortunately, most treatments produce only a partial alleviation of symptoms, with ~10% of patients unresponsive to any therapy. These numbers underscore the lack of truly effective treatment that can allow many patients to regain productive lives.

POSTTRAUMATIC STRESS DISORDER

Symptoms—PTSD is an anxiety disorder caused by an event in which there is fear of death. Symptoms are classified around three clusters: (1) hypervigilance that includes insomnia, exaggerated startle responses, and attentional problems; (2) reexperiencing that includes recurring vivid memories and dreams of the event; and (3) avoidance that involves avoiding places, people, and discussion of the event. About one in three combat veterans will develop the disorder, with epilepsy, concussion, or other brain insult as potential predisposing factors.

Etiology—The most important predictor of PTSD is the number of stressors (Bramsen et al., 2000), with childhood trauma and preexisting depression as likely factors (Breslau and Davis, 1992; Breslau et al., 1997). Women are twice as likely to develop PTSD as a result of equivalent trauma as men (Breslau et al., 1991), with lifetime prevalence of 5% and 10% for men and women, respectively.

EEG, reflexes, and P50 potential—Sleep patterns in PTSD are similar to those in schizophrenia and bipolar disorder, basically showing frequent awakenings, decreased SWS, increased REM sleep drive, and hypervigilance (Ross et al., 1989, 1999; Sandor and Shapiro, 1994; Woodward et al., 1996). In addition, PTSD patients have exaggerated startle responses and reflexes (Butler et al., 1990; Ornitz and Pynoos, 1989; Shalev et al., 2000). We described the manifestation of the P50 potential in PTSD. Briefly, we studied a group of combat-exposed veterans with PTSD and compared their P50 potential responses using a paired stimulus paradigm with those of normal age and gender matched controls, combat-exposed veterans without PTSD, and alcoholics in remission. PTSD patients tend to drink heavily, which may also represent an attempt at self-medication for the sensory gating abnormality produced by PTSD, so that alcoholics in remission and control subjects are essential. We found that the P50 potential was of significantly higher amplitude in PTSD after the second stimulus of a pair than in each of the other control groups (Gillette et al., 1997). Figure 11.8 shows the results of that study. Normally, the amplitude of the response to the second stimulus of a pair is 10–25%, which was evident in normal controls, combat-exposed veterans without PTSD, and alcoholic patients. However, in combat-exposed veterans with PTSD, habituation was reduced and the second response was ~60%. These results were best correlated with the reexperiencing symptom cluster in psychiatric testing instruments, suggesting that it is the recollection of the traumatic experience on a recurrent basis that leads to the sensory gating deficit.

We later replicated these findings in male combat veterans and also found that female rape victims suffered from the same decreased habituation (Skinner et al., 1999). We later found in a small sample of postmortem brain tissue from combat-exposed veterans with PTSD that there was a significant reduction in the number of neurons in the LC (Bracha et al., 2005). We proposed that the mechanism involved in PTSD is mainly a result of high-amplitude pulses of cortisol induced by the repeated recollection of the trauma that travel across the blood–brain barrier to bind in regions that have glucocorticoid receptors such as the hippocampus, cortex, and LC. All of these regions have been reported to be damaged in PTSD. The repeated exposure to cortisol would ultimately lead to neuronal cell death. Since the LC normally inhibits the PPN (see right side of Figure 11.7 PTSD), the loss of LC neurons would disinhibit the PPN and produce increased PPN output that would result in increased vigilance and REM sleep drive, in keeping with PPN functions. We also investigated the possibility that acute, high circulating levels of cortisol, possibly induced by the stress of reexperiencing the trauma, may lead to the LC cell death observed. A similar mechanism has been proposed for the effects of stress on hippocampal neurons (Sapolsky et al., 1984), which, like LC neurons, possess glucocorticoid receptors (Aronsson et al., 1988). In the rat,

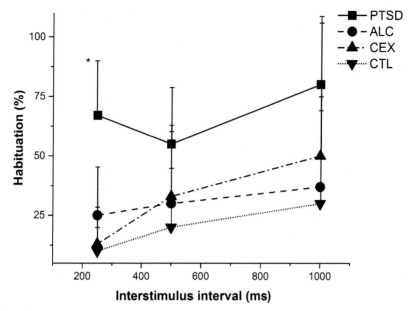

FIGURE 11.8 **Effects of PTSD on P50 potential habituation.** Graph of the mean percent (±S.E.) P50 potential habituation (sensory gating) using a paired stimulus paradigm in combat-exposed veterans with PTSD (filled squares), compared to alcoholic (ALC) patients in remission (filled circles), to combat-exposed (CEX) veterans without PTSD (upright triangles), and to control (CTL) (inverted triangles) groups. Three ISIs were tested, 250, 500, and 1000 ms. For each ISI, the amplitude of the P50 potential following the second stimulus was calculated as a percent of the amplitude of the P50 potential following the first stimulus. The percent habituation was significantly (*$p < 0.05$) higher in the DEP compared to all other groups at the 250 ms ISI. There was no statistically significant difference in the percent habituation of the groups at the 500 or 1000 ms ISI. Data from Gillette et al., 1997.

we found that neurotoxic lesions of LC led to decreased habituation of the P13 response, with a similar outcome resulting from daily administration of corticosterone (Miyazato et al., 2000).

These findings suggest that it is decreased, not increased, LC output to PPN that may be responsible for the symptoms of anxiety. The fact that yohimbine, an alpha-2 adrenergic receptor antagonist, induced anxiety in PTSD patients (Charney et al., 1987), while the adrenergic agonist clonidine reversed the effect (Charney and Heninger, 1986), suggests that the drug may be blocking not LC autoreceptors, but synaptic input from LC to PPN (as in Figure 11.7 PTSD). This would disinhibit cholinergic neurons and thus account for the wake-sleep symptoms observed and the changes in the P50 potential we reported. Along these lines, we determined that injections of yohimbine directly into the PPN increase the amplitude and decrease the habituation of the P13 potential in the rat, an effect reversed by injections of clonidine (Miyazato et al., 1999).

Treatment—Treatment of PTSD is limited to anxiolytic and therapy, with mixed success. We hypothesize that it may be possible to successfully treat and perhaps even prevent this disorder. For example, it would appear that the ideal anxiolytic compound would be one able to be an agonist at the α_{-2} receptor on PPN neurons (without the peripheral effects of agents such as clonidine). In addition, it may be possible to block the potentially toxic effects of cortisol on brain cells by using a glucocorticoid receptor blocker, such as RU-486, soon after the trauma. This would be a sort of preventive treatment in cases of recurrent increased levels of cortisol, presumably rescuing some neurons. Recent successes using propranolol, the β adrenergic blocker, as an amnestic agent soon after the trauma suggest that this is another option for the reexperiencing consequences of PTSD (Pittman et al., 2002; Vaiva et al., 2003).

ATTENTION DEFICIT HYPERACTIVITY DISORDER

Symptoms—Attentional deficits with hyperactivity and impulsive behavior are the hallmarks of ADHD. About 1–7% of children are diagnosed, with males three times more likely to be diagnosed. These symptoms lead to a number of consequences including learning disabilities, with vigilance being the primary process involved. One-third of ADHD patients develop depression, with comorbid anxiety, OCD, and sleep disorders being more common in this population.

Etiology—Twin studies suggest that as many as 75% of cases have a genetic component, with a number of different genes implicated. A problem arises with diagnosis since there are large differences in symptoms and behaviors between subjects labeled as ADHD.

EEG, reflexes, and P50 potential—The impulsive behavior and attentional problems are generally not accompanied by EEG sleep abnormalities (Greenhil et al., 1983). There appear to be large subgroups that have different sleep phenotypes, including hypoarousal similar to narcolepsy, delayed sleep onset as in bipolar disorder, sleep-disordered breathing, restless legs syndrome, and epilepsy (Miano et al., 2012). Only some ADHD patients have sleep problems, and many of those seem to be related to sleep-disordered breathing and periodic limb movements (Goraya et al., 2009; Sawyer et al., 2009), an effect that survives into adulthood (Philipsen et al., 2005). Studies of the P50 potential suggest that ADHD subjects do not exhibit a sensory gating deficit (Olincy et al., 2000), although other studies did find a deficit (Holstein et al., 2013). These results suggest that the diagnostic category includes different groups with opposite symptomatology. The attentional problems in the absence of sleep architecture abnormalities would tend to suggest that the RAS is not primarily involved in ADHD. Studies showed that ADHD patients have either decreased

(Barry et al., 2010) or increased (Herrmann and Demilrap, 2005) gamma band amplitude. This again points to a widely heterogeneous population. Future studies should target this measure, gamma band amplitude or power, as a tool for segregating patients similarly labeled with ADHD.

The impulsivity in ADHD in the absence of specific sleep architecture problems (such as hypervigilance and increased REM sleep drive that would implicate the PPN) would suggest that the frontal cortex is likely dysregulated. There appear to be imbalances in the right vs. left frontal lobe blood flow that persist into adulthood in some cases (Oner et al., 2005; Spaletta et al., 2001). These results lend further support for a cortical etiology in ADHD, with specific sleep disorders in only some subgroups. The attentional deficits may not be due to hypervigilance and increased REM sleep drive due to PPN overactivity, but perhaps only in a subpopulation.

Treatment—The use of stimulants has made an impact in some patients, with early use of amphetamine replaced by methylphenidate (Ritalin). The neurotoxic effect of amphetamine (Seiden et al., 1993) can be avoided with methylphenidate since amphetamine releases newly synthesized and stored pools of catecholamines, but methylphenidate releases only stored pools. The use of combined amphetamine and methylphenidate but at lower concentrations appears to mitigate these effects. Modafinil has been used with reported improvement (Kahbazi et al., 2009) or no effect (Arnold et al., 2014). Such results again point to large differences within the diagnostic category that have yet to be resolved.

AUTISM

Symptoms—Autism includes three types of symptoms, impaired social skills, impaired language skills, and repetitive behaviors (Jeste, 2011). A number of variations exist but those that share these core symptoms are referred to collectively as autism spectrum disorders. Onset is typically at 2–3 years of age, and as many as 1–2% of children may have some form of the disorder (Blumberg et al., 2013).

Etiology—Twin studies show that the risk of developing autism in identical twins is very high and quite high in fraternal twins (Devlin and Scherer, 2012; Hallmayer et al., 2011). Perinatal brain damage, especially to the cerebellum, led to autism in ~35% of patients (Becker and Stoodley, 2013). These numbers emphasize the genetic etiology of the disorder. The risk of developing autism from having older parents, SSRIs before birth, or vaccinations is negligible. However, many of the genes implicated in autism have also been implicated in schizophrenia, among them dysregulation of calcium channel function (Schmunk and Gargus, 2013) and FGF metabolism (Stevens et al., 2010). Recall that, as in schizophrenia, FGF dysregulation could affect the cerebellum and account for some of the movement disorder in these diseases. Dysregulation of other homeostatic systems is also evident, given the obesity associated with autism (Broder-Fingert et al., 2014; Curtin et al., 2014).

EEG, reflexes, and P50 potential—While sleep duration is decreased and fragmented sleep is common in autism (Hodge et al., 2014; Humphreys et al., 2014; Glodman et al., 2009), there are no major differences in sleep architecture (Giannotti et al., 2011), except for a decrease in REM sleep duration (Buckley et al., 2010). In addition, insomnia is present in over 80% of patients (Jeste, 2011; Richdale et al., 2014). Spectral analysis of the EEG in autistic subjects suggests a decrease in beta and an increase in theta frequencies (Daoust et al., 2004), reminiscent of thalamocortical dysrhythmia (Llinas et al., 1999). Finally, P50 potential studies using paired stimulus paradigm showed that there was indeed a decrease in habituation indicative of a sensory gating deficit in autism (Buchwald et al., 1992). However, there was also a decrease in amplitude of the P50 potential, suggestive of a decrement in the initial response of the system. Together, these findings suggest that the PPN is underactive, accounting for the decreased REM sleep duration, decreased beta frequency in the EEG, and decreased P50 potential amplitude. The prevalence of insomnia, therefore, may arise from a different source than PPN overactivity.

Treatment—Some relief from insomnia and prolonged sleep latency has been observed with melatonin (Jeste, 2011). Treatment generally involves intense behavior therapy ("applied behavioral analysis" and similar programs), and a small subset of patients actually recovers so that they no longer meet criteria for autism after years of such retraining. These results again suggest that the population is mixed and not enough is known about what distinguishes subgroups. The use of psychoactive agents is generally ineffective in treating the core symptoms. A great deal of research will be required in order to devise effective treatments for these core deficits in this complex disorder.

CLINICAL IMPLICATIONS

The National Institute of Mental Health (NIMH) at the National Institutes of Health (NIH) has embarked on a new approach termed Research Domain Criteria (RDoC). The idea is to transform diagnosis by incorporating scientific findings to lay the foundation for a new classification system. The plan is to transform psychiatric practice by bringing novel research to inform how to diagnose and treat mental disorders. This comes on the heels of failure of the current diagnostic approach to make sufficient headway in successful treatment. The RDoC approach is intended for clinical research and clinical trials to employ a matrix of dimensions of functions termed "constructs" grouped into major domains of functioning that constitute the unit of analysis and classes of variables to study these constructs. By shying away from classic diagnostic criteria and employing a more physiologically based approach, the failure of diagnostic criteria in accounting for a wide range of symptoms and behaviors can be avoided. The assumption is that many clinical trials fail because the diagnostic criteria accommodate a heterogeneous population that does not respond to the agent being used. The following discussion provides a series of variables that can be applied across classic diagnostic criteria to design treatments for domains of functioning.

The spectrum of symptoms in some of the disorders described here (schizophrenia, bipolar disorder, major depression, and PTSD) overlap greatly and can be divided into two categories. (1) The wake–sleep dysregulation, reflex changes, and P50 potential disturbances all suggest increased vigilance during waking in the form of hyperarousal and during sleep as increased REM sleep drive. The main center influencing arousal and REM sleep is the PPN and its cholinergic output. The overactive PPN output can be addressed with an agent like olanzapine, which increases serotonergic inhibition of the PPN, decreases muscarinic output to the SN, and also decreases dopaminergic output. (2) The other category of symptoms is decreased or interrupted gamma band activity. The stability of gamma band levels, which obviously is responsible for some of the cognitive and attentional deficits, can be addressed with an agent like the stimulant modafinil. Modafinil acts by increasing electrical coupling especially in GABAergic neurons (Garcia-Rill et al., 2007; Urbano et al., 2007), which decreases input resistance in GABAergic cells, thus decreasing firing and GABAergic output. This essentially disinhibits many transmitter systems normally suppressed by GABA. In addition, the added electrical coupling increases coherence and, because the system is activated through disinhibition, helps smoothly maintain gamma band activity. The individual or combined use of these two agents can address virtually all of the arousal and wake–sleep dysregulation in all of these disorders, at least in a large number of sufferers.

These suggestions cannot be considered lightly given the dangerous nature of suicidal ideation. The risks of combined therapy are unknown and must be monitored especially carefully. Moreover, the dosage remains questionable. Modafinil is prescribed for narcolepsy in 100 mg tablets, and some narcoleptic patients may take as much as 400 mg/day. However, the dosage for subtle maintenance of gamma band activity in disorders that already manifest hypervigilance is likely to be much lower. Certainly, 100 mg is too high a dose, so that formulations in the range of 25 mg could be titrated until therapeutic levels are achieved. A similar concern for olanzapine dosage remains, requiring great care in the combined use of these agents.

A number of avenues in the pursuit of better treatment for psychiatric disorders include developmental regulation and postpubertal development. In Chapter 5, we discussed the developmental decrease in REM sleep and how events that prevent it from taking place could lead to lifelong increases in hypervigilance and REM sleep drive. Since this decrease takes place from birth to puberty, gonadal steroids, which themselves have marked effects on arousal and vigilance, are worthy of added attention. After all, over three quarters of patients with schizophrenia, panic attacks, bipolar disorder, OCD, and narcolepsy manifest a postpubertal onset. Another avenue worth pursuing is the role of appetite control, specifically of leptin. Leptin is involved in brain development and can influence a number of proinflammatory cytokines. In the same chapter, we discussed leptin resistance, the increase in circulating levels due to the lack of activation of leptin receptors, which is present in obesity. Leptin dysregulation is also present in major depression and anxiety, in the treatment of schizophrenia and bipolar disorder (antipsychotics induce weight gain, as does lithium), and elevated levels are present in autism (Valleau and Sullivan, 2014). The role of this hormone in psychiatric disorder also needs investigation.

References

Akbarian, S., Bunney, W.E., Potkin, S.G., Wigal, S.B., Hagman, J.D., Sandman, C.A., Jones, E.G., 1993. Altered distribution of nicotinamide-adenine dinucleotide phosphate-diaphorase cells in frontal lobe of schizophrenics implies disturbances in cortical development. Arch. Gen. Psychiatry 50, 169–177.

Anderson, I.M., Haddad, P.M., Scott, J., 2012. Bipolar disorder. Brit. Med. J. 27, 345.e8508.

Andreasen, N.C., Flaum, M., 1991. Schizophrenia: the characteristic symptoms. Schizophr. Bull. 17, 27–49.

Arfken, C.l., Joseph, A., Sandhu, G.R., Roehrs, T., Douglass, A.B., Boutros, N.N., 2014. The status of sleep abnormalities as a diagnostic test for major depressive disorder. J. Affect. Disord. 156, 36–45.

Arnold, V.K., Feifel, D., Earl, C.Q., Yang, R., Adler, l.A., 2014. A 9-week, randomized, double-blind, placebo-controlled, parallel-group, dose-finding study to evaluate the efficacy and safety of modafinil as treatment for adults with ADHD. J. Atten. Disord. 18, 133–144.

Aronsson, M., Fuxe, K., Dong, Y., Agnati, L.F., Okret, S., Gustafsson, J.A., 1988. Localization of glucocorticoid receptor mRNA in the male rat brain by in situ hybridization. Proc. Natl. Acad. Sci. USA 85, 9331–9335.

Barry, R.J., Clarke, A.R., Hajos, M., McCarthy, R., Selikowitz, M., Dupuy, F.E., 2010. Resting-state EEG gamma activity in children with attention-deficit/hyperactivity disorder. Clin. Neurophysiol. 121, 1871–1877.

Becker, E.B., Stoodley, C.J., 2013. Autism spectrum disorder and the cerebellum. Int. Rev. Neurobiol. 113, 1–34.

Bergson, C., Levenson, R., Goldman-Rakic, P., Lidow, M.S., 2003. Dopamine receptor-interacting proteins: the Ca²⁺ connection in dopamine signaling. Trends Pharmacol. Sci. 24, 486–492.

Blumberg, S.J., Bramlett, M.D., Kogan, M.D., Schieve, L.A., Jones, J.R., Lu, M.C., 2013. Changes in prevalence of parent-reported autism spectrum disorder in school-aged

U.S. Children: 2007 to 2011–2012. National Health Statistics Reports no. 65. National Center for Health Statistics, Hyattsville, MD.

Bracha, H.S., Torrey, E.F., Bigelow, L.B., Lohr, J.B., Linington, B.B., 1991. Subtle signs of prenatal maldevelopment of the hand ectoderm in schizophrenia: a preliminary monozygotic twin study. Biol. Psychiatry 30, 719–725.

Bracha, H.S., Garcia-Rill, E., Mrak, R.E., Clothier, J., Karson, C.N., Skinner, R.D., 2005. Postmortem locus coeruleus neuron count: a case report of a veteran with probable combat-related posttraumatic stress disorder. J. Neuropsychiatry Clin. Neurosci. 17, 503–509.

Bramsen, I., Dirkzwager, A.J., Van der Ploeg, H.M., 2000. Predeployment personality traits and exposure to trauma as predictors of posttraumatic stress symptoms: a prospective study of former peacekeepers. Am. J. Psychiatry 157, 1115–1119.

Breslau, N., Davis, G.C., 1992. Posttraumatic stress disorder in an urban population of young adults: risk factors for chronicity. Am. J. Psychiatry 149, 671–675.

Breslau, N., Davis, G.C., Andreski, P., Peterson, E., 1991. Traumatic events and posttraumatic stress disorder in an urban population of young adults. Arch. Gen. Psychiatry 48, 216–222.

Breslau, N., Davis, G.C., Peterson, E.L., Schultz, L., 1997. Psychiatric sequelae of posttraumatic stress disorder in women. Arch. Gen. Psychiatry 54, 81–87.

Broder-Fingert, S., Brazauskas, K., Lindgren, K., Iannuzzi, D., Van Claeve, J., 2014. Prevalence of overweight and obesity in a large clinical sample of children with autism. Acad. Pediatr. 14, 408–414.

Brown, K.M., Tracy, D.K., 2013. Lithium: the pharmacodynamic actions of the amazing ion. Ther. Psychopharmacol. 3, 163–176.

Buchwald, J.S., Erwin, R., Van Lancker, D., Guthrie, D., Schwafel, J., Tanguay, P., 1992. Midlatency auditory evoked responses: P1 abnormalities in adult autistic subjects. Electroencephalogr. Clin. Neurophysiol. 84, 164–171.

Buckley, A.W., Rodriguez, A.J., Jennison, K., Buckley, J., Thurm, A., Sato, S., Swedo, S., 2010. REM sleep percentage in children with autism compared to children with developmental delay and typical development. Arch. Pediatr. Adolesc. Med. 164, 1032–1037.

Buckley, P.F., Miller, B.J., Lehrer, D.S., Castle, D.J., 2009. Psychiatric comorbidities and schizophrenia. Schizophr. Bull. 35, 383–402.

Butler, R.W., Braff, D.L., Pausch, J.L., Jenkins, M.A., Sprock, J., Geyer, M.A., 1990. Physiological evidence of exaggerated startle response in a subgroup of Vietnam veterans with combat-related PTSD. Am. J. Psychiatry 147, 1308–1312.

Caldwell, D.F., Domino, E.F., 1967. Electroencephalographic and eye movement patterns during sleep in chronic schizophrenic patients. Electroencephalogr. Clin. Neurophysiol. 22, 414–420.

Carroll, C.A., Vohs, J.L., O'Donnell, B.F., Shekhar, A., Hetrick, W.P., 2007. Sensorimotor gating in manic and mixed episode bipolar disorder. Bipolar Disord. 9, 221–229.

Caspi, A., Sugden, K., Moffitt, T.E., Taylor, A., Craig, I.W., Harrington, H., McClay, J., Mill, J., Martin, J., Braithwaite, A., Poulton, B., 2003. Influence of life stress on depression: moderation by a polymorphism in the 5-HTT gene. Science 301, 386–389.

Charney, D., Heninger, G., 1986. Abnormal regulation of noradrenergic function panic disorders. Arch. Gen. Psychiatry 43, 1042–1054.

Charney, D.S., Woods, S.W., Goodman, W.K., Heninger, G.R., 1987. Neurobiological mechanisms of panic anxiety: biochemical and behavioral correlates of yohimbine-induced panic attacks. Am. J. Psychiatry 144, 1030–1036.

Craddock, N., Owen, M.J., 2010. The Kraepelinian dichotomy—going, going… but still not gone. Br. J. Psychiatry 196, 92–95.

Curtin, C., Jojic, M., Bandini, L.G., 2014. Obesity in children with autism spectrum disorder. Harv. Rev. Psychiatry 22, 93–103.

Daoust, A.M., Limoges, E., Bolduc, C., Mottron, L., Godbout, R., 2004. EEG spectral analysis of wakefulness and REM sleep in high functioning autistic spectrum disorders. Clin. Neurophysiol. 115, 1368–1373.

Dement, W.C., 1967. Studies on the effects of REM deprivation in humans and animals. Res. Publ. Assoc. Res. Nerv. Ment. Dis. 43, 456–467.

Depue, R.A., Arbisi, P., Krauss, S., Iacono, W.G., Leon, A., Muir, R., Allen, J., 1990. Seasonal independence of low prolactin concentration and high spontaneous blink rates in unipolar and bipolar II seasonal affective disorder. Arch. Gen. Psychiatry 47, 356–364.

Devlin, B., Scherer, S.W., 2012. Genetic architecture in autism spectrum disorder. Curr. Opin. Genet. Dev. 22, 229–237.

D'Onofrio, S., Kezunovic, N., Hyde, J.R., Luster, B., Messias, E., Urbano, F.J., Garcia-Rill, E., 2015. Modulation of gamma oscillations in the pedunculopontine nucleus (PPN) by neuronal calcium sensor protein-1 (NCS-1): relevance to schizophrenia and bipolar disorder. J. Neurophysiol. 113, 709–719.

Etain, B., Henry, C., Bellivier, F., Mathieu, F., Leboyer, M., 2008. Beyond genetics: childhood affective trauma in bipolar disorder. Bipolar Disord. 10, 867–876.

Feinberg, I., 1969. Schizophrenia: caused by a fault in programmed synaptic elimination during adolescence? J. Psychiatry Res. 17, 319–334.

Feinberg, I., Braun, M., Koresko, R.L., Gottleib, F., 1969. Stage 4 sleep in schizophrenia. Arch. Gen. Psychiatry 21, 262–266.

Freedman, R., Adler, L.E., Waldo, M.C., Pachtman, E., Franks, R.D., 1983. Neurophysiological evidence for a defect in inhibitory pathways in schizophrenia: comparison of medicated and drug-free patients. Biol. Psychiatry 18, 537–551.

Gambino, F., Pavlowsky, A., Begle, A., Dupont, J.L., Bahi, N., Courjaret, R., Gardette, R., Hdjkacem, H., Skala, H., Poulain, B., Vitale, N., Humeau, Y., 2007. IL1-receptor accessory protein-like 1 (IL1RAPL1), a protein involved in cognitive functions, regulates N-type Ca^{2+}-channel and neurite elongation. Proc. Natl. Acad. Sci. USA 104, 9063–9068.

Ganguli, R., Reynolds, D.F., Kupfer, D.F., 1987. Electroencephalographic sleep in young, never-medicated schizophrenics. Arch. Gen. Psychiatry 44, 36–44.

Garcia-Rill, E., Davies, D., Skinner, R.D., Biedermann, J.A., McHalffey, C., 1991. Fibroblast growth factor-induced increased survival of cholinergic mesopontine neurons in culture. Dev. Brain Res. 60, 267–270.

Garcia-Rill, E., Skinner, R.D., Clothier, J., Dornhoffer, J., Uc, E., Fann, A., Mamiya, N., 2002. The sleep state-dependent midlatency auditory evoked P50 potential in various disorders. Thalamus Relat. Syst. 2, 9–19.

Garcia-Rill, E., Heister, D.S., Ye, M., Charlesworth, A., Hayar, A., 2007. Electrical coupling: novel mechanism for sleep–wake control. Sleep 30, 1405–1414.

Garcia-Rill, E., Kezunovic, N., D'Onofrio, S., Luster, B., Hyde, J., Bisagno, V., Urbano, F.J., 2014. Gamma band activity in the RAS-intracellular mechanisms. Exp. Brain Res. 232, 1509–1522.

Geyer, M.A., Braff, D.U., 1982. Habituation of the blink reflex in normals and schizophrenic patients. Psychophysiology 19, 1–6.

Giannotti, F., Cortesi, F., Cerquiglini, A., Vagnoni, C., Valente, D., 2011. Sleep in children with and without autistic regression. J. Sleep Res. 20, 338–347.

Gillette, G., Skinner, R.D., Rasco, L., Fielstein, E., Davis, D., Pawelak, J., Freeman, T., Karson, C.N., Boop, F.A., Garcia-Rill, E., 1997. Combat veterans with posttraumatic stress disorder exhibit decreased habituation of the P1 midlatency auditory evoked potential. Life Sci. 61, 1421–1434.

Glodman, S.E., Surdyka, K., Cuevas, R., Adkins, K., Wang, L., Malow, B.A., 2009. Defining sleep phenotype in children with autism. Dev. Neuropsychol. 34, 560–573.

Goraya, J.S., Cruz, M., Valencia, I., Kalevias, J., Khurana, D.S., Hardison, H.H., Marks, H., Legido, A., Kothare, S.V., 2009. Sleep study abnormalities in children with attention deficit hyperactivity disorder. Pediatr. Neurol. 40, 42–46.

Greenhil, L., Puig-Antich, J., Goetz, R., Hanlon, C., Davies, M., 1983. Sleep architecture and REM sleep measures in prepubertal children with attention deficit hyperactivity. Sleep 6, 91–101.

Hallmayer, J., Cleveland, S., Torres, A., Phillips, J., Cohen, B., Torigoe, T., Miller, J., Fedele, A., Collins, J., Smith, K., Lotspeich, L., Croen, L.A., Ozonoff, S., Lajonchere, C., Grether, J.K., Risch, N., 2011. Genetic heritability and shared environmental factors among twin pairs with autism. Arch. Gen. Psychiatry 68, 1095–1102.

Hashimoto, K., Shimizu, E., Komatsu, N., Nakazato, M., Okamura, N., Watanabe, H., Kumakiri, C., Shinoda, N., Okada, S., Takei, N., Iyo, M., 2003. Increased levels of serum basic fibroblast growth factor in schizophrenia. Psychiatry Res. 120, 211–218.

Heim, C., Newport, D.J., Mletzko, T., Miller, A.H., Nemeroff, C.B., 2008. The link between childhood trauma and depression: insights from HPA axis studies in humans. Psychoneuroendocrinology 33, 693–710.

Herrmann, C.S., Demiralp, T., 2005. Human EEG gamma oscillations in neuropsychiatric disorders. Clin. Neurophysiol. 116, 2719–2733.

Hirschfeld, R.M.A., 2001. The comorbidity of major depression and anxiety disorder: recognition and management in primary care. Prim. Care Companion J. Clin. Psychiatry 3, 244–254.

Hodge, D., Carollo, T.M., Lewin, M., Hoffman, C.D., Sweeney, D.P., 2014. Sleep patterns in children with and without autism spectrum disorder: developmental comparisons. Res. Dev. Disabil. 35, 1631–1638.

Holstein, D.H., Vollenweider, F.X., Geyer, M.A., Csomor, P.A., Belser, N., Eich, D., 2013. Sensory and sensorimotor gating in adult attention-deficit/hyperactivity disorder (ADHD). Psychiatry Res. 205, 117–126.

Holzman, P.S., Proctor, L.R., Hughes, D.W., 1973. Eye tracking patterns in schizophrenia. Science 181, 179–181.

Humphreys, J.S., Gringas, P., Blair, P.S., Scott, N., Henderson, J., Fleming, P.J., Emond, A.M., 2014. Sleep patterns in children with autistic spectrum disorders: a prospective cohort study. Arch. Dis. Child. 99, 114–118.

Itil, T.M., Hsu, W., Klingenberg, W., Saletu, B., Gannon, P., 1972. Digital computer-analyzed all-night sleep EEG patterns (sleep prints) in schizophrenics. Biol. Psychiatry 4, 3–16.

Janowsky, D.S., Davis, J.M., Huey, L., Judd, L.L., 1979. Adrenergic and cholinergic drugs as episode and vulnerability markers of affective disorders and schizophrenia. Psychopharmacol. Bull. 15, 33–34.

Jeste, S.S., 2011. The neurology of autism spectrum disorders. Curr. Opin. Neurol. 24, 132–139.

Jus, K., Bouchard, M., Jus, A., Villeneuve, A., Lachance, R., 1973. Sleep EEG studies in untreated long-term schizophrenic patients. Arch. Gen. Psychiatry 29, 286–290.

Kadrmas, A., Winokur, G., 1979. Manic depressive illness and EEG abnormalities. J. Clin. Psychiatry 40, 306–307.

Kahbazi, M., Ghoreishi, A., Rahimenijad, F., Mohammadi, M.R., Kamalipour, A., Akhondzadeh, S., 2009. A randomized, double-blind and placebo-controlled trial of modafinil in children and adolescents with attention deficit and hyperactivity disorder. Psychiatry Res. 168, 234–237.

Kaner, R.J., Baird, A., Mansukhani, A., Basilico, C., Summers, B.D., Florkiewicz, R.Z., Hajjar, D.P., 1990. Fibroblast growth factor receptor is a portal of cellular entry for herpes simplex virus type I. Science 248, 1410–1413.

Karson, C.N., Dykman, R.A., Paige, S.R., 1990. Blink rates in schizophrenia. Schizophr. Bull. 16, 345–354.

Kasri, N.N., Holmes, A.M., Bultynk, G., Parys, J.B., Bootman, M.D., Rietdorf, K., Missiaen, L., McDonald, F., Smedt, H., Conway, S.J., Holmes, A.B., Berridge, M.J., Roderick, H., 2004. Regulation of InsP$_3$ receptor activity by neuronal Ca^{2+}-binding proteins. EMBO J. 23, 312–321.

Kendler, K.S., Gatz, M., Gardner, C.O., Pedersen, N.L., 2006. A Swedish national twin study of lifetime major depression. Am. J. Psychiatry 163, 109–114.

Kessler, R.C., Nelson, C.B., McGonagle, K.A., Swartz, M., Blazer, D.G., 1996. Comorbidity of DSM-III-R major depressive disorder in the general population: results from the US National Comorbidity Survey. Br. J. Psychiatry Suppl. 30, 17–30.

King, L.J., 1974. A sensory-integrative approach to schizophrenia. Am. J. Occup. Ther. 28, 529–536.

Koh, P.O., Undie, A.S., Kabbani, N., Levenson, R., Goldman-Rakic, P., Lidow, M.S., 2003. Up-regulation of neuronal calcium sensor-1 (NCS-1) in the prefrontal cortex of schizophrenic and bipolar patients. Proc. Natl. Acad. Sci. USA 100, 313–317.

Kohl, S., Heekern, K., Klosterkotter, J., Kuhn, J., 2013. Prepulse inhibition in psychiatric disorders—apart from schizophrenia. J. Psychiatry Res. 47, 445–452.

Kovelma, J.A., Scheibel, A.B., 1984. A neurohistochemical correlate of schizophrenia. Biol. Psychiatry 19, 1601–1621.

Kudlow, P.A., Cha, D.S., Lam, R.W., McIntyre, R.S., 2013. Sleep architecture variation: a mediator of metabolic disturbance in individuals with major depressive disorder. Sleep Med. 14, 943–949.

Kupfer, D.J., Foster, F.G., Coble, P., McPartland, R.J., Ulrich, R.F., 1978. The application of EEG sleep for the differential diagnosis of affective disorders. Am. J. Psychiatry 135, 69–74.

Liu, T.Y., Hsieh, J.C., Chen, Y.S., Tu, P.C., Su, T.P., Chen, L.F., 2012. Different patterns of abnormal gamma oscillatory activity in unipolar and bipolar disorder patients during an implicit motor task. Neuropsychol. 50, 1514–1520.

Liu, X., Zhang, T., He, S., Hong, B., Chen, Z., Peng, D., Wu, Y., Wen, H., Lin, Z., Fang, Y., Jiang, K., 2014. Elevated serum levels of FGF-2, NGF and IGF-1 in patients with manic episode bipolar disorder. Psychiatry Res. 218, 54–60.

Llinas, R.R., Ribary, U., Jeanmonod, D., Kronberg, E., Mitra, P.P., 1999. Thalamocortical dysrhythmia: a neurological and neuropsychiatric syndrome characterized by magnetoencephalography. Proc. Natl. Acad. Sci. USA 96, 15222–15227.

Mamelak, A.N., Hobson, J.A., 1989. Dream bizarreness as the cognitive correlate of altered neuronal brain in REM sleep. J. Cogn. Neurosci. 1 (3), 201–222.

Manschrek, T.C., 1986. Motor abnormalities in schizophrenia. In: Nasrallah, H.A., Weinberger, D.R. (Eds.), Handbook of Schizophrenia. Elsevier, Amsterdam, pp. 65–96.

Mednick, S.A., Machon, R.A., Huttunen, M.O., Bonett, D., 1988. Adult schizophrenia following prenatal exposure to an influenza epidemic. Arch. Gen. Psychiatry 45, 189–192.

Merikangas, K.R., Jin, R., He, J.P., Kessler, R.C., Lee, S., Sampson, N.A., Viana, M.C., Andrade, L.H., Hu, C., Karam, E.G., Ladea, M., Medina-Mora, M.E., Ono, Y., Posada-Villa, J., Sagar, R., Wells, J.E., Zarkov, Z., 2011. Prevalence and correlates of bipolar spectrum disorder in the world mental health survey initiative. Arch. Gen. Psychiatry 68, 241–251.

Miano, S., Parisi, P., Villa, M.P., 2012. The sleep phenotypes of attention deficit hyperactivity disorder: the role of arousal during sleep and implications for treatment. Med. Hypotheses 79, 147–153.

Miyazato, H., Skinner, R.D., Garcia-Rill, E., 1999. Neurochemical modulation of the P13 mid-latency auditory evoked potential in the rat. Neuroscience 92, 911–920.

Miyazato, H., Skinner, R.D., Garcia-Rill, E., 2000. Locus coeruleus involvement in the effects of immobilization stress on the P13 midlatency auditory evoked potential in the rat. Prog. Neuro-Psychopharmacol. Biol. Psychiatry 24, 1177–1201.

Nasrallah, H.A., Schwarzkopf, S.B., Olson, S.C., Coffman, J.A., 1991. Perinatal brain injury and cerebellar vermal lobules I-X in schizophrenia. Biol. Psychiatry 29, 567–574.

Nordentoft, M., Mortensen, P.B., Pedersen, C.B., 2011. Absolute risk of suicide after first hospital contact in mental disorder. Arch. Gen. Psychiatry 68, 1058–1064.

Nutt, D.J., 2008. Relationship of neurotransmitters to the symptoms of major depressive disorder. J. Clin. Psychiatry 69, 4–7.

Olincy, A., Martin, L., 2005. Diminished suppression of the P50 auditory evoked potential in bipolar disorder subjects with a history of psychosis. Am. J. Psychiatry 162, 43–49.

Olincy, A., Ross, R.G., Harris, J.G., Young, D.A., McAndrews, M.A., Cawthra, E., McRae, K.A., Sullivan, B., Adler, L.E., Friedman, R., 2000. The P50 auditory event-evoked potential in adult attention-deficit disorder: comparison with schizophrenia. Biol. Psychiatry 47, 969–977.

Oner, O., Oner, P., Aysev, A., Kucuk, O., Ibis, E., 2005. Regional blood flow in children with ADHD: changes with age. Brain Dev. 27, 279–285.

Ornitz, E.M., Pynoos, R.S., 1989. Startle modulation in children with posttraumatic stress disorder. Am. J. Psychiatry 146, 866–870.

Ozerdem, A., Guntenkin, B., Atagun, I., Turp, B., Basar, E., 2011. Reduced long distance gamma (28–48 Hz) coherence in euthymic patients with bipolar disorder. J. Affect. Disord. 132, 325–332.

Perry, W., Minassian, A., Feifel, D., Braff, D.L., 2001. Sensorimotor gating deficits in bipolar disorder patients with acute psychotic mania. Biol. Psychiatry 50, 418–424.

Persson, M.L., Johansson, J., Vumma, R., Raita, J., Bjerkenstedt, L., Wiesel, F.A., Venizelos, N., 2009. Aberrant amino acid transport in fibroblasts from patients with bipolar disorder. Neurosci. Lett. 457, 49–52.

Philipsen, A., Feige, B., Hesslinger, B., Ebert, D., Carl, C., Hornyak, M., Lieb, K., Voderholzer, U., Riemann, D., 2005. Sleep in adults with attention-deficit/hyperactivity disorder: a controlled polysomnographic study including spectral analysis of the sleep EEG. Sleep 28, 877–884.

Picardi, A., Viroli, C., Tarsitani, L., Miglio, R., de Girolamo, G., Dell'Acqua, G., Biondi, M., 2012. Heterogeneity and symptom structure of schizophrenia. Psychiat. Res. 198, 386–394.

Picchioni, M.M., Murray, R.M., 2007. Schizophrenia. Br. Med. J. 335, 91–96.

Pittman, R.K., Sanders, K.M., Zusman, R.M., Healy, A.R., Cheema, F., Lasko, N.B., Cahill, L., Orr, S.P., 2002. Pilot study of secondary prevention of posttraumatic stress disorder with propranolol. Biol. Psychiatry 51, 189–192.

Richdale, A.L., Baker, E., Short, M., Gradisar, M., 2014. The role of insomnia, pre-sleep arousal and psychopathology symptoms in daytime impairment in adolescents with high-functioning autism spectrum disorder. Sleep Med. 15, 1082–1088.

Rodrigo, J., Suburu, A.M., Bentura, M.L., Fernandez, T., Nakada, S., Mikoshiba, K., Martinez-Murillo, R., Polak, J.M., 1993. Distribution of the inositol 1,4,5-triphosphate receptor, P400, in adult rat brain. J. Comp. Neurol. 337, 493–517.

Ross, R.J., Ball, W.A., Sullivan, K.A., Caroff, S.N., 1989. Sleep disturbance as the hallmark of posttraumatic stress disorder. Am. J. Psychiatry 146, 697–707.

Ross, R.J., Ball, W.A., Dinges, D.F., Kribbs, N.B., Morrison, A.R., Silver, S.M., Mulvaney, F.D., 1999. Rapid eye movement sleep changes during the adaptation night in combat veterans with posttraumatic stress disorder. Biol. Psychiatry 45, 938–941.

Sandor, P., Shapiro, C.M., 1994. Sleep patterns in depression and anxiety: theory and pharmacological effects. J. Psychosom. Res. 38, 125–139.

Sapolsky, R.M., Krey, L.C., McEwen, B.S., 1984. Prolonged glucocorticoid exposure reduces hippocampal neuron number. Implications for aging. J. Neurosci. 5, 1222–1227.

Sawyer, A.C., Clark, C.R., Keage, H.A., Moores, K.A., Clarke, S., Kohn, M.R., Gordon, E., 2009. Cognitive and electroencephalographic disturbances in children with attention-deficit/hyperactivity disorder and sleep problems: new insights. Psychiatry Res. 170, 183–191.

Schleker, C., Boehmerie, W., Jeromin, A., DeGray, B., Varshey, A., Sharma, Y., Szigeti-Buck, K., Ehrlich, B.E., 2006. Neuronal calcium sensor-1 enhancement of InsP$_3$ receptor activity is inhibited by therapeutic levels of lithium. J. Clin. Invest. 116, 1668–1674.

Schmunk, G., Gargus, J.J., 2013. Channelopathy pathogenesis in autism spectrum disorders. Front. Genet. 4 (222), 1–2.

Schulze, K.K., Hall, M., McDonald, C., Marshall, N., Walshe, M., Murray, R.M., Brannon, E., 2007. P50 auditory evoked potential suppression in bipolar disorder patients with psychotic features and their unaffected relatives. Biol. Psychiatry 62, 121–128.

Seiden, L.S., Sabol, K.E., Ricaurte, G.A., 1993. Amphetamine: effects of catecholamine systems and behavior. Annu. Rev. Pharmacol. Toxicol. 33, 639–677.

Seifritz, E., 2001. Contribution of sleep physiology to depressive pathophysiology. Neuropsychopharmacology 25, S85–S88.

Shalev, A.Y., Freedman, S., Peri, T., Brandes, D., Sahar, T., Orr, S.P., Pitman, R.K., 2000. Auditory startle response in trauma survivors with posttraumatic stress disorder: a prospective study. Am. J. Psychiatry 157, 255–261.

Skinner, R.D., Rasco, L., Fitzgerald, J., Karson, C.N., Matthew, M., Williams, D.K., Garcia-Rill, E., 1999. Reduced sensory gating of the P1 potential in rape victims and combat veterans with post traumatic stress disorder. Depress. Anxiety 9, 122–130.

Snyder, S., 2006. Dopamine receptor excess and mouse madness. Neuron 49, 484–485.

Spaletta, G., Pasini, A., Pau, F., Guido, G., Menghini, L., Caltagirone, C., 2001. Prefrontal blood flow dysregulation in drug naïve ADHD children without structural abnormalities. J. Neural Transm. 108, 1203–1216.

Spencer, K.M., Nestor, P.G., Niznikiewicz, M.A., Salisbury, D.F., Shenton, M.E., McCarley, R., 2003. Abnormal neural synchrony in schizophrenia. J. Neurosci. 23, 7407–7411.

Stevens, H.E., Smith, K.E., Rash, B.G., Vaccarino, F.M., 2010. Neural stem cell regulation, fibroblast growth factors, and the developmental origins of neuropsychiatric disorders. Front. Neurosci. 4 (59), 1–14.

Sullivan, P.F., Daly, M.J., O'Donovan, M., 2012. Genetic architectures of psychiatric disorders: the emerging picture and its implications. Nat. Rev. Genet. 13, 537–551.

Tandon, R., Greden, J.F., Haskett, R.F., 1993. Cholinergic hyperactivity and negative symptoms: behavioral effects of physostigmine in normal controls. Schizophr. Res. 9, 19–23.

Torres, K.C.L., Souza, B.R., Miranda, D.M., Sampiao, A.M., Nicolato, R., Neves, F.S., Barros, A.G., Dutra, W.O., Gollob, K.J., Roamno-Silva, M.A., 2009. Expression of neuronal calcium sensor-1 (NCS-1) is decreased in leukocytes of schizophrenia and bipolar disorder patients. Prog. Neuro-Pharmacol. Biol. Psychiat. 33, 229–234.

Uhlhaas, P.J., Singer, W., 2010. Abnormal neural oscillations and synchrony in schizophrenia. Nat. Rev. Neurosci. 11, 100–113.

Urbano, F.J., Leznik, E., Llinas, R.R., 2007. Modafinil enhances thalamocortical activity by increasing neuronal electronic coupling. Proc. Nat. Acad. Sci. 104, 12554–12559.

Vaiva, G., Ducrocq, F., Jezequel, K., Averland, B., Lestavel, P., Brunet, A., Marmar, C.R., 2003. Immediate treatment with propranolol decreases posttraumatic stress disorder two months after trauma. Biol. Psychiatry 54, 947–949.

Valleau, J.C., Sullivan, E.L., 2014. The impact of leptin on perinatal development psychopathology. J. Chem. Neuroanat. 61–62, 221–232.

Weinberger, D.R., 1982. Implications of normal brain development for the pathogenesis of schizophrenia. Arch. Gen. Psychiatry 44, 660–669.

Winokur, A., Gary, K.A., Rodner, S., Rae-Red, C., Fernando, A.T., Szuba, M.P., 2001. Depression, sleep physiology and antidepressant drugs. Depress. Anxiety 14, 19–28.

Woodward, S.H., Friedman, M.J., Bliwise, D.L., 1996. Sleep and depression in combat-related PTSD inpatients. Biol. Psychiatry 39, 182–192.

Yolken, R., 2004. Viruses and schizophrenia: a focus on herpes simplex virus. Herpes 11 (S2), 83A–88A.

Zarcone, V., Azumi, K., Dement, W., Gulevich, G., Kraimer, H., Pivik, R., 1975. REM phase deprivation and schizophrenia II. Arch. Gen. Psychiatry 32, 1431–1436.

12

Neurological Disorders and the RAS

Brennon Luster, BS, Erica Petersen, MD†, and Edgar Garcia-Rill, PhD**

*Center for Translational Neuroscience, Department of Neurobiology and Developmental Sciences, University of Arkansas for Medical Sciences, Little Rock, AR, USA
†Department of Neurosurgery, University of Arkansas for Medical Sciences, Little Rock, AR, USA

AROUSAL, WAKING, AND NEUROLOGICAL DISORDERS

Given the information outlined in the previous chapters, what aspects of reticular activating system (RAS) function will be involved in neurological disorders? (1) We know that the RAS participates in fight-or-flight responses; therefore, we would expect that responses to sudden alerting stimuli will be abnormal. For disorders in which the RAS is overactive, this would mean that such stimuli will produce exaggerated responses that would be manifested as *exaggerated startle responses or hyperactive reflexes* such as the blink reflex (see Chapter 7). (2) Another property of the RAS is its rapid habituation to repetitive stimuli. This is reflected in its lack of responsiveness to rapidly repeating stimuli, that is, its rapid habituation. This endows the RAS with its capacity for sensory gating, the property of decreasing responsiveness of repetitive events in favor of novel or different stimuli. For disorders in which this property is affected, we expect a *decrease in habituation or a sensory gating deficit* (see Chapter 6). (3) The RAS controls waking and sleep, so that sleep patterns would be dysregulated. If the RAS is downregulated by a disorder, we expect an inability to remain awake, the presence of *excessive daytime sleepiness*, and an excess of total sleep time, especially a decrease in slow-wave sleep (SWS). If, on the other hand, the RAS is upregulated, we expect difficulty in getting to sleep and maintaining sleep. This would be reflected in *insomnia* or disrupted sleep during the night, as well as *increased rapid eye movement (REM) sleep drive*, which is characterized by vivid nightmares and frequent awakenings (see Chapters 2 and 5). In Chapter 5, we considered what would happen if the developmental decrease in REM sleep drive did not occur. Such a condition would lead to increased REM sleep drive not only during sleep (resulting in intense dreaming) but also perhaps during waking, resulting in dreaming while awake or hallucinations, along with *hypervigilance.* (4) The RAS also modulates the maintenance of waking, a property ignored by many but one that affects a host of functions. The inability to maintain a steady waking state, in the form of *maintained gamma band activity*, will interfere with attention, learning, and memory, to name a few processes (see Chapter 9).

Another factor in all of these disorders is the level of frontal lobe blood flow. Decreased frontal lobe blood flow, or *hypofrontality*, is present in all of these disorders to some extent. This state is present during REM sleep and may account for the lack of critical judgment present in our dreams; however, such a state during waking would lead to reflexive reactions with lack of consideration of consequences. This condition is probably involved in the lack of habituation to repetitive stimuli, or a sensory gating deficit. Under the condition of decreased cortical modulation, fight-or-flight responses and reflexes would be exaggerated. Whether hypofrontality is a cause of RAS dysregulation or RAS dysregulation leads to hypofrontality remains to be determined. However, successful treatment of any of these disorders will be marked by normalization of wake-sleep rhythms, reflexes, and frontal lobe blood flow.

Therefore, in the assessment of the role of the RAS in neurological disorders, we need to consider (a) responses to reflexes such as the startle response and the blink reflex using electromyographic (EMG), (b) sensory gating using paired stimuli or prepulse inhibition paradigms such as the P50 potential, (c) sleep patterns or somnography using nighttime electroencephalogram (EEG), (d) the maintenance or interruption of gamma band activity during waking also using EEG or magnetoencephalography (MEG), and (e) the level of frontal lobe blood flow. These measures will allow us to determine in detail the role of the RAS in specific disorders.

PARKINSON'S DISEASE

Symptoms—Parkinson's disease (PD) is characterized by a variety of symptoms, which include 3–7 Hz resting tremor, rigidity, postural and gait abnormalities, akinesia, and bradykinesia. In addition, a number of other symptoms are present, including abnormal reflexes and higher-level impairments in frontal lobe function and cognition. Although many of the symptoms become manifest after a degenerative process has reduced the function of dopaminergic substantia nigra (SN) neurons below a certain threshold, there are a number of additional degenerative or functional changes in such areas as the locus coeruleus (LC), raphe nuclei, basal forebrain, and frontal cortex (Jellinger, 1991). These patients show decreased habituation of the blink and other reflexes (Ferguson et al., 1978; Kimura, 1973; Nakashima et al., 1993; Penders and Delwaide, 1971; Rothwell et al., 1983) and also exhibit anxiety disorder (including panic attacks) and depression (Cummings, 1992; Henderson et al., 1992; Menza et al., 1993; Stein et al., 1990; Vazquez et al., 1993). Moreover, cognitive impairments related to attentional deficits are present (Cooper et al., 1991; Jagust et al., 1992; Owen et al., 1993; Robbins et al., 1994) that, in general, correlate with decreased frontal lobe glucose utilization (Eidelberg et al., 1994; Holthoff et al., 1994; Jagust et al., 1992; Peppard et al., 1992; Pillon et al., 1991; Sawada et al., 1992).

Etiology—PD affects about 1% of people over the age of 60 and 4% of those over 80 and primarily involves the degeneration of the SN (Samil et al., 2004). PD has a mainly environmental etiology, with pollution from pulp and paper mills, pesticides, and heavy metals being the potential toxins; head trauma (boxing) and vascular (arteriosclerosis) injury represent fairly common causes, and there is a small genetic risk. As we discussed in Chapter 7, ~40% of patients with REM sleep behavior disorder (RBD) go on to develop PD (Schenck et al., 1996a,b), and many PD patients develop RBD in late stages (Boeve et al., 2003). RBD patients have serum anti-LC antibodies (Schenck et al., 1997), and we found that there was a marked cell loss in the LC of RBD patients (Schenck et al., 1996a,b), suggesting that in RBD, the degeneration of catecholaminergic cells may begin in the LC and progress to the SN, while in PD, the degeneration may begin in the SN and progress to the LC. However, recent studies suggest that early damage in both PD and RBD begins in the olfactory bulb and/or vagal afferents and may involve α-synuclein and *tau* pathologies (Doty, 2012; Kim, 2013). But it is not clear if these represent only a portion of patients.

EEG, reflexes, and P50 potential—The hyperactive reflexes of several kinds were described above (Ferguson et al., 1978; Kimura, 1973; Nakashima et al., 1993; Penders and Delwaide, 1971; Rothwell et al., 1983). PD patients show sleep disturbances that include increased REM sleep drive, decreased SWS, and frequent awakenings leading to daytime sleepiness, all resulting in insomnia (Jankovic, 2007). These observations suggest that the RAS, especially the pedunculopontine nucleus (PPN) that is in charge of waking and REM sleep, is overactive in PD. We carried out a study of the P50 potential in PD patients (Teo et al., 1997). We showed that the *amplitude* of the first P50 potential response of a pair of stimuli administered 250 ms apart was 40% higher in PD patients as a whole, but it was not statistically significant because there was a stage-dependent increase; that is, stage 3 patients had a 10% increase, stage 4 patients had a 42% increase, and stage 5 patients had a 55% increase. We concluded that this represented a trend towards increased P50 potential amplitude in late-stage PD. Moreover, P50 potential *habituation* (the amplitude of the second response as a percent of the first response of a pair) decreased significantly in the PD group as a whole. In addition, there was a statistically significant decrease in habituation with severity. Figure 12.1 shows on the left side the percent habituation with severity of PD compared to age- and gender-matched controls. The results showed that stage 5 PD patients (filled inverted triangles) exhibited decreased habituation at the 250 and 500 ms interstimulus intervals (ISIs) compared to controls (filled squares). These findings provided convincing evidence of a decrease in habituation (indicative of a decrease in sensory gating) and a trend towards increased amplitude (indicative of an increase in level of arousal) in a population of PD subjects. Also, we should note that animal studies subsequently confirmed that the PPN is overactive in animal models of PD (Orieux et al., 2001; Breit et al., 2000).

We then measured the P50 potential in PD patients who received bilateral pallidotomy that alleviated their motor symptoms, and we found that the amplitude and habituation of the P50 potential were within normal levels (Teo et al., 1998). Figure 12.1 shows on the right side that stage 5 PD patients (filled inverted triangles) before pallidotomy manifested decreased habituation of the P50 potential at 250 and 500 ms ISIs, as previously determined. However, soon after pallidotomy, P50 potential habituation in the

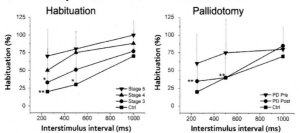

P50 potential in Parkinson's disease

FIGURE 12.1   **The P50 potential in PD.** Left side. Paired click stimuli were delivered at three ISIs, 250, 500, and 1000 ms, so that a recovery curve could be plotted. Age- and gender-matched control subjects (black squares) showed ~20%, 30%, and 60% habituation of the second response, respectively. PD patients at stage 3 showed ~30%, 50%, and 75% habituation, respectively. Stage 4 patients showed ~50%, 75%, and 80% respectively. Stage 5 patients showed decreased habituation at ~73%, 76%, and 95%, respectively. Stage 5 patients differed significantly form control subjects at the 250 ms (**$p<0.01$), and 500 ms (*$p<0.05$) ISIs, and from stage 3 patients at the 250 ms ISI (*$p<0.05$) (Teo et al., 1997). Right side. PD patients recorded before bilateral pallidotomy (inverted black triangles) showed habituation at ~60%, 75%, and 80%, respectively, for the three ISIs. After pallidotomy, the same patients showed ~35%, 35%, and 60%, respectively. The percent habituation differed significantly between the post- and presurgery results at the 250 and 500 ms ISIs (**$p<0.01$) and was not different from control subjects (black squares) (Teo et al., 1998).

same PD patients after surgery (filled circles) was not different from control (filled squares). These observations showed that, immediately upon executing the lesion of the internal pallidum, the patients began to show marked improvements in motor symptoms; for example, subjects who were so incapacitated that they could not speak before surgery began talking on the table. This means that the SN degeneration in PD sets up an imbalance that is immediately corrected by the surgery. No postsurgical delay is evident, suggesting that there is no reorganization in terms of axonal growth or even significant gene transcription that is required for recovery of function. We speculated that the increased PPN output in PD was instantly reinhibited or downregulated to normalize sensory gating. Interestingly, after surgery, these patients still needed L-dopa (Teo et al., 1998).

The immediate normalization of the P50 potential, which is generated by the PPN (see Chapter 6), shows that the RAS-related deficits in PD (e.g., sleep, arousal, and hyperactive reflexes) could be overcome by a surgical intervention that corrects for the overactive or disinhibited PPN output. Given that the RAS modulates arousal and preconscious awareness, agents or procedures that normalize RAS output could be effective in compensating for the effects of PD.

Treatment—The progressive nature of the disease makes treatment difficult since there is no cure. Prior to the use of L-dopa for the treatment of PD, lesions of the thalamus and pallidum were being used, but pharmacological treatment became the preferred form of therapy (Speakman, 1963). While the use of L-dopa has been effective for some of the symptoms of the disease, long-term use (>5 years) and increasing dosages can lead to dyskinesias. Other agents have been used but they provide less effective relief and can produce unwanted side effects. Surgical treatment includes the use of thalamotomy, pallidotomy, and subthalamic nucleus (STN) lesions. Nowadays, the most common surgical approach involves deep brain stimulation (DBS), with the most common site used being the STN.

DBS entails placing stimulating electrodes in the STN and delivering high-frequency (>80 Hz) short-duration (50–100 µs) pulses for as long as 20 of every 24 h. When stimulating nerve tissue, the rheobase is the lowest intensity or current amplitude that can activate the tissue, while the chronaxie is the pulse duration at which the threshold intensity is twice that of the rheobase (Irnich, 1980). The chronaxie of fibers is lower than that of cells (Nowak and Bullier, 1998), so that stimuli in the 50–100 µs range preferentially activate fibers, while durations of 400–1000 µs preferentially activate neurons. Stimulation frequencies in the range used for most DBS would not induce firing in cells but rather depolarize block neurons in the region. That is, the site being stimulated is essentially being inactivated, akin to a "physiological" lesion. The stimulating electrodes commonly used are ~1.2 mm in diameter, which represents a sizeable portion of the brain sites being stimulated. But it is not known if the introduction of these electrodes into a region represents a partial lesion. This requires that placement be very accurate, and several imaging tools are used to attempt to localize the site of stimulation, in addition to determining any palliative effects during the implantation procedure. STN DBS has been found effective in treating the motor symptoms of the disease and improving sleep quality and quality of life (Sobstyl et al., 2014) and appears to be effective in treating depression, but not anxiety (Cuoto et al., 2014). This is surprising since STN DBS appears to decrease P50 potential habituation (Gulberti et al., in press), which should represent an anxiolytic effect.

More recently, the PPN has become a target for DBS in PD. This potential use was predicted from animal studies using stimulation in the region of the PPN, which at the time was thought to represent in the mesencephalic locomotor region (MLR) (Garcia-Rill, 1986, 1991, 1997). The MLR was described as a region that could be stimulated by ramping up current using long-duration (0.5–1.0 ms) pulses at 40–60 Hz in the precollicular-postmamillary transected cat, to induce locomotion on a treadmill (Shik et al., 1966). The most effective site was the lateral cuneiform nucleus, dorsal to the lateral aspect of the superior cerebellar peduncle. As we saw in Chapters 2 and 7, this is precisely the location of the PPN *pars compacta*. Our discovery that every PPN neuron manifests gamma band activity subserved by N- and P-/Q-type calcium channels (Chapter 9) helps explain why the optimal stimulation frequency for stimulating the PPN is 40–60 Hz, and the parameters needed to activate these channels (ramps vs. steps to elicit oscillations in single neurons) also explain the need to use long-duration pulses and ramping up current steps to activate the nucleus as a whole (Garcia-Rill et al., 2014). Figure 12.2 is a three-dimensional reconstruction of the human brain showing sagittal sections of the midbrain and pons. Anterior is to the right, posterior to the left, with the basis pontis at the bottom and the inferior and superior colliculi at the top. The sections are stacked from lateral to medial and show the locations of individual PPN neurons as 100 μm spheres. Note the wedge shape of the PPN located ventrally to the inferior colliculus. The dashed line is the approximate approach used for the introduction of DBS electrodes for the treatment of PD. The small size of the *pars compacta* is a daunting target, requiring excellent precision in order to stimulate this region effectively.

A number of studies using PPN DBS for the treatment of PD have reported improvements in motor function (Mazzone et al., 2005, 2008, 2009), but not all groups reported positive effects (Ferraye et al., 2010; Moro et al., 2010). Ferraye et al. (2010) found that bilateral PPN stimulation at 15–25 Hz improved gait and decreased falls. Moro et al. (2010) used unilateral stimulation at 50 and 70 Hz to improve falls and motor scores. Stefani et al. (2007, 2013) used PPN stimulation at 10 and 25 Hz, with a significant improvement in sleep patterns and modest improvement in gait. Alessandro et al. (2010) used 25 Hz stimulation to show a significant amelioration in sleep scores and executive function. Thevanasathan et al. (2010, 2012) showed

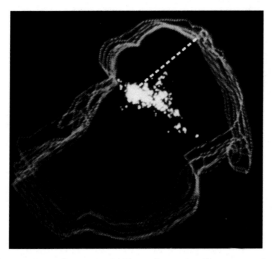

FIGURE 12.2 **Location of PPN DBS.** Three-dimensional reconstruction of the human brain showing sagittal sections of the midbrain and pons. Anterior is to the right, posterior to the left, with the basis pontis at the bottom and the inferior and superior colliculi at the top. The sections are stacked from lateral to medial and show the locations of individual PPN neurons as 100 μm spheres. Note the wedge shape of the PPN located ventrally to the inferior colliculus. The dashed line is the approximate approach used for DBS in PD.

that PPN stimulation at 20–35 Hz improved reaction time and fall scores. The latter study used double-blind analysis and established that bilateral stimulation was more effective than unilateral. One study performed sleep measures and found that PPN DBS improved not only nighttime sleep but also daytime sleepiness (Peppe et al., 2012). Others showed that PPN DBS may improve cognitive function (Tyckoki et al., 2011) and that low-frequency stimulation (5–30 Hz) may improve executive and higher functions (Stefani et al., 2013). One effect of continuous DBS is that the RAS responds well to novelty but habituates to repetitive or continuous input. This may be the reason why PPN DBS does not produce hyperarousing effects and indeed may help stabilize tonic RAS output. The latest findings show that PPN DBS using continuous 40 Hz stimulation (a frequency observed in PPN neurons, see Chapter 9) in an animal model of PD improved movement time in a delayed sensorimotor task and reduced 6-OHDA-induced rotational movements (Capozzo et al., 2014).

ALZHEIMER'S DISEASE

Symptoms—Alzheimer's disease (AD) is the most common form of dementia in which there are difficulty remembering events and decreases in short-term memory. As the disease progresses, there is confusion and long-term memory loss, along with decrements in executive function and attention. The lack of interaction with the external world leads to pronounced apathy, until patients disregard even themselves, failing to perform even the most basic hygienic functions. In some (20–45%) patients, there is sundowning agitation, in which a range of symptoms including confusion, fatigue, restlessness, and anger increase in the evening, suggestive of circadian dysregulation.

Etiology—AD occurs in people over 65 years of age and will soon affect 1 in 100 persons worldwide. Very few cases have a strictly genetic etiology, most cases involving environmental and genetic risk factors. The most common genetic risk factor is inherited apolipoprotein E ε4 allele, which may contribute to amyloid protein accumulation. *Tau* protein abnormalities may lead to microtubule breakdown and accumulation of neurofibrillary tangles. Head trauma and high cholesterol are additional risk factors. The exact mechanism that leads to cell death is controversial, with one suggestion being the disruption of intracellular calcium levels (Khachaturian, 1994; Mattson et al., 2000), with mutations in the presenilin genes of mice leading to increased intracellular calcium release from the endoplasmic reticulum (ER) (Stutzmann et al., 2006) (see Figure 12.3).

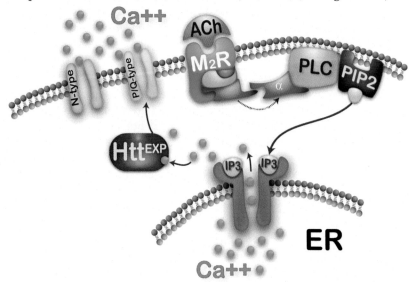

FIGURE 12.3   **Intracellular pathway for huntingtin.** In the case of muscarinic cholinergic receptors (M_2R blue structure in the membrane) activated by acetylcholine (ACh), as would occur in a PPN neurons, G proteins are activated to trigger the phospholipase C (PLC)-phosphatidylinositol bisphosphate (PIP2) pathway to release inositol trisphosphate (IP3) into the cytoplasm. The IP3 binds to the IP3 receptor in the ER to release calcium (red spheres) intracellularly. Expansion of huntingtin (Htt[exp] red cytoplasmic protein) binds calcium that flows from the ER and through P-/Q-type channels in the membrane. The Htt[exp] may increase intracellular calcium levels until they become toxic. Lithium may block the action of Htt[exp] to alleviate calcium concentrations, acting to protect neurons.

EEG, reflexes, and P50 potential—In general, the EEG findings in AD suggest an increase in lower frequencies such as delta and a decrease in higher frequencies such as beta and gamma (Horie et al., 1990; Kim et al., 2012; Koponen et al., 1989; Park et al., 2012). However, some studies point to increased gamma band EEG activity in some patients with AD (Koenig et al., 2005; van Deusen et al., 2008). The blink reflex and startle response are delayed and/or exaggerated in AD (Bonanni et al., 2007; Green et al., 1997; Ueki et al., 2006), indicative of decreased sensory gating. The P50 potential was found to be reduced in amplitude as well as decreased in habituation (Buchwald et al., 1989). This profile is similar to the P50 findings in autism (Buchwald et al., 1992) discussed in Chapter 11. Together, these findings suggest that the PPN is underactive in AD, accounting for the decreased REM sleep duration (Bliwise, 2004), decreased high frequencies in the EEG, and decreased P50 potential amplitude. The decreased habituation of the P50 potential may be explained by decreased descending cortical modulation of the RAS. Therefore, both ascending RAS output and descending cortical output are reduced, making it very difficult to reestablish appropriate levels of vigilance.

Treatment—There is also no cure for AD. Treatment is directed at specific symptoms, especially in later stages to address behavioral problems with the use of antidepressants and antipsychotics. Life expectancy after diagnosis is less than 10 years. Considering that there seems to be a decrease in arousal marked by apathy, it is surprising that stimulants, including modafinil, failed to induce improvements in apathy (Rea et al., 2014). One possibility is that cortical and subcortical cell loss is so advanced that there is not enough tissue to reactivate efficiently. The results to be discussed below on spatial neglect due to stroke suggest that decreased activity in large regions of the brain can be reactivated; however, in the case of AD, the cell loss may be so extreme that the RAS, even when stimulated pharmacologically, has little to activate. This devastating disease is still frustratingly difficult to treat effectively.

There is a potentially very effective new treatment. A drug developed in Japan called T-817MA is basically an inhibitor of *tau* protein phosphorylation, and its effects include restoration of transmitter function (Moreno et al., 2011) and cognitive decline in animals treated with amyloid protein (Kimura et al., 2009).

HUNTINGTON'S DISEASE

Symptoms—Huntington's disease (HD) is characterized by progressive motor (in the form of chorea), psychiatric (in the form of psychosis and related symptoms), and cognitive (in the form of executive function) impairments. Early symptoms include changes in mood and unsteady gait, which then develop into chorea and abnormal posture, as well as anxiety, depression, loss of affect (probably due to hypofrontality), and ultimately frank psychosis. Sleep disturbances include frequent awakenings, insomnia, daytime sleepiness, and RBD in some patients (Arnulf et al., 2008; Aziz et al., 2010; Cuturic et al., 2009; Morton et al., 2005).

Etiology—This rare late-onset (30s and 40s) disease is caused by an autosomal dominant mutation. The huntingtin gene (Htt) in HD undergoes CAG repeat expansion (Htt^{exp}) that leads to neurotoxicity and degeneration. The higher the number of CAG repeats, the higher the risk of developing HD. The Htt^{exp} leads to accumulation and formation of inclusions. The degeneration in HD is centered on medium spiny neurons of the striatum, but there is also cell loss in the cortex, hippocampus, and hypothalamus (Rosas et al., 2008; Vonsattel and DiFiglia, 1998).

EEG, reflexes, and P50 potential—The EEG in HD has been reported to show decreases in alpha and beta power (Painold et al., 2010) but conversely increases in delta and beta power (Bylsma et al., 1994). Surprisingly, changes in gamma band have not been described. However, in one of the most studied animal models of HD, the R6/2 mouse model, there is not only increased theta frequency but also massively increased gamma band activity (Fisher et al., 2013). This suggests that there may be overactivity of mechanisms mediating gamma band oscillations, which includes high-threshold, voltage-dependent calcium channels (see Chapter 9). The exaggerated influx of calcium through overactive N- and P-/Q-type calcium channels may contribute to the degeneration of cells in this disorder. Figure 12.3 is a variation of Figure 11.5, in which the presumed role of Htt^{exp} leads to excessive intracellular calcium release. Coupled with excessive influx of calcium through membrane channels, the increased intracellular calcium concentration could reach toxic levels. Basically, the excessive action of Htt^{exp} leads to the increased release of intracellular calcium due to its potentiation of InsP3 binding to the InsP3 receptor in the ER. In addition, Htt^{exp} promotes the action of calcium channels on the membrane that further increases calcium influx, until calcium concentrations become toxic. Lithium has also been proposed to have a neuroprotective effect in AD, and its site of action could well be the modulation of the activity of Htt^{exp}, thus downregulating the buildup of intracellular calcium.

The excessive activity of high-threshold voltage-dependent calcium channels that mediate gamma band activity in the PPN (Kezunovic et al., 2011) may be responsible for the manifestation of exaggerated levels of gamma band activity in the R6/2 mouse (Fisher et al., 2013). However, one would expect that the expression of gamma activity would ultimately decrease once enough neurons die. It remains to be determined if indeed this occurs in HD or in the R6/2 mouse, that is, an initially exaggerated level of EEG gamma band activity (that could signal the beginning of the degenerative process) followed by a marked decrease in gamma band levels once enough cells have died. If so, this would suggest that EEG gamma band levels could signal disease progression.

Blink, corneal, and jaw reflexes all manifest decreased habituation (Bollen et al., 1986), as does the auditory startle response (Swerdlow et al., 1995) in HD. We carried out a study of the P50 midlatency auditory evoked response in HD (Uc et al., 2003). We demonstrated decreased amplitude and prolonged latency of the P50 potential (in keeping with decreased arousal levels), as well as a lack of habituation of the P50 potential in a paired click paradigm, consistent with impairment of sensory gating in HD. These abnormalities were present in patients from all stages. There were no significant correlations with clinical features. Figure 12.4 is a graph of P50 potential habituation seen in HD, suggesting that there is a marked decrease in sensory gating in the disease. A similar study of the P50 potential attempting to correlate changes in this waveform that correlate to disease progression has not been carried out.

Treatment—There is no cure so that therapy is for specific symptoms, with tetrabenazine approved for the treatment of chorea and other agents used for the treatment of psychiatric symptoms. As mentioned in

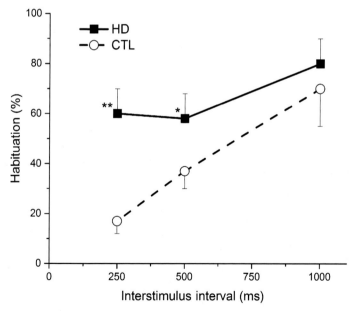

FIGURE 12.4 **The P50 potential in HD.** Paired click stimuli were delivered at three ISIs, 250, 500, and 1000 ms, so that a recovery curve could be plotted. Control subjects (open circles) showed ~15%, 35%, and 60% habituation of the second response, respectively. HD patients showed ~60%, 55%, and 75% habituation, respectively, which were significantly different from age- and gender-matched control subjects at the 250 ms (**$p<0.01$), and 500 ms (*$p<0.05$) ISIs (Uc et al., 2003).

Chapter 11, lithium has been used not only to stabilize mood but also as a neuroprotective agent. Some studies have shown prevention of the progression of HD with lithium (Danivas et al., 2013). The role of lithium in blocking the interaction of the excessive expression of NCS-1 with InsP3 and its receptor was discussed under bipolar disorder and schizophrenia in Chapter 11. The potential neuroprotective effect of lithium on that pathway, as proposed by Schlecker et al (2006), was shown in Figure 11.5. Figure 12.3 is a similar drawing showing the potential action of lithium in blocking the interaction of Htt^{exp} with InsP3 and its receptor to release calcium from the ER intracellularly, as well as the role of Htt^{exp} with membrane calcium channels to increase calcium influx. That is, lithium may counteract the stimulatory role of Htt^{exp} so that this may be a mechanism underlying the neuroprotective role of lithium in HD.

INSOMNIA

Symptoms—This so-called sleep disorder is actually a "waking disorder," in which the waking system is overactive, intruding deeply into sleep time. It is termed "primary" if it is not related to some other medical or psychiatric condition. We saw in Chapter 11 how insomnia is a hallmark of a number of psychiatric disorders such as schizophrenia, bipolar disorder, and major depression, in which insomnia is termed "secondary." Insomnia is also present in a number of neurological diseases described above. Insomnia includes difficulty falling asleep (prolonged sleep latency), frequent awakenings (difficulty maintaining sleep), and shortened sleep duration (resulting in daytime sleepiness, irritability, and fatigue), all of which lead to impairments in daytime functioning. Almost any condition that affects arousal and vigilance can induce insomnia.

Etiology—About half of the population experiences insomnia at least half the nights, and about 5% meets clinical criteria. Aside from comorbid neurological and psychiatric disorders, risk factors include principally exposure to continued stress. This makes it fairly common in modern society, but sleeping less than 4 hours per night does increase mortality (Kripke et al., 2002). There may be a genetic component since over one-half of patients diagnosed with primary insomnia reported familial insomnia (Dauvilliers et al., 2005). A number of imaging studies suggest decreased frontal lobe function and/or hypofrontality (for reviews, see Bonnet and Arand, 2010; Riemann et al., 2010).

EEG, reflexes, and P50 potential—The EEG characteristics of insomnia do not show major differences with good sleepers, with some studies reporting an increase in low beta and decrease in high beta frequency power (Cervena et al., 2014), as well as decreases in REM sleep (Baglioni et al., 2014). In general, however, the differences in the EEG are subtle but do suggest intrusion of higher frequency during typically low-frequency states, such as the incidence of higher beta activity during SWS (Feige et al., 2013; Perlis et al., 2001; Spiegelhalder et al., 2012). Experts in the field agree that primary insomnia patients show hyperarousal not only at night but also during the day (Bonnet and Arand, 2010; Riemann et al., 2010). This particular spectrum suggests that there is high-frequency activity during SWS as well as decreased REM sleep output but the hyperarousal persists during waking. As we saw in Chapter 9, the high-frequency activity generated by the RAS during waking and REM sleep is controlled by N- and P-/Q-type calcium channels, with N-type channels being permissive and P-/Q-type channels being essential (Kezunovic et al., 2011). One potential avenue of research in insomnia involves these channels. As described in Figures 11.2 and 11.3, NCS-1 at low concentrations potentiates beta/gamma oscillations in PPN neurons (D'Onofrio et al., in press; Garcia-Rill et al., in press). In insomnia, there may be a dysregulation in this protein that leads to increased high-frequency activity. The use of lithium to alleviate the hypervigilance in schizophrenia and bipolar disorder may also modulate the hyperarousal in insomnia. However, dosage may need to be low in order to only partially reduce high-frequency activity. The metabolic pathway described in Figure 11.5 shows the potential site of the downregulation by lithium of NCS-1 interactions with InsP3, as proposed previously (Schlecker et al., 2006). The same pathway may be partially modulated by low-dosage lithium to reduce the hyperarousal in insomnia. Hopefully, future off-label and clinical trials may prove beneficial in this difficult to treat condition.

Another potential avenue of treatment is the modulation of the CaMKII "waking" pathway (see Chapter 9). The increased vigilance and decreased REM sleep may indicate that the CaMKII pathway is overactivated in relation to the cAMP/PKA pathway. This pathway can be inhibited by KN-93 (see Figure 9.14), so that mild downregulation of CaMKII may provide beneficial results in insomnia. This is clinically difficult because of the many functions of CaMKII, so that other approaches may be needed. A different approach would be partial blockade of P-/Q-type calcium channels to reduce gamma oscillations and, therefore, arousal. New agents must be developed to elicit such effects, mimicking the specificity of agatoxin. Such results would suggest that insomnia may be due to P-/Q-type channel over expression or over activity.

Treatment—Treatments for insomnia are palliative and include benzodiazepines and nonbenzodiazepine hypnotics, but the risk of physical dependence is high when used chronically. Melatonin can be effective but it is not sufficiently well studied. For example, in hibernating animals, melatonin not only helps induce sleep but also retracts the gonads, so that any reduction in libido should be seriously considered.

NEGLECT

Symptoms—A deficit in attention to or awareness of one side of space results in the inability to process and perceive stimuli that is not a result of a loss of sensation. Most commonly, hemispatial neglect can result from strokes or trauma to the contralateral hemisphere but can also result from thalamic or RAS lesions. Lesions to the left hemisphere result in only transient neglect, while right side lesions lead to permanent deficits. The neglect can result not only in inattention or perceptual disregard but also in motor or intentional neglect. We consider this symptom cluster or neglect syndrome here due to the involvement of arousal in the resolution of the disorder. The accompanying flattened affect is suggestive of hypofrontality.

Etiology—Most causes involve trauma or stroke, but it has been reported to result from a number of diseases such as frontal lobe lesions, parietal and frontal cortex strokes, thalamic or RAS stroke or lesion, and all three neurodegenerative disorders discussed above, AD, PD, and HD. Unilateral lesions of the cortex or thalamus produce unilateral neglect (Orem et al., 1973; Watson et al., 1981), but bilateral lesions of the intralaminar thalamus produce akinetic mutism (Mills and Swanson, 1978), while bilateral RAS lesions produce coma (Watson et al., 1974, 1981).

EEG, reflexes, and P50 potential—The EEG in neglect is generally depressed, with overall slowing, increased delta band activity, and inability to generate fast activity (Coslett et al., 1987; Demeurisse et al., 1988; Watson et al., 1977). In general, reflexes and reaction times are increased in neglect patients (Ptak and Schnider, 2006). The P50 potential is somewhat reduced in amplitude and habituation, but these effects are not significant, perhaps because recordings are done with a single midline electrode and sources in the two hemispheres may be summating algebraically (Woods et al., 2006).

Treatment—Prism adaptation is a fairly effective therapy for hemispatial neglect, probably because the effort required to carry out the tasks requires affective incitement of the downregulated hemisphere. Hypoarousal exacerbates, if not directly causes, symptoms of neglect (Coslett et al., 1987; Demeurisse et al., 1988; Storrie-Baker et al., 1997; Watson et al., 1977). Therefore, measures that increase arousal should relieve the symptoms of neglect. For example, the cold pressor test can be used to transiently reduce neglect (Storrie-Baker et al., 1997). We recorded the P50 potential from left spatial neglect patients who showed a nonsignificant decrease in amplitude before, along with an increase after cold pressor testing. In addition, the stimulant modafinil eliminated the neglect and restored P50 potential amplitude and habituation to normal levels (Woods et al., 2006). These findings suggest that indeed, as has been proposed previously, neglect represents a decrement in arousal levels and restoration of those levels will decrease or eliminate the symptom. Therefore, where indicated, stimulants such as amphetamine, methylphenidate, or modafinil may be effective, although the low abuse potential of modafinil, because it does not activate the dopamine pathway like amphetamine and methylphenidate, may be the best option.

CLINICAL IMPLICATIONS

DBS—The results of DBS discussed in Chapter 11 demonstrate that DBS of the STN and PPN are now effective for certain measures, are surgically fairly safe, and are well tolerated. As far as the PPN is concerned, stimulation at gamma frequencies appears to improve function in posture and movement, perhaps because the preferred frequency of these cells is being imposed by DBS. Moreover, the use of continuous application of DBS may induce habituation and establish a stable level of activation, essentially helping maintain gamma band frequencies. If this is the case, and appropriate studies on these patients are still necessary, this method may be amenable for the treatment of other disorders involving dysregulation in PPN output, due to either overactivity as in PD or lack of maintenance of gamma band activity or interrupted gamma activity. That is, increased PPN output may be tractable by DBS in schizophrenia, posttraumatic stress disorder (PTSD), major depression, as well as AD, PD, HD, and perhaps even insomnia. Obviously, the use of DBS would be called for only in unresponsive and intractable cases, in which all other options have been exhausted. Disorders with downregulation of PPN function such as autism and neglect may be more difficult to address. Much more testing in animals and patients is required, along with investigation of physiological mechanisms at the cellular level using patch clamp recordings. The fact is that such physiological measures are absolutely essential in order to demonstrate that manipulations are having a physiologically relevant effect that is indeed altering symptomatology.

In Chapter 1, we discussed thalamocortical dysrhythmia (TCD), in which decreased activation or inhibition leads to increased activation of low-threshold spikes (LTSs) mediated by T-type calcium channels and slowing of thalamocortical rhythms (Llinas et al., 2005). A number of disorders become "neurogenic" so that perceived symptoms persist despite resolution of the peripheral damage, whether due to hair cell loss in tinnitus, chronic pain in the case of somatic injury, or dysregulation of perception in psychosis.

However, simply using systemic administration of LTS channel blockers would be inappropriate since the same "pacemaker" currents of I_H and LTS that mediate slow oscillations also control the rhythm of the heart. Intralaminar thalamic lesions for a number of TCD disorders

(Jeanmonod et al., 2001) are rarely used now, but one alternative in order to drive thalamocortical oscillations to a higher frequency may be the use of PPN DBS.

Calcium—Calcium is one of the most regulated ions in the brain, working in a narrow physiological window above which calcium levels can be neurotoxic and below which a number of processes will not ensue (Burgoyne, 2007). A number of calcium-sensing proteins have evolved to buffer, transport, and sense calcium levels (Berridge et al., 2003). Disturbances in these proteins have been implicated in a number of disorders, especially as the final common pathway in cell death in neurodegenerative disorders like AD, PD, and HD and in functional distortions in mental retardation, schizophrenia, and bipolar disorder (Braunewell, 2005; Mattson et al., 2000). For example, the interleukin 1 receptor accessory protein-like 1 (IL1RAPL1) interacts with NCS-1 and regulates N-type calcium channels (Gambino et al., 2007), and mutations in this gene result in a form of X-linked mental retardation (Carrie et al., 1999). As far as the RAS is concerned, the essential calcium channels involved in PPN and Pf function are P-/Q-type high-threshold, voltage-dependent channels (Kezunovic et al., 2011, 2012). The "P" was coined from the discovery of this current in Purkinje cells (Llinas et al., 1989), followed by identification of similar "Q" currents in cerebellar granule cells (Randall and Tsien, 1995). These channels are also referred to as $Ca_v2.1$ and are encoded by the CACNA1A gene. Mutations in this gene can lead to familial hemiplegic migraine (Plomp et al., 2001), while decreased P-/Q-type channel activity can lead to epilepsy (synchronization of lower frequencies; see Chapter 8) and ataxia (due to cerebellar cell malfunction) (Ophoff et al., 1998). The generation of gamma band activity by these channels in PPN and Pf cells implicates them in a large number of functions. A recent review discusses a number of therapeutic avenues being pursued that could result in new and improved treatments for several of the diseases described here (Nimmrich and Gross, 2012).

References

Alessandro, S., Ceravolo, R., Brusa, L., Pierantozzi, M., Costa, A., Galati, S., Placidi, F., Romigi, A., Iani, C., Marzetti, F., Peppe, A., 2010. Non-motor functions in parkinsonian patients implanted in the pedunculopontine nucleus: focus on sleep and cognitive problems. J. Neurol. Sci. 289, 44–48.

Arnulf, I., Nielsen, J., Lohmann, E., Schiefer, J., Schieffer, J., Wild, E., Jennum, P., Konofal, E., Walker, M., Oudiette, D., Tabrizi, S., Durr, A., 2008. Rapid eye movement sleep disturbances in Huntington disease. Arch. Neurol. 65, 482–488.

Aziz, N.A., Anguelova, G.V., Marinus, J., Lammers, G.J., Roos, R.A., 2010. Sleep and circadian rhythm alterations correlate with depression and cognitive impairment in Huntington's disease. Parkinsonism Relat. Disord. 16, 345–350.

Baglioni, C., Regen, W., Teghen, A., Spiegelhalder, K., Feige, B., Nissen, C., Riemann, D., 2014. Sleep changes in the disorder of insomnia: a meta-analysis of polysomnographic studies. Sleep Med. Rev. 18, 195–213.

Berridge, M.J., Bootman, M.D., Roderick, L., 2003. Calcium signaling: dynamics, homeostasis and remodeling. Nat. Rev. 4, 517–529.

Bliwise, D.L., 2004. Sleep disorders in Alzheimer's disease and other dementias. Clin. Cornerstone 6 (Suppl. 1A), S16–S28.

Boeve, B.F., Silber, M.H., Parisi, J.E., Dickson, D.W., Ferman, T.J., Benarroch, E.E., Schmeichel, A.M., Smith, G.E., Petersen, R.C., Ahlskog, J.E., Matsumoto, J.Y., Knopman, D.S., Schenck, C.H., Mahowald, M.W., 2003. Synucleinopathy pathology and REM sleep behavior disorder plus dementia or parkinsonism. Neurology 61, 40–45.

Bollen, E., Arts, R.J., Roos, R.A., van der Velde, E.A., Buruma, O.J., 1986. Brainstem reflexes and brainstem auditory evoked responses in Huntington's chorea. J. Neurol. Neurosurg. Psychiatry 49, 313–315.

Bonanni, L., Anzellotti, F., Varanese, S., Thomas, A., Manzoli, L., Onofrij, M., 2007. Delayed blink reflex in dementia with Lewy bodies. J. Neurol. Neurosurg. Psychiatry 78, 1137–1139.

Bonnet, M.H., Arand, D.L., 2010. Hyperarousal and insomnia: state of the science. Sleep Med. Rev. 14, 9–15.

Braunewell, K.H., 2005. The darker side of Ca^{2+} signaling by neuronal Ca^{2+}-sensor proteins: from Alzheimer's disease to cancer. Trends Pharmacol. Sci. 26, 345–351.

Breit, S., Bouali-Benazzouz, R., Benabid, A., Benazzouz, A., 2000. Unilateral lesion of the nigrostriatal pathway induces an increase of neuronal activity of the pedunculopontine nucleus, which is reversed by the lesion of the subthalamic nucleus in the rat. Eur. J. Neurosci. 14, 1833–1842.

Buchwald, J.S., Erwin, R., Read, S., Lancker, V., Cummings, J.L., 1989. Midlatency auditory evoked responses: differential abnormality of P1 in Alzheimer's disease. Electroencephalogr. Clin. Neurophysiol. 74, 378–384.

Buchwald, J.S., Erwin, R., Van Lancker, D., Guthrie, D., Schwafel, J., Tanguay, P., 1992. Midlatency auditory evoked responses: P1 abnormalities in adult autistic subjects. Electroencephalogr. Clin. Neurophysiol. 84, 164–171.

Burgoyne, R.D., 2007. Neuronal calcium sensor proteins: generating diversity in neuronal Ca^{2+} signaling. Nat. Rev. Neurosci. 8, 182–193.

Bylsma, F.W., Peyser, C.E., Folstein, S.E., Ross, C., Brandt, J., 1994. EEG power spectra in Huntington's disease: clinical and neuropsychological correlates. Neuropsychology 32, 137–150.

Capozzo, A., Vitale, F., Mattei, C., Mazzone, P., Scarnati, E., 2014. Continuous stimulation of the pedunculopontine tegmental nucleus at 40 Hz affects preparative and executive control in a delayed sensorimotor task and reduces rotational movements induced by apomorphine in the 6-OHDA parkinsonian rat. Behav. Brain Res. 27, 333–342.

Carrie, A., Jun, L., Bienvenu, T., Vinet, M.C., McDonell, N., Couvert, P., Zemni, R., Cardona, A., Van Buggenhout, G., Frints, S., Hamel, B., Moraine, C., Ropers, H.H., Strom, T., Howell, G.R., Whittaker, A., Ross, M.T., Kahn, A., Fryns, J.P., Beldjord, C., Marynen, P., Chelly, J., 1999. A new member of the IL-1 receptor family highly expressed in hippocampus and involved in X-linked mental retardation. Nat. Genet. 23, 25–31.

Cervena, K., Espa, F., Perogamvros, L., Perrig, S., Merica, H., Ibanez, V., 2014. Spectral analysis of the sleep onset period in primary insomnia. Clin. Neurophysiol. 125, 979–987.

Cooper, J.A., Sagar, H.J., Jordan, N., Harvey, N.S., Sullivan, E.V., 1991. Cognitive impairment in early, untreated Parkinson's disease and its relationship to motor disability. Brain 114, 2095–2122.

Coslett, H.B., Bowers, D., Heilman, K.M., 1987. Reduction of cerebral activation after right hemisphere stroke. Neurology 37, 957–962.

Cummings, J.L., 1992. Depression and Parkinson's disease: a review. Am. J. Psychiatry 149, 443454.

Cuoto, M.I., Monteiro, A., Oliveira, A., Lunet, N., Massano, J., 2014. Depression and anxiety following deep brain stimulation in Parkinson's disease; systematic review and meta-analysis. Acta Med. Port. 27, 372–382.

Cuturic, M., Abramson, R.K., Vallini, D., Frank, E.M., Shamsnia, M., 2009. Sleep patterns in patients with Huntington's disease and their unaffected first-degree relatives: a brief report. Behav. Sleep Med. 7, 245–254.

Danivas, V., Moily, N.S., Thimmaiah, R., Muralidharan, K., Puroshotham, M., Muthane, U., Jain, S., 2013. Off label use of lithium in the treatment of Huntington's disease: a case series. Indian J. Psychiatry 55, 81–83.

Dauvilliers, Y., Morin, C., Cervena, K., Carlander, B., Touchon, J., Besset, A., Billiard, M., 2005. Family studies in insomnia. J. Psychosom. Res. 58, 271–278.

Demeurisse, G., Hublet, C., Paternot, J., 1988. Quantitative EEG in subcortical neglect. Neurophysiol. Clin. 28, 259–265.

D'Onofrio, S., Kezunovic, N., Hyde, J.R., Luster, B., Messias, E., Urbano, F.J., Garcia-Rill, E., 2015. Modulation of gamma oscillations in the pedunculopontine nucleus (PPN) by neuronal calcium sensor protein-1 (NCS-1): relevance to schizophrenia and bipolar disorder. J. Neurophysiol. 113, 709–719.

Doty, R.L., 2012. Olfaction in Parkinson's disease and related disorders. Neurobiol. Dis. 46, 527–552.

Eidelberg, D., Moeller, J.R., Dhawan, V., Spetsieris, P., Takikawa, S., Ishikawa, T., Chaly, T., Robeson, W., Margouleff, D., Przedborski, S., Fahn, S., 1994. The metabolic topography of parkinsonism. J. Cereb. Blood Flow Metab. 14, 783–801.

Feige, B., Baglioni, C., Spiegelhalder, K., Hirscher, V., Nissen, C., Riemann, D., 2013. The microstructure of sleep in primary insomnia: an overview and extension. Int. J. Psychophysiol. 89, 171–180.

Ferguson, I.T., Lenman, J.A.R., Johnston, B.B., 1978. Habituation of the orbicularis oculi reflex in dementia and dyskinetic states. J. Neurol. Neurosurg. Psychiatry 41, 824–828.

Ferraye, M.U., Debu, B., Fraix, V., Goetz, L., Ardouin, C., Yelnik, J., Henry-Lagrange, C., Seigneuret, E., Piallat, B., Krack, P., Le Bas, J.F., Benabid, A.L., Chabardes, S., Pollak, P., 2010. Effects of pedunculopontine nucleus area stimulation on gait disorders in Parkinson's disease. Brain 133, 205–214.

Fisher, S.P., Black, S.W., Schwartz, M.D., Wilk, A.J., Chen, T.M., Lincoln, W.U., Liu, H.W., Kilduff, T.S., Morairty, S.R., 2013. Longitudinal analysis of the electroencephalogram and sleep phenotype in the R6/2 mouse model of Huntington's disease. Brain 136, 2159–2172.

Gambino, F., Pavlowsky, A., Begle, A., Dupont, J.L., Bahi, N., Courjaret, R., Gardette, R., Hdjkacem, H., Skala, H., Poulain, B., Vitale, N., Humeau, Y., 2007. IL1-receptor accessory protein-like 1 (IL1RAPL1), a protein involved in cognitive functions, regulates N-type Ca^{2+}-channel and neurite elongation. Proc. Natl. Acad. Sci. USA 104, 9063–9068.

Garcia-Rill, E., 1986. The basal ganglia and the locomotor regions. Brain Res. Rev. 11, 47–63.

Garcia-Rill, E., 1991. The pedunculopontine nucleus. Prog. Neurobiol. 36, 363–389.

Garcia-Rill, E., 1997. Disorders of the reticular activating system. Med. Hypotheses 49, 379–387.

Garcia-Rill, E., Kezunovic, N., D'Onofrio, S., Luster, B., Hyde, J., Bisagno, V., Urbano, F.J., 2014. Gamma band activity in the RAS-intracellular mechanisms. Exp. Brain Res. 232, 1509–1522.

Garcia-Rill, E., Hyde, J., Kezunovic, N., Urbano, F.J., Petersen, E., in press. The physiology of the pedunculopontine nucleus: implications for deep brain stimulation. J. Neural Transm.

Green, J.B., Burba, A., Freed, D.M., Elder, W.W., Xu, W., 1997. The P1 component of the middle latency auditory potential may differentiate a brainstem subgroup of Alzheimer's disease. Alzheimer Dis. Assoc. Disord. 11, 153–157.

Gulberti, A., Hamel, W., Buhmann, C., Bolemans, K., Zittel, S., Gerloff, C., Westphal, M., Engel, K., Schneider, T.R., Moll, C.K., in press. Subthalamic deep brain stimulation improves auditory sensory gating deficit in Parkinson's disease. Clin. Neurophysiol.

Henderson, R., Kurlan, R., Kersun, J.M., Como, P., 1992. Preliminary examination of the comorbidity of anxiety and depression in Parkinson's disease. J. Neuro-Psychiatry Clin. Neurosci. 4, 257–264.

Holthoff, V.A., Vieregge, P., Kessler, J., Pietrzyk, U., Herholz, K., Bonner, J., Wagner, R., Wienhard, K., Pawlik, G., Heiss, W.D., 1994. Discordant twins with Parkinson's disease: positron emission tomography and early signs of impaired cognitive circuits. Ann. Neurol. 36, 176–182.

Horie, T., Koshino, Y., Murata, T., Omori, M., Isaki, K., 1990. EEG analysis in patients with senile dementia and Alzheimer's disease. Jpn. J. Psychiatry Neurol. 44, 91–98.

Irnich, W., 1980. The chronaxie time and its practical implications. Pacing Clin. Electrophysiol. 3, 292–301.

Jagust, W.J., Reed, B.R., Martin, E.M., Eberlingm, J.L., Nelson-Abbott, R.A., 1992. Cognitive function and regional blood flow in Parkinson's disease. Brain 115, 521–537.

Jankovic, J., 2007. Parkinson's disease: clinical features and diagnosis. J. Neurol. Neurosurg. Psychiatry 79, 368–376.

Jeanmonod, D., Magnon, M., Morel, A., Siegemund, M., Cancro, A., Lanz, M., Llinas, R., Ribary, U., Kronberg, E., Schulman, J., Zonenshayn, M., 2001. Thalamocortical dysrhythmia II. Clinical and surgical aspects. Thalamus Relat. Syst. 1, 245–254.

Jellinger, K.A., 1991. Pathology of Parkinson's disease: changes other than the nigrostriatal pathway. Mol. Chem. Neuropathol. 14, 153–197.

Kezunovic, N., Urbano, F.J., Simon, C., Hyde, J., Smith, K., Garcia-Rill, E., 2011. Mechanism behind gamma band activity in the pedunculopontine nucleus (PPN). Eur. J. Neurosci. 34, 404–415.

Kezunovic, N., Hyde, J., Simon, C., Urbano, F.J., Garcia-Rill, E., 2012. Gamma band activity in the developing parafascicular nucleus (Pf). J. Neurophysiol. 107, 772–784.

Khachaturian, Z.S., 1994. Calcium hypothesis of Alzheimer's disease and brain aging. Ann. N. Y. Acad. Sci. 747, 1–11.

Kim, H.J., 2013. Alpha-synuclein expression in Parkinson's disease: a clinician's perspective. Exp. Neurobiol. 22, 77–83.

Kim, J.S., Lee, S.H., Park, G., Kim, S., Bae, S.M., Kim, D.W., Im, C.H., 2012. Clinical implications of quantitative electroencephalography and current source density in patients with Alzheimer's disease. Brain Topogr. 25, 461–474.

Kimura, J., 1973. Disorder of interneurons in parkinsonism: the orbicularis oculi reflex to paired stimuli. Brain 96, 87–96.

Kimura, T., Nguyen, P.T.H., Ho, S.A., Tran, A.H., Ono, T., Nishino, H., 2009. T-817MA, a neurotrophic agent, ameliorates the deficits in adult neurogenesis and spatial memory in rats infused i.c.v. with amyloid-β peptide. Br. J. Pharmacol. 157, 451–463.

Koenig, T., Prichep, L., Dierks, T., Hubl, D., Wahlund, L.O., John, E.R., Jelic, V., 2005. Decreased EEG synchronization in Alzheimer's disease and mild cognitive impairment. Neurobiol. Aging 26, 165–171.

Koponen, H., Partanen, J., Paakonen, A., Mattila, E., Riekkinen, P.J., 1989. EEG spectral analysis in delirium. J. Neurol. Neurosurg. Psychiatry 52, 980–985.

Kripke, D.F., Garfinkel, L., Wingard, D.L., Klauber, M.R., Marler, M.R., 2002. Mortality associated with sleep duration and insomnia. Arch. Gen. Psychiatry 59, 131–136.

Llinas, R.R., Sugimori, M., Lin, J.W., Cherksey, B., 1989. Blocking and isolation of a calcium channel from neurons in mammals and cephalopods utilizing a toxin fraction (FTX) from funnel-web spider poison. Proc. Natl. Acad. Sci. USA 86, 1689–1693.

Llinas, R.R., Urbano, F.J., Leznik, E., Ramirez, R.R., van Marle, H.J.F., 2005. Rhythmic and dysrhythmic thalamocortical dynamics: GABA systems and the edge effect. Trends Neurosci. 28, 325–333.

Mattson, M., LaFerla, F., Chan, S., Leissring, M., Shepel, P., Geiger, J., 2000. Calcium signaling in the ER: its role in neuronal plasticity and neurodegenerative disorders. Trends Neurosci. 23, 222–229.

Mazzone, P.S.P., Lozano, A., Sposato, S., Scarnati, E., Stefani, A., 2005. Brain stimulation and movement disorders: where we going? In: Proceedings of 14th Meeting of the World Society of Stereotactic and Functional Neurosurgery (WSSFN). Monduzzi, Bologna.

Mazzone, P., Sposato, S., Insola, A., Dilazzaro, V., Scarnati, E., 2008. Stereotactic surgery of nucleus tegmenti pedunculopontine [corrected]. Br. J. Neurosurg. 22 (Suppl. 1), S33–S40.

Mazzone, P., Insola, A., Sposato, S., Scarnati, E., 2009. The deep brain stimulation of the pedunculopontine tegmental nucleus. Neuromodulation 12, 191–204.

Menza, M.A., Robertson-Hoffman, D.E., Bonapace, A.S., 1993. Parkinson's disease and anxiety: comorbidity with depression. Biol. Psychiatry 34, 465–470.

Mills, R.P., Swanson, P.D., 1978. Vertical oculomotor apraxia and memory loss. Arch. Neurol. 4, 149–153.

Moreno, H., Choi, S., Yu, E., Brusco, J., Mopreira, J.E., Sugimore, M., Llinas, R.R., 2011. Blocking effects of human tau on squid giant synapse transmission and its prevention by TB1T MA. Frontiers Neurosci. 3, 3.

Moro, E., Hamani, C., Poon, Y.Y., Al-Khairallah, T., Dostrovsky, J.O., Hutchison, W.D., Lozano, A.M., 2010. Unilateral pedunculopontine stimulation improves falls in Parkinson's disease. Brain 133, 215–224.

Morton, A.J., Wood, N.I., Hastings, M.H., Hurelbrink, C., Barker, R.A., Maywood, E.S., 2005. Disintegration of the sleep–wake cycle and circadian timing in Huntington's disease. J. Neurosci. 25, 157–163.

Nakashima, K., Shimoyama, R., Yokoyama, Y., Takahashi, K., 1993. Auditory effects on the electrically elicited blink reflex in patients with Parkinson's disease. Electroencephalogr. Clin. Neurophysiol. 89, 108–112.

Nimmrich, V., Gross, G., 2012. P/Q-type calcium channel modulators. Br. J. Pharmacol. 167, 741–759.

Nowak, L.G., Bullier, J., 1998. Axons, but not cell bodies, are activated by electrical stimulation in cortical gray matter. Exp. Brain Res. 118, 477–488.

Ophoff, R.A., Terwindt, G.M., Frants, R.R., Ferrari, M.D., 1998. P/Q-type Ca^{2+} channel defects in migraine, ataxia, and epilepsy. Trends Pharmacol. Sci. 19, 121–127.

Orem, J., Schlag-Rey, M., Schlag, J., 1973. Unilateral visual neglect and thalamic intralaminar lesions in the cat. Exp. Neurol. 40, 784–797.

Orieux, G., Francois, C., Feger, J., Yelnik, J., Vila, M., Ruberg, M., Agid, Y., Hirsch, E., 2001. Metabolic activity of excitatory parafascicular and pedunculopontine inputs to the subthalamic nucleus in a rat model of Parkinson's disease. Neuroscience 97, 79–88.

Owen, A.M., Roberts, A.C., Hodges, J.R., Summers, B.A., Polkey, C.E., Robbins, T.W., 1993. Contrasting mechanisms of impaired attentional set-shifting in patients with frontal lobe damage or Parkinson's disease. Brain 116, 1159–1175.

Painold, A., Anderer, P., Holl, A.K., Saleu-Zhylarz, G.M., Saletu, B., Bonelli, R.M., 2010. Comparative EEG mapping studies in Huntington's disease patients and controls. J. Neural Transm. 117, 1307–1318.

Park, J.Y., Lee, S.K., An, S.K., Lee, S.J., Kim, J.J., Kim, K.H., Namkoong, K., 2012. Gamma oscillatory activity in relation to memory ability in older adults. Int. J. Psychophysiol. 86, 58–65.

Penders, C.A., Delwaide, P.J., 1971. Blink reflex studies in patients with parkinsonism before and during surgery. J. Neurol. Neurosurg. Psychiatry 34, 674–678.

Peppard, R.F., Martin, W.R.W., Carr, G.D., Grochowski, E., Schulzer, M., Guttman, M., McGeer, P.L., Tsui, A.G., Caine, D.B., 1992. Cerebral glucose metabolism in Parkinson's disease with and without dementia. Arch. Neurol. 49, 1262–1268.

Peppe, A., Pierantozzi, M., Baiamonte, V., Moschella, V., Caltagirone, C., Stanzione, P., Stefani, A., 2012. Deep brain stimulation of pedunculopontine tegmental nucleus: role of sleep modulation in advanced Parkinson disease patients—one-year follow-up. Sleep 35, 1637–1642.

Perlis, M.L., Smith, M.T., Andrews, P.J., Orff, H., Giles, D.E., 2001. Beta/gamma EEG activity in patients with primary and secondary insomnia and good sleeper controls. Sleep 24, 110–117.

Pillon, B., Dubois, B., Polska, A., Agid, Y., 1991. Severity and specificity of cognitive impairment in Alzheimer's, Huntington's, and Parkinson's diseases and progressive supranuclear palsy. Neurology 41, 634–643.

Plomp, J.J., van den Maagdenberg, A.M., Molenaar, P.C., Frants, R.R., Ferrari, D., 2001. Mutant P/Q-type calcium channel electrophysiology and migraine. Curr. Opin. Investig. Drugs 2, 1250–1260.

Ptak, R., Schnider, A., 2006. Reflexive orienting in spatial neglect is biased towards behaviorally salient stimuli. Cereb. Cortex 16, 337–346.

Randall, A., Tsien, R.W., 1995. Pharmacological dissection of multiple types of Ca^{2+} channel currents in rat cerebellar granule neurons. J. Neurosci. 15, 2995–3012.

Rea, R., Carotenuto, A., Fasanaro, A.M., Traini, E., Amenta, F., 2014. Apathy in Alzheimer's disease: any effective treatment? Sci. World J. 2014, 1–9, Article ID 421385.

Riemann, D., Spiegelhalder, K., Feige, B., Vodenholzer, U., Berger, M., Perils, M., Nissen, C., 2010. The hyperarousal model of insomnia: a review of the concept and its evidence. Sleep Med. Rev. 14, 19–31.

Robbins, T.W., James, M., Owen, A.M., Lange, K.W., Lees, A.J., Leigh, P.N., Marsden, C.D., Quinn, N.P., Summers, B.A., 1994. Cognitive deficits in progressive supranuclear palsy, Parkinson's disease, and multiple system atrophy in tests sensitive to frontal lobe dysfunction. J. Neurol. Neurosurg. Psychiatry 57, 79–88.

Rosas, H.D., Salat, D.H., Lee, S.Y., Zaleta, A.K., Hevelone, N., Hersch, S.M., 2008. Complexity and heterogeneity: what drives the ever-changing brain in Huntington's disease? Ann. N. Y. Acad. Sci. 1147, 196–205.

Rothwell, J.C., Obeso, J.A., Traub, M.M., Marsden, C.D., 1983. The behavior of the long-latency stretch reflex in patients with Parkinson's disease. J. Neurol. Neurosurg. Psychiatry 46, 35–44.

Samil, A., Nutt, J.G., Ransom, B.R., 2004. Parkinson's disease. Lancet 363, 1783–1793.

Sawada, H., Udaka, F., Kameyama, M., Seriu, N., Nishinaka, K., Shindou, K., Kodama, M., Nishitani, N., Okumira, K., 1992. SPECT findings in Parkinson's disease associated with dementia. J. Neurol. Neurosurg. Psychiatry 55, 960–963.

Schenck, C.H., Bundlie, S.R., Mahowald, M.W., 1996a. Delayed emergence of a parkinsonian disorder in 38% of 29 older men initially diagnosed with idiopathic rapid eye movement sleep behavior disorder. Neurology 46, 388–393.

Schenck, C., Garcia-Rill, E., Skinner, R.D., Anderson, M., Mahowald, M.W., 1996b. A case of REM sleep behavior disorder with autopsy-confirmed Alzheimer's disease: post mortem brainstem histochemical analyses. Biol. Psychiatry 40, 422–425.

Schenck, C.H., Ullevig, C.M., Mahowald, M.W., Dalmau, J., Posner, J.B., 1997. A controlled study of serum anti-locus coeruleus antibodies in REM sleep behavior disorder. Sleep 20, 349–351.

Schlecker, C., Boehmerie, W., Jeromin, A., DeGray, B., Varshey, A., Sharma, Y., Szigeti-Buck, K., Ehrlich, B.E., 2006. Neuronal calcium sensor-1 enhancement of $InsP_3$ receptor activity is inhibited by therapeutic levels of lithium. J. Clin. Invest. 116, 1668–1674.

Shik, M.L., Severin, F.V., Orlovskii, G.N., 1966. Control of walking and running by means of electric stimulation of the midbrain. Biofizika 11, 659–666.

Sobstyl, M., Zabek, M., Gorecki, W., Mossakowski, Z., 2014. Quality of life in advanced Parkinson's disease after bilateral subthalamic stimulation: 2 years follow-up study. Clin. Neurol. Neurosurg. 124C, 161–165.

Speakman, T.J., 1963. Results of thalamotomy for Parkinson's Disease. Can. Med. Assoc. J. 89, 652–656.

Spiegelhalder, K., Regen, W., Feige, B., Holz, J., Piosczyk, H., Baglioni, C., Riemann, D., Nissen, C., 2012. Increased EEG sigma and beta power during NREM sleep in primary insomnia. Biol. Psychol. 91, 329–333.

Stefani, A., Lozano, A.M., Peppe, A., Stanzione, P., Galati, S., Troppei, D., Pierantozzi, M., Brusa, L., Scarnati, E., Mazzone, P., 2007. Bilateral deep brain stimulation of the pedunculopontine and subthalamic nuclei in severe Parkinson's disease. Brain 130, 1596–1607.

Stefani, A., Peppe, A., Galati, S., Stampanoni, A., Bassi, M., D'Angelo, V., Pierantozzi, M., 2013. The serendipity case of the pedunculopontine nucleus low-frequency brain stimulation: chasing a gait response, finding sleep, and cognitive improvement. Front. Neurol. 4 (68), 1–12.

Stein, M.B., Heuser, I.J., Juncos, J.L., Uhde, T.W., 1990. Anxiety disorders in patients with Parkinson's disease. Am. J. Psychiatry 147, 217–220.

Storrie-Baker, H.J., Segalowitz, S.J., Black, S.E., McLean, J.A., Sullivan, N., 1997. Improvement of spatial neglect with cold-water calorics: an electrophysiological test of the arousal hypothesis of neglect. J. Int. Neuropsychol. Soc. 3, 394–402.

Stutzmann, G.E., Smith, I., Caccamo, A., Oddo, S., LaFerla, F.M., Parker, I., 2006. Enhanced ryanodine receptor recruitment contributes to Ca^{2+} disruptions in young, adult, and aged Alzheimer's disease mice. J. Neurosci. 26, 5180–5189.

Swerdlow, N.R., Paulsen, J., Braff, D.L., Butters, N., Geyer, M.A., Swanson, M.R., 1995. Impaired prepulse inhibition of acoustic and tactile startle response in patients with Huntington's disease. J. Neurol. Neurosurg. Psychiatry 58, 192–200.

Teo, C., Rasco, L., Al-Mefty, K., Skinner, R.D., Garcia-Rill, E., 1997. Decreased habituation of midlatency auditory evoked responses in Parkinson's disease. Mov. Disord. 12, 655–664.

Teo, C., Rasco, L., Skinner, R.D., Garcia-Rill, E., 1998. Disinhibition of the sleep state-dependent P1 potential in Parkinson's disease-improvement after pallidotomy. Sleep Res. Online 1, 62–70.

Thevanasathan, W., Silburn, P.A., Brooker, H., Coyne, T.J., Kahn, S., Gill, S.S., Aziz, T.Z., Brown, P., 2010. The impact of low-frequency stimulation of the pedunculopontine nucleus region on reaction time in Parkinsonism. J. Neurol. Neurosurg. Psychiatry 81, 1099–1104.

Thevanasathan, W., Cole, M.H., Grapel, C.L., Hyam, J.A., Jenkinson, N., Brittain, J.S., Coyne, T.J., Silburn, P.A., Aziz, T.Z., Kerr, G., Brown, P., 2012. A spatiotemporal analysis of gait freezing and the impact of pedunculopontine nucleus stimulation. Brain 135, 1446–1454.

Tyckoki, T., Mandat, T., Nauman, P., 2011. Pedunculopontine nucleus deep brain stimulation in Parkinson's disease. Arch. Med. Sci. 7, 555–564.

Uc, E.Y., Skinner, R.D., Rodnitzky, R.L., Garcia-Rill, E., 2003. The midlatency auditory evoked potential P50 is abnormal in Huntington's disease. J. Neurol. Sci. 212, 1–5.

Ueki, A., Goto, K., Sato, N., Iso, H., Morita, Y., 2006. Prepulse inhibition of acoustic startle response in mild cognitive impairment and mild dementia of Alzheimer type. Psychiatry Clin. Neurosci. 60, 55–62.

Van Deusen, J.A., Vuurman, E.F.M.P., Verhey, F.R.J., van Kranen-Mastenbroek, V.H.J.M., Riedel, W.J., 2008. Increased EEG gamma band activity in Alzheimer's disease and mild cognitive impairment. J. Neural Transm. 115, 1301–1311.

Vazquez, A., Jimenez-Jimenez, F.J., Garcia-Ruiz, P., Garcia-Urra, D., 1993. "Panic attacks" in Parkinson's disease: a long-term complication of levodopa therapy. Acta Neurol. Scand. 87, 14–18.

Vonsattel, J.P., DiFiglia, M., 1998. Huntington disease. J. Neuropathol. Exp. Neurol. 57, 369–384.

Watson, R.T., Heilman, K.M., Miller, B.D., King, F.A., 1974. Neglect after mesencephalic reticular formation lesions. Neurology 24, 294–298.

Watson, R.T., Andriola, M., Heilman, K.M., 1977. The electroencephalogram in neglect. J. Neurol. Sci. 34, 343–348.

Watson, R.T., Valenstein, E., Heilman, K.M., 1981. Thalamic neglect. Possible role of the medial thalamus and reticularis in behavior. Arch. Neurol. 38, 501–506.

Woods, A.J., Mennemeier, M., Garcia-Rill, E., Meythaler, J., Mrak, V.W., Jewel, G.R., Murphy, H., 2006. Bias in magnitude estimation following left hemisphere injury. Neuropsychology 44, 1406–1412.

Drug Abuse and the RAS

Francisco J. Urbano, PhD[*], *Veronica Bisagno, PhD*[*], *Edgar Garcia-Rill, PhD*[‡]

[*]Laboratorio Fisiologia y Biologia Molecular, Facultad Ciencias Exactas, Ciudad Universitaria, Buenos Aires, Argentina

[‡]Center for Translational Neuroscience, Department of Neurobiology and Developmental Sciences, University of Arkansas for Medical Sciences, Little Rock, AR, USA

WAKING AND DRUG ABUSE

Drug abuse only takes place during waking. We only experience withdrawal or relapse to drug abuse when we are awake. Furthermore, higher motor activation while awake has been associated with greater addiction liability. The ventral tegmental area (VTA), a key neural substrate for the modulation of psychostimulant abuse, is modulated by cholinergic reticular activating system (RAS) output from the laterodorsal tegmental (LDT) and pedunculopontine (PPN) nuclei.

The RAS modulates oscillating rhythms between the thalamus, hypothalamus, basal forebrain, and cortex that are characterized in the EEG during wake–sleep states (Garcia-Rill, 2009). Importantly, psychostimulant abuse is characterized by several neurobiological phenomena that happen during waking. All three main nuclei in the RAS, the PPN and LDT that are mainly cholinergic (Koyama and Sakai, 2000; Garcia-Rill, 2009); the locus coeruleus (LC), which is mainly noradrenergic; and the raphe nucleus (RN), which is mainly serotonergic, have been described as affecting neural substrates related to addiction like the VTA. This is important because the basal activity of VTA neurons can be functionally associated to individual vulnerability to psychostimulant abuse (Marinelli and White, 2000).

The PPN nucleus contains cholinergic and noncholinergic neurons that have an excitatory effect on LC and RN neurons (Egan and North, 1986; Takakusaki et al., 1996). It is important to note that one of the targets of PPN

Waking and the Reticular Activating System in Health and Disease
http://dx.doi.org/10.1016/B978-0-12-801385-4.00013-6

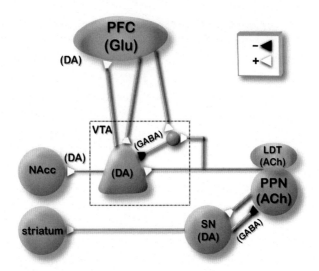

FIGURE 13.1   **RAS projections to abuse and motor pathways.** Schematic diagram showing projections from the PPN/LDT to key dopaminergic nuclei underlying psychostimulant
effects. The PPN and LDT project to the ventral tegmental area (VTA) and substantia nigra
(SN). Dopaminergic VTA neurons in turn project to the medial prefrontal cortex (mPFC) and
nucleus accumbens (NAcc). The PPN/LDT also project to the SN, which in turn projects to
the striatum. Upper left box: (+) represents excitatory glutamatergic or cholinergic synapses,
and (−) GABAergic inhibitory synapses.

noncholinergic neurons is the VTA (Maskos, 2008). Cholinergic efferents
from the PPN to the VTA form a loop that includes the medial prefrontal
cortex (mPFC) (Figure 13.1). This loop is composed of mPFC glutamatergic
efferents to dopaminergic (DA) and GABAergic neurons in the VTA and
to the nucleus accumbens (NAcc) through a polysynaptic circuit that includes the PPN/LDT nuclei. In addition, the VTA sends dopaminergic
and GABAergic efferents to the NAcc (Figure 13.1). Activation of the PPN
increases VTA dopaminergic output and increases extracellular DA levels in the NAcc and mPFC (Floresco et al., 2003), which suggests that the
PPN in part regulates the reward and motivational functions of the VTA
(Good and Lupica, 2009). Higher glutamatergic efferent activation from
the mPFC would in turn reduce VTA dopaminergic output through its direct activation of local GABAergic interneurons in the VTA (Figure 13.1).
Recent optogenetic experiments confirmed that the LDT-VTA pathway
stimulation can elicit psychostimulant-like behavior in the absence of
drug administration (Lammel et al., 2012).

The noradrenergic input from the LC to the PPN is inhibitory, presumably
via α-$_2$ adrenergic receptors (Williams and Reiner, 1993). Furthermore, the
RN sends inhibitory serotonergic projections to both the PPN and the LC
(Leonard and Llinas, 1994). Other inhibitory inputs to the PPN come from
GABAergic neurons in the substantia nigra (SN), an area also involved in
the locomotor activation produced by psychostimulant administration.

Since midbrain dopaminergic neurons originating in the VTA and SN *pars compacta* (SNc) have been previously described as the neural substrates underlying individual vulnerability to psychostimulant addiction (Koob, 1998; White and Kalivas, 1998; Wise, 1998), understanding the functional modulation of the VTA and SNc by the RAS is key to understanding how reinforcing, drug craving, and psychomotor stimulant effects are modulated by a wake-promoting nucleus such as the PPN.

STIMULANTS

Amphetamines

Amphetamine, and its most addictive derivative methamphetamine, exerts its actions through an increase in DA extracellular levels in the terminal and cell body regions of midbrain DA neurons, by causing reverse transport of DA and preventing its uptake via the DA transporter (Seiden et al., 1993; Sulzer et al., 1995). The basic behavioral effect is to increase locomotor activity, or hyperlocomotion. This higher locomotor activity in response to novelty has been described as a predictor of future propensity to amphetamine self-administration (Pierre and Vezina, 1997).

Recently, intranuclear PPN or LDT injections of muscarinic M_2 agonists have helped clarify the role of acetylcholine (ACh) on methamphetamine-induced locomotion (Dobbs and Mark, 2012). The activation of inhibitory cholinergic muscarinic autoreceptors in the LDT, but not in the PPN, can attenuate methamphetamine-induced ACh and DA release in the VTA (and its upstream targets like the NAcc), which would reduce the expected hyperlocomotion induced by that psychostimulant (Dobbs and Mark, 2012).

Not only do the PPN and LDT send out cholinergic projections, but also these nuclei themselves are modulated by ACh input to both muscarinic (mAChR) and nicotinic (nAChR) receptors. These cholinergic projections arise in the contralateral PPN and LDT (Semba and Fibiger, 1992). Although the mechanisms for nicotinic modulation of the PPN are clearer (see below), muscarinic modulation of the PPN has been associated with the degree of psychostimulant-mediated effects in the whole individual.

Muscarinic agonists have been shown to reduce amphetamine-induced hyperlocomotion activation (Shannon and Peters, 1990), while muscarinic antagonists mediated facilitation of amphetamine-induced rotation through M_2 receptors (Hagan et al., 1987) (Figure 13.2). Unfortunately, all of these experiments were performed using systemic administration of muscarinic modulators, making it difficult to determine the exact neural substrate of these actions in the brain (e.g., it is possible that muscarinic receptor modulation might be affecting the activity of another cholinergic nucleus such as the nucleus basalis located in the basal forebrain that projects to the cortex and the thalamic reticular nucleus) (Asanuma, 1997).

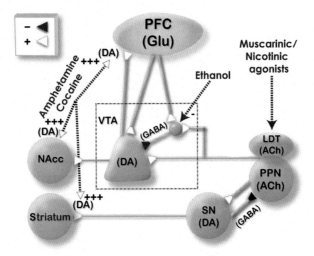

FIGURE 13.2   **Schematic diagram showing psychostimulant and neuromodulator effects on PPN/LDT dopaminergic circuits.** Amphetamine/cocaine (two stimulants that exert their effects by drastically increasing extracellular DA concentration) self-administration can be modulated by LDT/PPN efferents to the VTA. Moreover, VTA local GABAergic interneurons (inside dashed lines) are inhibited by ethanol, disinhibiting VTA dopaminergic neurons. However, cholinergic modulation of PPN/LDT or VTA can also be altered by ethanol self-administration.

Direct interactions between the PPN and VTA were revealed using excitotoxic lesions of the PPN (Alderson et al., 2004). PPN-lesioned animals showed a clear reduction in the response to a progressive ratio self-administration schedule of amphetamine (Alderson et al. 2004). In addition, a previous study showed that lesions of the PPN blocked the motivational (measured using a conditioned place preference paradigm) effects of amphetamine (Bechara and van der Kooy, 1989).

Our group recently described how the stimulant modafinil can prevent neurotoxic effects of methamphetamine on striatal circuits (Raineri et al., 2011, 2012). Modafinil acts by increasing electrical coupling especially in GABAergic neurons (Garcia-Rill et al., 2007a,b; Urbano et al., 2007), which decreases input resistance in GABAergic cells not only throughout the brain but also at the level of the PPN. Such effects would disinhibit glutamatergic and dopaminergic interactions between the PPN and VTA nuclei. In addition, modafinil was found to improve methamphetamine-mediated behavioral impairment described in the mPFC (Gonzalez et al. 2014). In conclusion, PPN activation has been implicated in the animal's voluntary search for amphetamine. Modafinil has shown a promising profile for the treatment of cognitive deficits mediated by methamphetamine.

Cocaine

The rewarding properties and abuse potential of cocaine derives in part from elevated synaptic levels of DA neurotransmission in limbic circuits (Wilson et al., 1976; Wise and Bozarth, 1987) (Figure 13.2). Cocaine can also increase serotonin (5-HT) neurotransmission via inhibition of its reuptake (Ross and Renyi, 1969). Individual vulnerability to cocaine self-administration has been associated with changes in neuronal intrinsic properties, in particular with higher action potential frequency and bursting of VTA neurons and to a lesser extent SNc neurons (Anderson and Pierce, 2005). In addition, animals that manifested higher self-administration rates also exhibited higher locomotor responses to a novel environment prior to psychostimulant administration (Anderson and Pierce, 2005; Grimm and See, 1997), strongly suggesting that the basal modulation of VTA neurons by the PPN/LDT nuclei can be considered critical to future psychostimulant abuse liability.

However, there is a discrepancy about the role of the PPN in cocaine reward mechanisms. After a bilateral PPN lesion, some authors have shown no blocking effects on the reinforcing effects of cocaine in a place conditioning paradigm (Parker and van der Kooy, 1995). Local injection of carbachol (a nonspecific muscarinic and nicotinic agonist) into the PPN reduced cocaine self-administration (Corrigal et al. 2002). Carbachol-induced local inhibition of the PPN might be mediated by G protein modulation of P/Q-type calcium currents (Kezunovic et al., 2013), suggesting a cocaine-mediated mechanism centered on the importance of voltage-dependent calcium channels, rather than only on DA-dependent reinforcement mechanisms.

More recent work further supports the role of the PPN/LDT in cocaine self-administration (Schmidt et al., 2009) or conversely rules it out (Steidl et al., 2014). While Schmidt et al. (2009) showed that enhanced glutamatergic transmission in the PPN/LDT nuclei promoted cocaine priming-induced reinstatement of cocaine seeking, others (Steidl et al., 2014) used the infusion of diphtheria toxin conjugated to urotensin-II into the PPN (which resulted in the loss of >95% of PPN cholinergic neurons) to report no significant alteration in cocaine self-administration.

A possible explanation for the discrepancies described in recent years may be related to the strong serotonergic effects of cocaine (Ross and Renyi, 1969). Indeed, cocaine-mediated blocking effects of both DA and serotonin reuptake could alter the activity of PPN efferents to the VTA as well as the degree of activation of serotonergic afferents from the RN to the PPN. Such a possibility has been recently proposed to explain the differential actions of cocaine and methylphenidate on GABAergic transmission in mouse ventrobasal thalamic nucleus (Goitia et al., 2013). Therefore, PPN activation is important in the modulation of cocaine self-administration, although the mechanisms underlying such an effect need further characterization.

Tobacco

We investigated the effects of nicotine on PPN neurons in vitro and found that nicotinic receptor agonists induced a depolarization early in development (~day 12) that switched across days 15–17 until they elicited hyperpolarization by day 21 (Good et al., 2007). Some receptors were found to be postsynaptic on PPN neurons, others presynaptic. We also studied the effects of nicotine in vivo and found that nicotinic agonists decreased the P13 potential, a vertex-recorded auditory response generated by the PPN (Chapter 6) in a dose-dependent manner (Mamiya et al., 2005). These in vitro and in vivo results both suggest that nicotine has an inhibitory effect on the PPN, at least initially reducing arousal (Figure 13.2). This may explain the initial anxiolytic effects of cigarette smoke (Benowitz, 1996; Simosky et al., 2002). Inhaled nicotine is known to permeate the lungs where it is absorbed into the bloodstream. The short delivery time and elimination half-lives (8 min and 2 h, respectively) assure that, within a short time, the effect can be reproduced by smoking another cigarette (Benowitz, 1996). After absorption into the blood, nicotine readily crosses the blood–brain barrier. In rats, concentrations of nicotine in the brain are 5–7 times higher than blood concentrations (Gosheh et al., 2001). Although nicotine is generally considered a stimulant, smokers assert that, in addition to its stimulatory effects on concentration and attention, the primary effect of smoking is that it calms and relaxes (Frith, 1972; Spielberger, 1986). Therefore, nicotine in cigarette smoke could initially reduce PPN output and decrease arousal levels, at least briefly. This would explain why patients with disorders of hypervigilance, such as schizophrenia, anxiety disorder, and depression (see Chapter 11), tend to overindulge as a form of self-medication (Goff et al., 1992). This initial anxiolytic response is followed by an anxiogenic one, since smoking has been reported to produce such effects (Benowitz, 1996; Fleming and Lombardo, 1987; Frith, 1972). Anxiogenic effects may be mediated by nicotinic activation of sites other than the PPN. If true, this push–pull (anxiolytic–anxiogenic) effect would explain the difficulty in decreasing dependence to the smoking habit in the adult.

Nicotine in the fetus may have long-term consequences on RAS activity. Nicotine from cigarette smoke can cross into the placenta, so fetuses of mothers who smoke are exposed to higher nicotine concentrations in both amniotic fluid and umbilical vein than maternal vein serum (Luck et al., 1985). Coupled with an increased propensity for young women to begin smoking during the childbearing ages of 18–25, prenatal exposure to cigarette smoke represents a critical health problem. Toxicological effects of perinatal cigarette smoke exposure include lower birth weight (Eskanazi et al., 1995), higher rate of spontaneous abortion (Kline et al., 1997), and increased incidence of sudden infant death syndrome (Bulterys, 1990). Maternal smoking during pregnancy can lead to increased aggression (Weissman et al., 1999) and problems with sustained attention and impulsivity in adolescent offspring (Fried et al., 1992). Children of smoking mothers are at increased risk for ADHD (Milberger et al., 1997; Wasserman et al., 1999), conduct disorders (Fergusson et al., 1998; Wakschlag et al., 1997), and drug abuse (Weissman et al., 1999), and they are responsible for high rates of violent and persistent criminal offenses (Rasanen et al. 1999; Weissman et al., 1999).

Behavioral deficits related to arousal and attentional problems in humans have been identified in rats exposed to nicotine prenatally. These animal models show deficits in attention and memory in maze performance (Levin et al., 1993; Sorenson et al., 1991), learning (Levin et al., 1993), and operant behaviors (Martin and Becker, 1971). We studied the effects of prenatal exposure to cigarette smoke on the physiology of PPN cells postnatally (Garcia-Rill et al., 2007a,b). We found that PPN neurons exhibited lowered resting membrane potential and lowered action potential threshold (both would tend to increase PPN firing), both of which could be related to an increase in hyperpolarization-activated I_H current (that could also increase firing). All of these effects would increase tonic arousal levels during waking and REM sleep. We speculated that hyperpolarization-activated cyclic nucleotide-gated proteins may be affected by prenatal cigarette smoke exposure and that such changes could be responsible for the arousal and attentional deficits (Milberger et al., 1997), as well as impulse control problems (exaggerated fight-or-flight responses) reported in the children of mothers who smoked during pregnancy (Fried et al., 1992; Wasserman et al., 1999).

DEPRESSANTS

Ethanol

Our group initially described that ethanol suppressed the P13 potential (i.e., a vertex-recorded midlatency auditory-evoked potential generated by PPN efferents) in a dose-dependent manner (Miyazato et al., 1999). This work suggested that ethanol's effects on arousal might be mediated by direct inhibition of the PPN. Recent studies have focused mainly on GABAergic interneurons in the PPN and VTA nuclei in order to explain the observed changes in ethanol self-administration. PPN lesions have been shown to block ethanol-conditioned place preference in ethanol-dependent and ethanol-withdrawn mice (Ting-A-Kee et al., 2009), suggesting that ethanol's positive reinforcing effects are $GABA_A$ receptor-dependent. Moreover, local injections of muscimol (a $GABA_A$ agonist) into the PPN reduced ethanol self-administration in rats (Samson and Chappell, 2001). However, the PPN is composed of nonoverlapping populations of cholinergic, glutamatergic, and GABAergic neurons (Wang and Morales, 2009). Thus, other neuronal populations might also be engaged when GABAergic interneurons are affected by ethanol. Interestingly, injecting muscarinic cholinergic receptor agonists (e.g., carbachol) into the PPN, or cholinergic antagonists (e.g., methylscopolamine) into the VTA, was able to reduce ethanol self-administration in rats (Katner et al., 1997). Muscarinic activation of the PPN was proposed to reduce the activation of only cholinergic cells (Katner et al., 1997), centering the effect of ethanol on the alteration of normal PPN-VTA cholinergic interactions (Figure 13.2).

Partial effects of ethanol on only GABAergic or cholinergic neural populations cannot fully explain its drastic effect on P13 amplitude, a measure of PPN output through the intralaminar thalamus (Miyazato et al., 1999). Recent publications from our group have described the fact that P/Q-type, and to a lesser extent N-type, calcium channel-mediated mechanisms underlie beta/gamma band oscillatory activity of *all PPN* neuronal groups (Garcia-Rill et al., 2013, 2014; Urbano et al., 2012). In addition, we have also shown that muscarinic modulation of PPN gamma band activity was mediated by the long-lasting modulation of P/Q-type calcium channels (Kezunovic et al., 2013). Thus, with all of this new information in mind, the effects of alcohol on PPN physiology could be explained as a calcium channel blockade, as previously described in other systems (Solem et al., 1997).

CLINICAL IMPLICATIONS

Almost 25 million people in the United States need treatment for drug or alcohol abuse, but less than 12% actually receive treatment at a special facility. Of those who recover from drug abuse, over one-half will relapse. Just like chronic illness, drug abuse requires repeated episodes of therapy. The odds of remaining drug-free within one year are only one-third but are over 80% after three years of abstinence. Therefore, it is clear that the main factor in the increasing incidence of drug and alcohol abuse is relapse. The main factors in relapse are drug-related cues that must be avoided especially within the first ninety days, which is the period of most likely relapse.

The symptoms of relapse are similar to those in depression and anxiety, which may require pharmacological therapy in addition to cognitive therapy and support groups. The triggers to relapse involve the context of our sensory world, for example, socializing with drug users, being in a place associated with drug abuse, and being in stressful or emotional situations, that is, in situations in which relief from overwhelming input requires the use of a drug. These functions all involve the RAS and its function in preconscious awareness. For example, smokers manifest more impaired sleep during withdrawal, with insomnia-like symptoms (Jaehne et al. 2012, 2014). Alcoholics experience persistent insomnia, which increases the risk of relapse (Brower et al., 2011). Similar symptoms are present in heroin, cocaine, methamphetamine, and addicted users of other drugs (Hasler et al. 2012). Basically, disturbed sleep predicts relapse to alcohol and psychoactive drug abuse (Brower and Perron, 2010). However, treatment of insomnia by itself does not predict recovery from drug abuse (Friedmann et al., 2008). It is also problematic that treatment for insomnia involves the use of cross-dependent drugs such as benzodiazepines (Arendt et al., 2007). However, it is clear that normalization of sleep homeostasis is essential for preventing relapse. Just as with psychiatric and neurological disorders, rebalancing the wake–sleep homeostatic system is key to treatment response and in fact signals successful alleviation of symptoms. The P50 potential may be a tool for monitoring recovery of arousal-related functions since alcoholics in remission show no sensory gating deficits of the P50 potential (Gillette et al., 1997) (see Figure 11.7).

Acknowledgments

This work is supported by grants from FONCyT Agencia Nacional de Promoción Científica y Tecnológica; BID 1728 OC.AR. PICT 2007–1009; PICT 2008–2019 and PICT-2012-1769 (to Dr. Urbano); and CONICET-PIP 2011-2013-11420100100072 and CONICET-PICT-2012-0924 (to Dr. Bisagno).

References

Alderson, H.L., Latimer, M.P., Blaha, C.D., Phillips, A.G., Winn, P., 2004. An examination of d-amphetamine self-administration in pedunculopontine tegmental nucleus-lesioned rats. Neurosci. 125, 349–358.

Anderson, S.M., Pierce, R.C., 2005. Cocaine-induced alterations in dopamine receptor signaling: implications for reinforcement and reinstatement. Pharmacol. Ther. 106, 389–403.

Arendt, J.T., Conroy, D.A., Brower, K.J., 2007. Treatment options for sleep disturbances during alcohol recovery. J. Addict. Res. 26, 41–45.

Asanuma, C., 1997. Distribution of neuromodulatory inputs in the reticular and dorsal thalami nuclei. In: Steriade, M., Jones, E.G., McCormick, D.A. (Eds.), Thalamus, Vol II: Experimental and Clinical Aspects. Elsevier, Amsterdam, pp. 93–154.

Bechara, A., van der Kooy, D., 1989. The tegmental pedunculopontine nucleus: a brainstem output of the limbic system critical for the conditioned place preferences produced by morphine and amphetamine. J. Neurosci. 9, 3400–3409.

Benowitz, N.L., 1996. Pharmacology of nicotine: addiction and therapeutics. Annu. Rev. Pharmacol. Toxicol. 36, 597–613.

Brower, K.J., Perron, B.E., 2010. Sleep disturbances as a universal risk factor for relapse in addictions to psychoactive substances. Med. Hypotheses 74, 928–933.

Brower, K.J., Krentzman, A., Robinson, E.A.R., 2011. Persistent insomnia, abstinence, and moderate drinking in alcohol-dependent individuals. Am. J. Addict. 20, 435–440.

Bulterys, M., 1990. High incidence of sudden infant death syndrome among northern Indians and Alaska natives compared with southwestern Indians: possible role of smoking. J. Community Health 15, 185–194.

Corrigall, W.A., Coen, K.M., Zhang, J., Adamson, L., 2002. Pharmacological manipulations of the pedunculopontine tegmental nucleus in the rat reduce self-administration of both nicotine and cocaine. Psychopharmacology 160, 198–205.

Dobbs, L.K., Mark, G.P., 2012. Acetylcholine from the mesopontine tegmental nuclei differentially affects methamphetamine induced locomotor activity and neurotransmitter levels in the mesolimbic pathway. Behav. Brain Res. 226, 224–234.

Egan, T.M., North, R.A., 1986. Actions of acetylcholine and nicotine on rat locus coeruleus neurons in vitro. Neuroscience 19, 565–571.

Eskanazi, B., Prehn, A.W., Cristianson, R.E., 1995. Passive and active maternal smoking as measured by serum cotinine: the effect on birthweight. Am. J. Public Health 85, 395–398.

Fergusson, D.M., Woodward, L.J., Horwood, L.J., 1998. Maternal smoking during pregnancy and psychiatric adjustment in late adolescence. Arch. Gen. Psychiatry 55, 721–727.

Fleming, S.E., Lombardo, T.W., 1987. Effects of cigarette smoking on phobic anxiety. Addict. Behav. 12, 195–198.

Floresco, S.B., West, A.R., Ash, B., Moore, H., Grace, A.A., 2003. Afferent modulation of dopamine neuron firing differentially regulates tonic and phasic dopamine transmission. Nat. Neurosci. 6, 968–973.

Fried, P.A., O'Connell, C.M., Watkinson, B., 1992. 60 and 72-month follow-up of children prenatally exposed to marijuana, cigarettes and alcohol: cognitive and language assessments. J. Dev. Behav. Pediatr. 13, 383–391.

Friedmann, P.D., Rose, J.S., Swift, R., Stout, R.L., Millman, R.P., Stein, M.D., 2008. Trazodone for sleep disturbance after alcohol detoxification: a double-blind, placebo-controlled trial. Alcohol. Clin. Exp. Res. 32, 1652–1660.

Frith, C., 1972. Smoking behavior and its relation to the smoker's immediate experience. Br. J. Clin. Psychol. 10, 73–78.

Garcia-Rill, E., 2009. Reticular activating system. In: Stickgold, R., Walker, M. (Eds.), The Neuroscience of Sleep. Elsevier, Oxford, pp. 133–139.

Garcia-Rill, E., Buchanan, R., McKeon, K., Skinner, R.D., Wallace, T., 2007a. Smoking during pregnancy: postnatal effects on arousal and attentional brain systems. NeuroToxicology 28, 915–923.

Garcia-Rill, E., Heister, D.S., Ye, M., Charlesworth, A., Hayar, A., 2007b. Electrical coupling: novel mechanism for sleep–wake control. Sleep 30, 1405–1414.

Garcia-Rill, E., Kezunovic, N., Hyde, J., Beck, P., Urbano, F.J., 2013. Coherence and frequency in the reticular activating system (RAS). Sleep Med. Rev. 17, 227–238.

Garcia-Rill, E., Kezunovic, N., D'Onofrio, S., Luster, B., Hyde, J., Bisagno, V., Urbano, F.J., 2014. Gamma band activity in the RAS-intracellular mechanisms. Exp. Brain Res. 232, 1509–1522.

Gillette, G., Skinner, R.D., Rasco, L., Fielstein, E., Davis, D., Pawelak, J., Freeman, T., Karson, C.N., Boop, F.A., Garcia-Rill, E., 1997. Combat veterans with posttraumatic stress disorder exhibit decreased habituation of the P1 midlatency auditory evoked potential. Life Sci. 61, 1421–1434.

Goff, D.C., Henderson, D.C., Amico, E., 1992. Cigarette smoking in schizophrenia: Relationship to psychopathology and medication side effects. Am. J. Psychol. 149, 1189–1194.

Goitia, B., Raineri, M., González, L.E., Rozas, J.L., Garcia-Rill, E., Bisagno, V., Urbano, F.J., 2013. Differential effects of methylphenidate and cocaine on GABA transmission in sensory thalamic nuclei. J. Neurochem. 124, 602–612.

Gonzalez, B., Raineri, M., Cadet, J.L., Garcia-Rill, E., Urbano, F.J., Bisagno, V., 2014. Modafinil improves methamphetamine-induced object recognition deficits and restores prefrontal cortex ERK signaling in mice. Neuropharmacology 87, 188–197.

Good, C.H., Lupica, C.R., 2009. Properties of distinct ventral tegmental area synapses activated via pedunculopontine or ventral tegmental area stimulation in vitro. J. Physiol. 587, 1233–1247.

Good, C.H., Bay, K.D., Buchanan, R., Skinner, R.D., Garcia-Rill, E., 2007. Muscarinic and nicotinic responses in the developing pedunculopontine nucleus (PPN). Brain Res. 1129, 147–155.

Gosheh, O.A., Dwoskin, L.P., Miller, D.K., Crooks, P.A., 2001. Accumulation of nicotine and its metabolites in rat brain after intermittent or continuous peripheral administration of [2′-¹⁴C] nicotine. Drug Metab. Dispos. 29, 645–651.

Grimm, J.W., See, R.E., 1997. Cocaine self-administration in ovariectomized rats is predicted by response to novelty, attenuated by 17-β estradiol, and associated with abnormal vaginal cytology. Physiol. Behav. 61, 755–761.

Hagan, J.J., Tonnaer, J.A.D.M., Rijk, H., Broekkamp, C.L.E., van Delft, A.M.L., 1987. Facilitation of amphetamine-induced rotation by muscarinic antagonists is correlated with M2 receptor affinity. Brain Res. 410, 69–73.

Hasler, B.P., Smith, L.J., Cousins, J.C., Bootzin, R.R., 2012. Circadian rhythms, sleep, and substance abuse. Sleep Med. Rev. 16, 67–81.

Jaehne, A., Unbehaun, T., Feige, B., Lutz, U.C., Batra, A., Riemann, G., 2010. How smoking affects sleep: a polysomnographical analysis. Sleep Med. 13, 1286–1292.

Jaehne, A., Unbehaun, T., Feige, B., Cohrs, S., Rodenbeck, A., Riemann, D., 2014. Sleep changes in smokers before, during and 3 months after nicotine withdrawal. Addict. Biol, in press.

Katner, S.N., McBride, W.J., Lumeng, L., Li, T.K., Murphy, J.M., 1997. Alcohol intake of P rats is regulated by muscarinic receptors in the pedunculopontine nucleus and VTA. Pharmacol. Biochem. Behav. 58, 497–504.

Kezunovic, N., Hyde, J., Goitia, B., Bisagno, V., Urbano, F.J., Garcia-Rill, E., 2013. Muscarinic modulation of high frequency oscillations in pedunculopontine neurons. Front. Neurol. 4 (176), 1–13.

Kline, J., Stein, Z.A., Susser, M., Warburton, D., 1997. Smoking: a risk factor for spontaneous abortion. N. Engl. J. Med. 297, 793–796.

Koyama, Y., Sakai, K., 2000. Modulation of presumed cholinergic mesopontine tegmental neurons by acetylcholine and monoamines applied iontophoretically in unanesthetized cats. Neurosci. 96, 723–733.

Koob, G.F., 1998. Circuits, drugs, and drug addiction. Adv. Pharmacol. 42, 978–982.

Lammel, S., Lim, B.K., Ran, C., Huang, K.W., Betley, M.J., Tye, K.M., Deisseroth, K., Malenka, R.C., 2012. Input-specific control of reward and aversion in the ventral tegmental area. Nature 491, 212–217.

Leonard, C.S., Llinas, R., 1994. Serotonergic and cholinergic inhibition of mesopontine cholinergic neurons controlling REM sleep: an in vitro electrophysiological study. Neuroscience 59, 309–330.

Levin, E.D., Briggs, S.J., Christopher, N.C., Rose, J.E., 1993. Prenatal nicotine exposure and cognitive performance in rats. Neurotoxicol. Teratol. 15, 251–260.

Luck, W., Nau, H., Hansen, R., Steldinger, R., 1985. Extent of nicotine and cotinine transfer to the human fetus, placenta and amniotic fluid of smoking mothers. Dev. Pharmacol. Ther. 8, 384–395.

Mamiya, N., Buchanan, R., Wallace, T., Skinner, R.D., Garcia-Rill, E., 2005. Nicotine suppresses the P13 auditory evoked potential by acting on the pedunculopontine nucleus in the rat. Exp. Brain Res. 164, 109–119.

Marinelli, M., White, F.J., 2000. Enhanced vulnerability to cocaine self-administration is associated with elevated impulse activity of midbrain dopamine neurons. J. Neurosci. 20, 8876–8885.

Mark, G.P., Shabani, S., Dobbs, L.K., Hansen, S.T., 2011. Cholinergic modulation of mesolimbic dopamine function and reward. Physiol. Behav. 104, 76–81.

Martin, J.C., Becker, R.F.T., 1971. The effects of maternal nicotine absorption or hypoxic episodes upon appetitive behavior of rat offspring. Dev. Psychobiol. 4, 133–147.

Maskos, U., 2008. The cholinergic mesopontine tegmentum is a relatively neglected nicotinic master modulator of the dopaminergic system: relevance to the drugs of abuse and pathology. Br. J. Pharmacol. 1, 438–445.

Milberger, S., Biederman, J., Farraone, S.V., Chen, L., Jones, J., 1997. ADHD is associated with early initiation of cigarette smoking in children and adolescents. J. Am. Acad. Child Adolesc. Psychiatry 36, 37–44.

Miyazato, H., Skinner, R.D., Cobb, M., Andersen, B., Garcia-Rill, E., 1999. Midlatency auditory-evoked potentials in the rat: effects of interventions that modulate arousal. Brain Res. Bull. 48, 545–553.

Parker, J.L., van der Kooy, D., 1995. Tegmental pedunculopontine nucleus lesions do not block cocaine reward. Pharmacol. Biochem. Behav. 52, 77–83.

Pierre, P.J., Vezina, P., 1997. Predisposition to self-administer amphetamine: the contribution of response to novelty and prior exposure to the drug. Psychopharmacology 129, 277–284.

Raineri, M., Peskin, V., Goitia, B., Taravini, I.R., Giorgeri, S., Urbano, F.J., Bisagno, V., 2011. Attenuated methamphetamine induced neurotoxicity by modafinil administration in mice. Synapse 65, 1087–1098.

Raineri, M., Gonzalez, B., Goitia, B., Garcia-Rill, E., Krasnova, I.N., Cadet, J.L., Urbano, F.J., Bisagno, V., 2012. Modafinil abrogates methamphetamine-induced neuroinflammation and apoptotic effects in the mouse striatum. PLoS ONE 7 (10), e46599.

Rasanen, P., Hakko, H., Isohanni, M., Hodgins, S., Jarvelin, M.R., Tiihonen, J., 1999. Maternal smoking during pregnancy and risk of criminal behavior among adult male offspring in the Northern Finland 1966 Birth Cohort. Am. J. Psychiatry 156, 857–862.

Ross, S.B., Renyi, A.L., 1969. Inhibition of the uptake of tritiated 5-hydroxytryptamine in brain tissue. Eur. J. Pharmacol. 7, 270–277.

Samson, H.H., Chappell, A., 2001. Injected muscimol in pedunculopontine tegmental nucleus alters ethanol self-administration. Alcohol 23, 41–48.

Schmidt, H.D., Famous, K.R., Pierce, R.C., 2009. The limbic circuitry underlying cocaine seeking encompasses the PPTg/LDT. Eur. J. Neurosci. 30, 1358–1369.

Seiden, L.S., Sabol, K.E., Ricaurte, G.A., 1993. Amphetamine: effects on catecholamine systems and behavior. Annu. Rev. Pharmacol. Toxicol. 32, 639–677.

Semba, K., Fibiger, H.C., 1992. Afferent connections of the laterodorsal and the pedunculopontine tegmental nuclei in the rat: a retro- and antero-grade transport and immunohistochemical study. J. Comp. Neurol. 323, 387–410.

Shannon, H.E., Peters, S.C., 1990. A comparison of the effects of cholinergic and dopaminergic agents on scopolamine-induced hyperactivity in mice. J. Pharmacol. Exp. Ther. 255, 549–553.

Simosky, J.K., Stevens, K.E., Freedman, R., 2002. Nicotinic agonists and psychosis. Curr. Drug Targets CNS Neurol. Disord. 1, 177–190.

Solem, M., McMahon, T., Messing, R.O., 1997. Protein kinase A regulates regulates inhibition of N- and P/Q-type calcium channels by ethanol in PC12 cells. J. Pharmacol. Exp. Ther. 282, 1487–1495.

Sorenson, C.A., Raskin, L.A., Suh, Y., 1991. The effects of prenatal nicotine on radial-arm maze performance in rats. Pharmacol. Biochem. Behav. 40, 991–993.

Spielberger, C.D., 1986. Psychological determinants of smoking behavior. In: Tollison, R.D. (Ed.), Smoking and Society. Lexington Books, Lexington, MA, pp. 89–134.

Steidl, S., Wang, H., Wise, R.A., 2014. Lesions of cholinergic pedunculopontine tegmental nucleus neurons fail to affect cocaine or heroin self-administration or conditioned place preference in rats. PLoS ONE 9, e84412.

Sulzer, D., Chen, T.-K., Lau, Y.Y., Kristensen, H., Rayport, S., Ewing, A., 1995. Amphetamine redistributes dopamine from synaptic vesicles to the cytosol and promotes reverse transport. J. Neurosci. 15, 4102–4108.

Takakusaki, K., Shiroyama, T., Yamamoto, T., Kitai, S.T., 1996. Cholinergic and noncholinergic tegmental pedunculopontine projection neurons in rats revealed by intracellular labeling. J. Comp. Neurol. 7, 2353–2356.

Ting-A-Kee, R., Dockstader, C., Heinmiller, A., Grieder, T., van der Kooy, D., 2009. GABA(A) receptors mediate the opposing roles of dopamine and the tegmental pedunculopontine nucleus in the motivational effects of ethanol. Eur. J. Neurosci. 29, 1235–1244.

Urbano, F.J., Leznik, E., Llinas, R.R., 2007. Modafinil enhances thalamocortical activity by increasing neuronal electrotonic coupling. Proc. Natl. Acad. Sci. USA 104, 12554–12559.

Urbano, F.J., Kezunovic, N., Hyde, J., Simon, C., Beck, P., Garcia-Rill, E., 2012. Gamma band activity in the reticular activating system (RAS). Front. Neurol. Sleep Chronobiol. 3 (6), 1–16.

Wakschlag, L.S., Lahey, B.B., Loeber, R., Green, S.M., Gordon, R.A., Leventhal, B.L., 1997. Maternal smoking during pregnancy and the risk of conduct disorders in boys. Arch. Gen. Psychiatry 54, 670–676.

Wang, H.L., Morales, M., 2009. Pedunculopontine and laterodorsal tegmental nuclei contain distinct populations of cholinergic, glutamatergic and GABAergic neurons in the rat. Eur. J. Neurosci. 29, 340–358.

Wasserman, R.C., Kelleher, K.J., Bocian, A., Baker, A., Childs, G.E., Indacochea, F., Stulp, C., Gardner, W.P., 1999. Identification of attentional and hyperactivity problems in primary care: a report from pediatric research in office settings and the ambulatory sentinel practice network. Pediatrics 103, E38.

Weissman, M.M., Warner, V., Wickramaratne, P.J., Kandel, D.B., 1999. Maternal smoking during pregnancy and psychopathology in offspring followed to adulthood. J. Am. Acad. Child Adolesc. Psychiatry 38, 892–899.

White, F.J., Kalivas, P.W., 1998. Neuroadaptations involved in amphetamine and cocaine addiction. Drug Alcohol Depend. 51, 141–153.

Williams, J.A., Reiner, P.B., 1993. Noradrenaline hyperpolarizes identified rat mesopontine cholinergic neurons in vitro. J. Neurosci. 13, 3878–3883.

Wilson, M.C., Bedford, J.A., Buelke, J., Kibbe, A.H., 1976. Acute pharmacological activity of intravenous cocaine in the rhesus monkey. Psychopharmacol. Commun. 2, 251–261.

Wise, R.A., 1998. Drug-activation of brain reward pathways. Drug Alcohol Depend. 51, 13–22.

Wise, R.A., Bozarth, M.A., 1987. A psychomotor stimulant theory of addiction. Psychol. Rev. 94, 469–492.

The Science of Waking and Public Policy

Edgar Garcia-Rill, PhD

Center for Translational Neuroscience, Department of Neurobiology
and Developmental Sciences, University of Arkansas for Medical Sciences,
Little Rock, AR, USA

THREE STATES

We have three states. We are awake, asleep, and asleep and dreaming. We spend on average about 16h awake, 7h asleep, and 1h asleep and dreaming. The amount of life we devote to each of these states is a good indication of the importance that we should assign to them. It is during waking that we accomplish things great and mundane, that we carry on relationships fulfilling and shallow, and that we think, wish, invent, and enjoy. We saw how waking is a complex process to arrive at and even more complicated to maintain. But it is that state that provides the background for our lives, the stream of preconsciousness, the context of sensory experience. It is the waking state that provides preconscious awareness, the continuous flow of information on which we base our decisions, our wishes, and our survival. It is the process that underlies attention, learning, and memory, without which we would be unable to pay attention, remember, or learn.

While much research is devoted to attention, learning, and memory, without the process of maintained waking, none of these functions would be possible. Moreover, without considering the process of waking and preconscious awareness and without controlling for the context of sensory experience (see Chapter 2, the Gorilla in the Room), the study of these three functions is on shaky ground. The old adage, "life is what happens when you are not paying attention," is what the reticular activating system (RAS) is about. The adage has negative connotations implying that if we do not pay attention, we are likely to succumb to accidents. However, preconscious awareness is a process that allows us to survive and to not get into trouble; basically, it is what life is about.

Waking and the Reticular Activating System in Health and Disease
http://dx.doi.org/10.1016/B978-0-12-801385-4.00014-8

When we are asleep, there is no sensation, no perception, and no content or context to the world. We are not in play and there is nobody home. The lack of lateral inhibition across cortical columns makes sensation impossible until higher-frequency oscillations reestablish coincident firing and lateral inhibition. However, it is during slow-wave sleep (SWS) that the brain regains much of the used energy stores that are spent during the waking hours. It is not clear if this state has additional functions, for example, in memory consolidation. For example, some studies suggest that sleep helps the formation of dendritic spines after learning (Yang et al., 2014). SWS appears to be essential for survival, although more research on sleep deprivation without a stress component (most studies on sleep deprivation are marred by inducing stress) is needed.

When we dream, the internal world is manufacturing a fictitious play; it is a neurogenic epiphenomenon, a psychedelic world without the benefit of fully functioning frontal lobes. Psychedelic because the brain is not in touch with the external world. As discussed in Chapter 1, the internal gear is not meshing with the external gear. That is, the absence of sensory information leads the brain to manufacture a "sensory" content. This dissociation between the external world and internal brain states may also be present during sensory deprivation, in which the absence of the stream of worldly sensations leads to distorted perception. These effects were encountered in the 1950s by high-altitude pilots who wore pressure suits that led cutaneous input to habituate, who had white noise continuously running in their headsets, and who had a uniform blue visual display. Some pilots reported "out-of-body" phenomena, such as seeing themselves sitting on the wing watching themselves flying the plane. The lack of change in the content of sensory experience, to which the brain habituated rapidly, led to sensory deprivation, to a lack of external input. The result was a neurogenic manufacture of a distorted "external" world. It is not a far reach to believe that the sensory deprivation induced by remaining in a static posture for prolonged periods, such as during meditation, could provide similar effects.

Recall that decreased frontal lobe blood flow during rapid eye movement (REM) sleep may account for the lack of critical judgment during dreaming. The brain paralyzes us so that we do not act out our dreams, which could be dangerous if we do act them out. The brain ensures that only our eye muscles act out our dreams; they are the only muscles not subjected to decreased muscle tone. If the atonia of REM sleep were absent, we would suffer from REM sleep behavior disorder. We should take note of the fact that the brain does not want us to act out our dreams before we place too much weight on dream content. Moreover, we dream during each REM sleep episode throughout the night, but usually can recall the content of only the last episode. While we can recall or reconstruct events during the last 8h of waking, we find it very difficult to reconstruct every dream episode during the night. We can recall the latest dream only if we are awakened at its end. If dream content is so important, why is it not easier to recall? REM sleep, unlike SWS, is not essential since drugs such as monoamine oxidase inhibitors can eliminate it without major long-term effects (although this needs more study).

The functions of REM sleep are not entirely known, although some work suggests that it participates in certain types of memory. This is puzzling since memories would need to be cemented without fully functioning frontal lobes, but perhaps that is why our memories for complex events, for example, witnessing an accident, are so notoriously unreliable. Other theories suggest that REM sleep is a state when unnecessary memories are culled (Jouvet, 2001). Another theory is that the eye movements during REM sleep serve to stir the aqueous humor to bring oxygen to the cornea during sleep (Maurice, 1998). For now, the most parsimonious explanation is that dreams are an epiphenomenon of a process of brain activation during sleep.

Much weight is paid to the process of lucid dreaming. Such states probably represent intermediate levels of consciousness. We saw in Chapter 10 that transitions between states can last for minutes, and in Chapter 3, we saw that the process of regaining consciousness may be stepwise. Moreover, as mentioned in Chapter 10, upon waking, blood flow to the brain increases first in the thalamus and brain stem and only later in the frontal lobes. Such processes as lucid dreaming may represent transition states especially marked by the lack of cortical activation. The absence of complexity and variety to the marginally conscious brain may actually provide a period of simplicity that provides options unencumbered by multiple considerations. We see the "big picture" instead of getting lost in the details. The absence of "interference" from the cortex is evident, for example, in athletic performance.

WAKEFULNESS AND PERFORMANCE

Peak performance in athletics is characterized by "clearing the mind" or being "unconscious." Perhaps, the absence of multiple cortical options allows more exact subcortical control of movement. Preconscious awareness and the RAS, as we saw in Chapter 10, modulate our movements. If humans had evolved only to see, hear, and feel, we would be stuck on a rock like a sea anemone and look like a stalk with one eye. There would be no need for muscles, and thus, muscles would not have evolved. But we do have muscles and we use them to move and to gain control of our environment. Human beings have evolved the capacity to carry out very sophisticated and rapid movements. Our most sophisticated and rapid movements involve our fingers and tongue. Humans perform movements up to about $10\,s^{-1}$ (Hz), so that we do have a limit, as we saw in Chapter 8. Most of our repetitive movements, however, are slower in frequency. How our movements are controlled has been of great concern to brain researchers. Most studies have been directed to the presumed linear properties of movement. That is, many researchers have attempted to relate the activity of single neurons to the action of single muscles, an undertaking that has proven to be elusive. Other researchers have spent years dissecting the internal organization of rhythmically active groups of neurons, describing the nature of each synapse in locomotion and respiration oscillators (rhythmically active groups of neurons; see Chapter 7).

The manner the brain triggers and controls movement is thought to be through resonance. Nicholas Bernstein, a Russian physiologist and mathematician early in the twentieth century, was first responsible for describing movements in terms of the sums of three or four harmonic oscillations (Bernstein, 1967). These oscillations are presumably provided by specific regions of the brain and carried by several different output pathways to neurons that drive muscles. That is why damage to one pathway or brain region will typically produce a rhythmic disturbance, a lack of smooth control, like tremor (see Chapter 8, "The 10 Hz Fulcrum"). Tremor may be the manifestation of one specific resonant frequency unsmoothed due to the absence of some harmonics. That is, the 6–8 Hz tremor in PD is missing the additional 2–4 Hz normally present in physiological tremor. At the same time, rhythmic information is received from our muscles, tendons, and joints, all in the form of resonant frequencies carried by several different input pathways, especially those to the cerebellum. These inputs provide an oscillating background of activity, a dynamic starting point, from which we start a movement. This means that, when the body moves, it represents a nonlinear, complex dynamical system already in nonequilibrium.

Bernstein's analysis showed that complex movements are sensitive to initial conditions, that a small section of the movement represents the movement as a whole (is self-similar), and that movement is the result of coordinated (self-organized) activity. The rhythmic properties of movement, which arise from harmonic resonance, were proposed to be induced by the repetition, reiteration, of a formula or "motor engram." To demonstrate these points, draw a five-pointed star free hand several times. The stars will look similar to each other but never identical. However, each individual's stars will differ from everyone else's. A person's signature, for example, is characteristic of that individual and it looks similar whether written large or small, that is, it is indifferent to the scale of the movement of writing. Each individual's signature will be similar whether written with a pen on paper or with a piece of chalk on a blackboard. Yet, the sequence and selection of muscle contractions is markedly different whether writing horizontally with a pen while sitting or writing vertically with a piece of chalk while standing. In other words, the end result is indifferent to the position of the movement. These examples serve to emphasize the nonlinear, probabilistic nature of motor control by the brain.

Bernstein was ahead of his time. Unfortunately, when he formulated these ideas, the mathematics of the science of complexity, of nonlinear dynamic systems, had not been invented. Bernstein believed that it would someday be possible to describe the dynamics of voluntary movement in mathematical terms that would also reflect the underlying brain activity. The problem with movement is that the brain's commands are designed to affect the biomechanical relationships between forces, that is, the weight of the bones and muscles that are influenced by gravity and by changes in acceleration during movement. The forces and the changes in forces are in continuous flux as the movement proceeds, meaning that the motor command from the brain continuously compensates for these forces and those of the environment. Therefore, the motor command may be considerably different for apparently similar movements, such as in walking, in which one step looks like the next but the terrain and the biomechanical forces of the legs are constantly varying with every step. Under these conditions, the brain's motor commands, the activity of particular cells, will differ significantly from one step to the next. This is perhaps why single-cell activity in central brain regions is only generally correlated, rather than exhibiting exact firing patterns, with each successive movement. The best way to think about movements is that, rather than being "programmed," they actually "develop" over time; they evolve, like a "complex melody of scores of dynamic waves," as Bernstein put it. Practice and training are, therefore, very much like natural selection, the selection of a viable movement that will accomplish the aim successfully, perhaps not the absolute best option, but one with a good probability of success.

When a movement is performed, it is a statistical event, the best option available given the circumstances. The actual movement was dependent on many, many factors, and the final product may not have been the absolutely best option, but as good an option as could be managed given those factors. Practice and training basically narrow down the options, weeding out the grossest mistakes, cutting down the possibilities of not performing more or less the best option. Sometimes, the best possible movement is generated, and it is right and is the perfect movement. When that perfect movement occurs, the brain instantly realizes it. The moment is savored as heartily as a great idea. The brain has accomplished the ideal, the optimal use of one of our faculties. And all of this perfection is best carried out without interference from the cortex. The cooperation between gamma band activity in the cerebellum and gamma band activity in the RAS probably provides the activity that leads to peak performance.

When performing a complex, rapid movement, such as a layup in basketball or a golf swing, the least interference, such as a touch or the sound of a camera shutter, will disturb the evolution of the sequence ever so slightly. The RAS is activated sufficiently by these wayward stimuli to lead to considerable distortions in the end result. For example, the golf swing of a professional is in the order of 160 km/h with rotations about three axes, at the end of which the face of the club must meet the ball at right angles. The slightest deflection of that angle, say, one degree, could send the ball 30 m off the fairway at the end of a drive. The slightest sound, the smallest twitch of a startle response, will easily cause such a deflection. A baseball pitcher has a 10 ms window during a throw to release the ball so that it will arrive in the strike zone. The least deviation beyond that window will cause the ball to hit the ground or fly over the catcher's head. The least distraction will lead to such a deviation.

This is why human beings empathize with the movements of others, why they resonate with athletic performance. People instantly recognized the impeccability of the movements of Michael Jordan driving for the basket, floating suspended in the air, seeming to defy gravity. How did Michael Jordan do that? It was perfection in movement. It was the best option, similar to the movements of other professionals, and perhaps even similar to our own movements, but Jordan's were at a higher level of faultlessness. Tiger Woods' golf swing once reached that gold standard, and the excellence, the perfection, of these movements can be appreciated. Instantly, others recognize the faultlessness, the impeccability of these motions, and, every now and then, we golf hackers also get one right. Our brain immediately understands the completeness of the optimal golf swing, as long as the cortex does not get in the way. From the moment these athletes were conceived, from the initial melding of genes from their parents, through their prenatal and postnatal development, from the experiences that shaped their brains, their actions evolved into a unique probability. The perfection of their movements is a testament to that history, one not so different from our own in some ways, alien in others, but a history that led to a more optimal choice, a more perfect option.

Why do people show such interest in sports? Why are humans willing to pay considerable amounts to see top athletes perform? Perhaps because these trained athletes are capable of perfection of movement. Our brains perceive athletic performance as no less perfect than a masterpiece in music, oil, or stone. Humans empathize equally with the flawless performance of a master pianist, a prima ballerina, or a Rudolf Nureyev, who seemed to hang suspended in the air and defy gravity, much like Michael Jordan. Our brains collectively experience the same awe at a concert hall when a master guitarist finishes a solo encore, the last perfect note hanging suspended in the air. The movements of an athlete, like the motions of a musician, form an integrated whole, a prime example of our reason for being a manifestation of what we have become, of what our development as individuals and our evolution as a species has provided. The modulation of voluntary movement by the RAS has additional implications.

INTENT

In Chapter 10, we discussed the role of waking and the RAS in volition and free will. These considerations have further implications. For example, legal liability for a crime is a two-part inquiry: *actus reus* (the voluntary act) and *mens rea* (the intent necessary to commit the act). The legal meaning of intent, however, is controversial. In the penal code, purpose and knowledge are seen as descriptive terms. It is defined willfully as being synonymous with "knowingly" and knowingly as acting with "awareness" of engaging in the conduct under the defined circumstances of the offense and to be "practically certain" that the conduct will result in the proscribed act. Despite this attempt at clarity, however, defining intent remains a juridical quagmire.

Enter the concept advanced by Libet, and discussed in Chapter 10, that volition is "subconscious" because the readiness potential begins long before there is a "conscious" recognition that the movement was willed. The natural extension is that there is no such thing as free will. This creates a massive legal problem since impulsive acts would be classified as "subconscious," therefore, presumably unintentional. Our interpretation of these results, that the readiness is evidence that the movement was "preconscious," not "subconscious," not only restores the concept of free will but also eliminates the legal conundrum and renders some sanity to the potential legal considerations of the definition of intent. Given our alternative conclusions on Libet's experiment, we are still ultimately responsible for our actions in a legal sense. An upcoming law review will address this issue.

There are exceptions, such as in the absence of the atonia of REM sleep, as can occur, fortunately rarely, when some patients take zolpidem as a sleep aid. Carlos Schenck has described parasomnias, sleep disorders involving involuntary movements and behaviors, that carry legal implications. Sleep sex can involve verbal and physical behaviors, and violent behaviors also can occur while asleep (Ohayon and Schenck, 2010; Schenck et al., 2007). In some cases, violent behaviors may create a legal problem for the "unconscious" perpetrator. In such cases, expert testimony will be critical. However, the onus for determining who can provide expert testimony now rests on the judge. Previously, the jury was expected to determine the validity of expert testimony, but since the Supreme Court decision on *Daubert v. Merrell Dow Pharmaceuticals, Inc.*, it is the judge who makes that decision. Judges now have a pretrail session on the admissibility of expert testimony. The Supreme Court *Daubert* decision used the philosophy of Sir Karl Popper, the preeminent philosopher of the twentieth century, to explain why the judge can and should carry out such determinations and why the expert witness must now justify his or her opinion much more thoroughly. That is, the expert witness must explain the scientific validity so that the judge can understand it, not simply rest on personal opinion. After all, Popper stated:

> Never aim at more precision than is required by the problem in hand. Thus I have no faith in precision: I believe that simplicity and clarity are values in themselves, but not that precision or exactness is a value in itself. Clarity and precision are different and sometimes incompatible aims.... What can be said can and should always be said more and more simply and clearly.
>
> Popper, 1956/1983

We described the concept in a law review (Beecher-Monas and Garcia-Rill, 1999). This review discussed how these Supreme Court decisions apply to a modern world of nondeterminism and nonlinear dynamics, especially as it concerns brain function and human behavior.

Despite the requirement to filter such testimony, judges have been reluctant to apply the *Daubert* decision. Nowhere is this phenomenon more evident than in the admissibility of post-traumatic stress disorder (PTSD) evidence. Despite the well-established science (see Chapter 11) behind the disorder, PTSD testimony is often excluded from evidence, generally without any analysis of its scientific validity. In sharp contrast, psychological syndrome testimony (such as "battered woman syndrome"), which rests on very shaky ground indeed and cannot meet standards of scientific validity, is widely admitted (also largely without any analysis of its scientific merit). A subsequent law review addressed the specific case of the science and admissibility of expert testimony in PTSD (Garcia-Rill and Beecher-Monas, 2001).

GENETICS

Perhaps the most interesting legal issue is the genetics behind violence. For nearly 20 years, we have known that psychiatrists cannot predict whether a person who has committed a violent act will be violent in the future (Webster et al., 1994) and neither can lay people (Quinsey et al., 1998). The very best anyone can do is speculate. Even the most scientific predictions based on thorough examination, diagnosis of mental symptoms, past patterns of behavior, and probabilistic assessments are wrong nearly as often as they are right. The most common courtroom predictions, frequently based solely on hypotheticals, are wrong twice as often as they are right (Monahan, 2000). Nevertheless, legislatures and courts often demand expert risk assessments, which themselves are subject to large margins for error. These predictions are known as "future dangerousness."

Two additional law reviews discussed the relationship of genetics and neuroscience in the context of future dangerousness predictions (Beecher-Monas and Garcia-Rill, 2003, 2006). At least since the late nineteenth century, courts and prisons have attempted to discriminate between the innately criminal and those who acted merely by force of circumstance (whose crimes, being caused by circumstance rather than nature, would not pose a future danger to society). In order to distinguish the dangerous criminals from the merely circumstantial ones, predictions of "future dangerousness" became vital to the criminal justice system and continue as a pervasive influence in death penalty adjudications and sex offender civil commitment hearings. These articles discussed the fallacy of genetic determinism and explained the complex interactions between genes, environment, and developmental forces in generating behavior.

However, there is a more fundamental problem that involves genetics and neuroscience. The overwhelming emphasis on genetic models has provided what could turn out to be a money pit for research funding. Throughout this book, we discussed how high-frequency activity, especially in the $40\,s^{-1}$ range, underlies our waking processes, from awareness, to attention, to learning and memory. At that speed, the role of genetic transcription is negligible. Proteins simply are not made at sufficiently fast rates to influence how we carry out a conversation or play a game of soccer. The genome spoke long before in providing the essential elements for brain processes and as such is an important area of study. However, as Jouvet posited, "the genome speaks only to the unconscious" (Jouvet, 2001). Our behavior, thoughts, and ideas are a result of high-frequency activity far too fast for genetic transcription. Nevertheless, the current fad is to develop a genetic model for every facet of brain function. This involves the use of knockout or knockin technology. Briefly, the genetic code is transfected and recombined so that, after a controlled breeding schedule, the gene in question is eliminated or a new one is inserted into the genome of an animal, usually a mouse.

While this is a very powerful technique to study, for example, the action of drugs, it has a major overlooked and unaddressed weakness. When you delete a gene from an animal, the resulting progeny do not manifest simply the absence of that gene. The fact is that, when you remove a gene, you upregulate hundreds of other genes and downregulate almost as many others. That is, the resulting animal would probably never exist in nature and would probably not represent an accurate model for a human genetic disorder. That is, knockout and knockin studies do not control for the *other* genes that are misregulated by the deletion or insertion. Such controls would be prohibitively difficult and are thus ignored. Moreover, which gene is removed is critical. For example, a gene coding for a structural protein will affect many more other genes and functions than a gene coding for a limited use protein. Take the case of gap junctions. These are important for electrical coupling and the much-needed coherence of brain rhythms (see Chapters 1 and 4). In mice with connexin 43 gene deletion, large numbers of other genes are statistically altered, and these encode a large number of proteins with multiple functions (Iacobas et al., 2004). That is, the gene deletion altered a large number of other functions not controlled for. Moreover, the gene deletion can affect a multitude of downstream partners, producing alterations in entire genetic and therefore functional networks (Iacobas et al., 2007). The best that can be said is that this technology may not be producing fully transparent models.

FUNCTIONAL MAGNETIC RESONANCE IMAGING

Another technology that overreaches its scientific foundations is functional magnetic resonance imaging (fMRI). While the MRI and fMRI methods are excellent clinical tools that can reveal details of the anatomical distribution of tumors or lesions, it is a poor real-time functional tool. For example, the fMRI is being used to estimate such processes as attention and perception. However, the fMRI actually measures changes in oxygenated blood, not neuronal activity. The peak of the MRI signal occurs seconds after the brain activity (Bandettini, 2009). For a system that functions at 40 Hz, this is an eternity and thus represents a very indirect measure of neuronal activity. The method simply does not have the temporal resolution to detect changes in neuronal activity on a real-time basis or on a single-trial basis since images are the result of considerable averaging. The functional aspect of the fMRI is generally oversold.

One example is the Brain Initiative of the National Institutes of Health (NIH). The Brain Initiative has nine major research domains that are targeted and all of them involve the study of circuits. While this is an encouraging topic, those who framed these domains neglected the other half of brain function, the intrinsic properties of nerve cells, which control firing patterns (see Chapter 1). The brain works through circuits that maintain certain firing patterns thanks to the role of membrane channels and intrinsic oscillations. By merely studying the circuits, all that is gained is a static picture, a still photograph, of the orchestra. The technique of fMRI cannot hope to measure the second half of the control of neuronal activity. The music is missing. The initiative as it is thus appears a subsidy to imaging studies that will generate beautiful pictures, attractive still photographs of the orchestra. However, we will not be able to know what music the orchestra is playing.

There are only two technologies that measure brain activity in real time, the electroencephalogram (EEG) and the magnetoencephalogram (MEG). The EEG provides electrical signals that are unfortunately distorted by the scalp and skull, but the MEG provides magnetic signals undistorted by tissue. This is the closest methodology currently available for measuring real-time brain activity, but few neuroscientists are familiar with it, and the cost of the MEG is prohibitive without subsidy from clinical use. Only such technology can provide access to high-frequency brain rhythms, and researchers and the NIH need to wake up to that fact. Such technologies are critical for the performance of sophisticated studies on humans. The absolutely essential need, methods for facilitating it, and necessary policies for translational neuroscience were addressed in a previous book (Garcia-Rill, 2012).

THE FUTURE OF SCIENCE

How did we get here? A recent book proposes that the posing of ambitious questions leads to innovation (Berger, 2014). This is largely true for scientific research, but that is not what is practiced by those who support science. Unfortunately, reviewers of grant applications now demand perfection before funding a study. That means safe, predictable, "well-controlled" research. It also means that the opportunity for leaps of understanding and innovation is quashed. We speak from experience as our findings, first on electrical coupling and then on gamma band activity in the RAS, were supported by a limited number of scientists with true vision.

As someone with almost 30 years' experience reviewing grant applications, there is a definite difference between earlier review attitudes and current ones. We used to read an application with this question foremost: Should this science be done and will it advance brain research? The methods could be improved or corrected where needed, and advice was rendered by reviewers on how to best obtain the results. It was the "big idea" that mattered. With the doubling of the NIH funding in the 1990s, there was a tripling of the number of scientists and of subsequent grant applications. That created increased research and increased competitiveness. With leveling and even decrements in funding after 2001, reviewers have become more conservative. The retrenchment to a "safe" and incremental project has become the norm. NIH has created some of the problems, first by cutting the time during which applications are discussed. In the interest of saving money, review committees that typically lasted 2.5 days now last 1.5 days or less. In the interest of recruiting more reviewers, they simplified the scoring system such that the discrimination between applications has decreased.

The current system not only fails to ensure that the most innovative science is performed but also fails to discriminate between the top ranked applications. The use of bullets instead of full sentences in the reviews, along with shortened discussion time, allows unjustified opinions to determine an application's future. The use of premeeting uploading of reviews and the requirement to not discuss one-half of the applications at the meeting create an environment in which applications are dismissed without discussion by more than a couple of reviewers. The brief reviews during the meetings protect unsupported criticisms and do not allow sufficient time for others to question them. Reviewers are now younger and more critical, and many do not think about what are the big questions that should be asked. Despite lip service by NIH that innovation should be a critical factor, it is only one of several equally weighted factors, so that it can easily become lost in the noise. One factor that remains in the likelihood that an application will be funded is the "popularity" of the method in vogue, whether or not it is the best way to advance science. Above, we discussed two techniques in vogue that are strongly supported, but while extremely valuable as diagnostic tools, in research, they are limited in advancing the field as currently practised.

The origins of this condition are multiple, but a major one that has been proposed is the assumption of never-ending growth in biomedical science, and that has created the potential for long-term decline (Alberts et al., 2014). A number of specific recommendations have been proposed, which need to be heeded by the NIH and by the research community as a whole. Without new and progressive policies for research, especially neuroscience research, the field will become less and less attractive. Fundamental changes in the policies governing grant applications, training, and the administration of awards are needed.

WAKING AND PUBLIC POLICY

The persistence of waking leading to insomnia and of sleep deprivation, as well as drug-induced sleepiness, is a major health hazard. Almost a third, over 100 million, of automobile drivers in the United States have fallen asleep at the wheel. About a quarter of pilots admit to making errors due to sleepiness. As such, policies to curb sleep deprivation also need to be addressed. Fatigue due to sleepiness is the cause of almost one-fifth of all transport accidents in the world (Arendt, 2000). Sleep deprivation not only is a major cause of accidents but also leads to significant losses in productivity with serious consequences to the economy. Sleep deprivation leads not only to daytime sleepiness but also to decreased motor and sensory performance (Jackson et al., 2008). David Dinges is one of the pioneers in the field of sleep deprivation and performance and has developed a number of fatigue management technologies (Abe et al., 2014). Technologies that predict and detect sleepiness and fatigue may prevent operator errors and accidents. Amazingly, one of the smallest agencies of the government is in charge of investigating all air, highway, rail, marine, and pipeline accidents.

The National Transportation Safety Board (NTSB) is an independent federal agency numbering less than 100 employees that has no legislative power and can only make recommendations. However, it is so well respected that most (>80%) of the recommendations it makes are implemented. Some of the regulations previously and currently proposed involve the setting of time limits for operators after the consumption of alcohol, sleep aids, and other drugs. Sleep deprivation in transportation accidents has been found to involve degraded decision-making and to lead to visual or cognitive fixation, poor communication and coordination, and slowed reaction time. The problem requires multiple solutions, including better scheduling policies and education. However, one factor that has been overlooked is the effect on waking of a variety of agents that affect arousal that are not usually regulated.

The most common sleep-inducing agents that are not regulated are allergy and antimotion sickness medications, whether in the form of antihistamines or anticholinergics. Mefloquine is a drug used for the treatment of malaria but has a serious side effect, sleepiness. As discussed in Chapter 4, we use mefloquine to block gap junctions, which decreases coherence, especially at high frequencies. That is, it leads to slowed EEG rhythms and sleepiness. Carbenoxolone is used in the United Kingdom for the treatment of ulcers since it blocks gap junctions in acid-secreting cells. It has the significant side effect of inducing sleepiness and, in fact, is a major factor in noncompliance by patients. These are just some examples of agents that also need to be regulated. But perhaps the most impact on operator error would be requiring operators to pass a test of alertness, for example, a short psychomotor vigilance task (Abe et al., 2014), before being able to start the engine.

CHANGING DIRECTIONS

Throughout this book are indications of the need for additional research in the field of waking and the RAS. For example, there is a great need to incorporate the context of sensory experience into studies on attention, memory, and learning, perhaps using priming or masking methods. Why is this important? Imagine the sheer number of applications to which controlling preconscious awareness can be put. A child with painful burns can be debrided only if distracted. Chronic pain disappears if we pay attention to something we find important. Distraction is another way of redirecting preconscious awareness and has a massive number of clinical uses. Methods that would accomplish such control would decrease the need for analgesics and anesthetics.

In terms of basic research, a more thorough description of the activity of cell clusters within RAS nuclei will tell us about the algorithm used to generate various frequencies and if different frequencies modulate specific functions. For example, in the cortex, conscious thought is limited in capacity so that we can express only 2–3 thoughts at a time, mainly because we run out of different frequency bands (Buschman and Miller, 2010). Does the RAS have similar limitations? Do different cell clusters generate different frequencies of activity and does that determine whether we are awake or in REM sleep? Is gamma activity during waking reciprocally activated from gamma activity during REM sleep? Is there a limit to preconscious awareness? It would seem so given how much information we miss in a complex environment. Nevertheless, this is critical information not only for basic knowledge about the context of our sensory world but also for processes related to awareness and attentional demands during complex acts, like flying an airplane. This system is at the root of our ability to improve our performance, to decrease distraction, and to avoid accidents.

This area of research is wide open but admittedly difficult to study. The challenge is there for the next generations of neuroscientists to embrace. Hopefully, they will enjoy the challenge as much as I have, and funding for science can be repaired sufficiently to entice our best and brightest, for it is a wonderfully fulfilling calling in life.

BRAIN MUSIC

Coupled with the responsibility to perform research at the highest ethical standards and assuming a philosophy based on the fact that, as a scientist, it is unethical to believe in something for which there is no evidence (Clifford, 1877) is the need for the scientist to inform the public. Without public support, research funding will grow increasingly difficult. One parallel that we have proposed over the years in public lectures, based on the considerable evidence that the brain works through rhythmic activity, is *Brain Music*. The idea is that the brain does not work like a computer; it is not digital or all or none like the action potential because most of the computing is analog, in waveforms, and performed by dendritic potentials and intrinsic oscillations. The idea to be explored in an upcoming book for laypersons is that the brain works like an orchestra.

This idea suggests that the various regions of the brain can be likened to the sections of an orchestra. When several brain regions are active simultaneously, they generate frequencies of activity in harmony with each other, very much in the way the different sections of the orchestra produce notes, frequencies of sound, to yield musical harmony. This process is akin to thinking. The expressed result of this process is movement. The sequential activity of the brain over time is what dictates function, for example, in generating a complex series of movements, such as in writing. The sequential activity of the orchestra over time is the series of notes, the melody, of the music. A well-performed sequence of movements, such as those involved in walking up steps, inserting a key in a lock, or opening a door, involves a large number of neurons in several brain regions generating firing patterns changing over time in a dynamic manner. Similarly, a well-performed melody involves a number of instruments playing a series of notes in a harmonious manner over time. The neuronal algorithm for a hierarchically organized movement can be likened to sheet music. The function of the brain is to generate thought and movement, just as the function of the orchestra is to generate music.

Music does not come from outside, it is a mental construct, a product of the brain, a "perception" of ordered sound. It is a product of the characteristics of the human mind. Although many animals have ears similar to ours and can perceive sound, none can generate melodies. Only brains as complex as ours can manipulate and order sounds well enough to make music. Music is organized sound, the sound of our brains. Rodolfo Llinás, an eminent neuroscientist at New York University, suggested that "music is machine language for the brain." Machine language, as you know, is a language composed of a set of numbers and symbols that can direct computer operations without the need for translation. Similarly, music is proposed to be a language composed of a set of notes and frequencies that can direct brain operations without the need for translation. While there is much work to be done in supporting this suggestion with experimental evidence, if even partly correct, this may explain why music is so universal, so natural. Music may well be the language of the brain.

References

Abe, T., Mollicone, D., Basner, M., Dinges, D., 2014. Sleepiness and safety: where biology needs technology. Sleep Biol. Rhythms 12, 74–84.

Akerstedt, T., 2000. Consensus statement: fatigue and accidents in transport operations. J. Sleep Res. 9, 395.

Alberts, B., Kirschner, M.W., Tilghman, S., Varmus, H., 2014. Rescuing US biomedical research from its systemic flaws. Proc. Natl. Acad. Sci. USA 111, 5773–5777.

Arendt, J., 2002. Melatonin, circadian rhythms, and sleep. New Engl. J. Med. 343, 1114–1116.

Bandettini, P.A., 2009. Functional MRI limitations and aspirations. In: Kraft, E., Gulyas, B., Poppel, E. (Eds.), Neural Correlates of Thinking. Springer, New York, NY, pp. 15–38.

Beecher-Monas, E., Garcia-Rill, E., 1999. The law and the brain: judging scientific evidence of intent. J. Appell. Pract. Proc. 1, 243–277.

Beecher-Monas, E., Garcia-Rill, E., 2003. Danger at the edge of chaos: future dangerousness and predicting violent behavior. Cardozo Law Rev. 24, 1845–1901.

Beecher-Monas, E., Garcia-Rill, E., 2006. Genetic predictions of future dangerousness: is there a blueprint for violence? Law Contemp. Probl. 69, 301–341.

Berger, W., 2014. The Power of Inquiry to Spark Breakthrough Ideas. Bloomsbury Press, New York, NY.

Bernstein, N., 1967. The Coordination and Regulation of Movement. Pergamon Press, New York, NY, 196 pp.

Buschman, T.J., Miller, E.K., 2010. Shifting the spotlight of attention: evidence for discrete computations in cognition. Frontiers Human Neurosci. 4, 194.

Buschman, T.J., Siegel, M., Roy, J.E., Miller, E.K., 2010. Neural substrates of cognitive capacity limitations. Proc. Natl. Acad. Sci. USA 108, 11252–11255.

Clifford, W.C., 1877. The Ethics of Belief and Other Essays. Prometheus Books, New York, NY, 140 pp.

Garcia-Rill, E., 2012. Translational Neuroscience: A Guide to a Successful Program. Wiley-Blackwell, New York, NY, 152 pp.

Garcia-Rill, E., Beecher-Monas, E., 2001. Gatekeeping stress: the science and admissibility of post-traumatic stress disorder. UALR Law Rev. 24, 9–40.

Iacobas, D.A., Scemes, E., Spray, D.C., 2004. Gene expression alterations in connexin null mice extend beyond the gap junction. Neurochem. Int. 45, 243–250.

Iacobas, D.A., Iacobas, S., Spray, D.C., 2007. Connexin-dependent transcellular transcriptomic networks in mouse brain. Prog. Biophys. Mol. Biol. 94, 169–185.

Jackson, M.L., Croft, R.J., Owens, K., Pierce, R.J., Kennedy, G.A., Crewther, D., Howard, M.E., 2008. The effect of acute sleep deprivation on visual evoked potentials in professional drivers. Sleep 31, 1261–1269.

Jouvet, P., 2001. The Paradox of Sleep, the Story of Dreaming. MIT Press, Cambridge, MA, 227 pp.

Maurice, D.M., 1998. The Von Sallmann lecture 1996: an ophthalmological explanation of REM sleep. Exp. Eye Res. 66, 139–146.

Monahan, J., 2000. Violence risk assessment: scientific validity and evidentiary admissibility. Wash Lee Law Rev. 57, 901–918.

Ohayon, M.M., Schenk, C.H., 2010. Violent behavior during sleep: prevalence, comorbidity and consequences. Sleep Med. 11, 941–946.

Popper, K.R., 1956/1983. Realism and the Aim of Science. Rowan and Littlefield, Lanham, MD, 420 pp.

Quinsey, V.L., Harris, G.T., Rice, M.E., Cormeir, C.A., 1998. Violent Offenders: Appraising and Managing Risk. American Psychological Association, Washington, DC, 356 pp.

Schenck, C.H., Amulf, I., Mahowald, M.W., 2007. Sleep and sex: what can go wrong? A review of the literature on sleep related disorders and abnormal sexual behaviors and experiences. Sleep 30, 683–702.

Webster, C.D., Harris, G.T., Rice, M.E., Cormier, C., Quinsey, V.L., 1994. The Violence Prediction Scheme: Assessing Dangerousness in High Risk Men. University of Toronto Center of Criminology, Toronto 92 pp.

Yang, G., Lai, C.S.W., Cichon, J., Ma, L., Li, W., Gan, W.B., 2014. Sleep promotes branch-specific formation of dendritic spines after learning. Science 344, 1173–1178.

Index

Note: Page numbers followed by *f* indicate figures.

Printed in the United States
By Bookmasters